Other Titles in This Series

170 S. P. Novikov, Editor, Topics in Topology and Mathematical Physics
169 S. G. Gindikin and E. B. Vinberg, Editors, Lie Groups and Lie Algebras: E. B. Dynkin's Seminar
168 V. V. Kozlov, Editor, Dynamical Systems in Classical Mechanics
167 V. V. Lychagin, Editor, The Interplay between Differential Geometry and Differential Equations
166 O. A. Ladyzhenskaya, Editor, Proceedings of the St. Petersburg Mathematical Society, Volume III
165 Yu. Ilyashenko and S. Yakovenko, Editors, Concerning the Hilbert 16th Problem
164 N. N. Uraltseva, Editor, Nonlinear Evolution Equations
163 L. A. Bokut', M. Hazewinkel, and Yu. G. Reshetnyak, Editors, Third Siberian School "Algebra and Analysis"
162 S. G. Gindikin, Editor, Applied Problems of Radon Transform
161 Katsumi Nomizu, Editor, Selected Papers on Analysis, Probability, and Statistics
160 K. Nomizu, Editor, Selected Papers on Number Theory, Algebraic Geometry, and Differential Geometry
159 O. A. Ladyzhenskaya, Editor, Proceedings of the St. Petersburg Mathematical Society, Volume II
158 A. K. Kelmans, Editor, Selected Topics in Discrete Mathematics: Proceedings of the Moscow Discrete Mathematics Seminar, 1972–1990
157 M. Sh. Birman, Editor, Wave Propagation. Scattering Theory
156 V. N. Gerasimov, N. G. Nesterenko, and A. I. Valitskas, Three Papers on Algebras and Their Representations
155 O. A. Ladyzhenskaya and A. M. Vershik, Editors, Proceedings of the St. Petersburg Mathematical Society, Volume I
154 V. A. Artamonov et al., Selected Papers in K-Theory
153 S. G. Gindikin, Editor, Singularity Theory and Some Problems of Functional Analysis
152 H. Draškovičová et al., Ordered Sets and Lattices II
151 I. A. Aleksandrov, L. A. Bokut', and Yu. G. Reshetnyak, Editors, Second Siberian Winter School "Algebra and Analysis"
150 S. G. Gindikin, Editor, Spectral Theory of Operators
149 V. S. Afraĭmovich et al., Thirteen Papers in Algebra, Functional Analysis, Topology, and Probability, Translated from the Russian
148 A. D. Aleksandrov, O. V. Belegradek, L. A. Bokut', and Yu. L. Ershov, Editors, First Siberian Winter School "Algebra and Analysis"
147 I. G. Bashmakova et al., Nine Papers from the International Congress of Mathematicians, 1986
146 L. A. Aĭzenberg et al., Fifteen Papers in Complex Analysis
145 S. G. Dalalyan et al., Eight Papers Translated from the Russian
144 S. D. Berman et al., Thirteen Papers Translated from the Russian
143 V. A. Belonogov et al., Eight Papers Translated from the Russian
142 M. B. Abalovich et al., Ten Papers Translated from the Russian
141 H. Draškovičová et al., Ordered Sets and Lattices
140 V. I. Bernik et al., Eleven Papers Translated from the Russian
139 A. Ya. Aĭzenshtat et al., Nineteen Papers on Algebraic Semigroups
138 I. V. Kovalishina and V. P. Potapov, Seven Papers Translated from the Russian
137 V. I. Arnol'd et al., Fourteen Papers Translated from the Russian
136 L. A. Aksent'ev et al., Fourteen Papers Translated from the Russian
135 S. N. Artemov et al., Six Papers in Logic
134 A. Ya. Aĭzenshtat et al., Fourteen Papers Translated from the Russian
133 R. R. Suncheleev et al., Thirteen Papers in Analysis
132 I. G. Dmitriev et al., Thirteen Papers in Algebra
131 V. A. Zmorovich et al., Ten Papers in Analysis

(*Continued in the back of this publication*)

Topics in Topology and Mathematical Physics

American Mathematical Society

TRANSLATIONS

Series 2 • Volume 170

Advances in the Mathematical Sciences – 27

(*Formerly Advances in Soviet Mathematics*)

Topics in Topology and Mathematical Physics

S. P. Novikov
Editor

American Mathematical Society
Providence, Rhode Island

ADVANCES IN THE MATHEMATICAL SCIENCES EDITORIAL COMMITTEE

V. I. ARNOLD
S. G. GINDIKIN
V. P. MASLOV

Translation edited by A. B. SOSSINSKY

1991 *Mathematics Subject Classification.* Primary 35Q40, 57Rxx, 58E05, 58F05, 58F07; Secondary 16W30.

ABSTRACT. The papers in this collection grew out of talks recently presented at S. P. Novikov's seminar on topology and mathematical physics in Moscow. They are devoted to various problems in the theory of completely integrable systems and relations to topology, algebra, and mathematical physics. The book will be of interest to researchers and graduate students working in the corresponding areas of mathematics.

Library of Congress Catalog Card Number: 91-640741
ISBN 0-8218-0455-3
ISSN 0065-9290

Copying and reprinting. Material in this book may be reproduced by any means for educational and scientific purposes without fee or permission with the exception of reproduction by services that collect fees for delivery of documents and provided that the customary acknowledgment of the source is given. This consent does not extend to other kinds of copying for general distribution, for advertising or promotional purposes, or for resale. Requests for permission for commercial use of material should be addressed to the Assistant Director of Production, American Mathematical Society, P. O. Box 6248, Providence, Rhode Island 02940-6248. Requests can also be made by e-mail to reprint-permission@math.ams.org.
Excluded from these provisions is material in articles for which the author holds copyright. In such cases, requests for permission to use or reprint should be addressed directly to the author(s). (Copyright ownership is indicated in the notice in the lower right-hand corner of the first page of each article.)

© Copyright 1995 by the American Mathematical Society. All rights reserved.
The American Mathematical Society retains all rights
except those granted to the United States Government.
Printed in the United States of America.

⊚ The paper used in this book is acid-free and falls within the guidelines
established to ensure permanence and durability.
♻ Printed on recycled paper.

10 9 8 7 6 5 4 3 2 1   00 99 98 97 96 95

Contents

The S. P. Novikov Seminar
V. M. BUCHSTABER and S. P. NOVIKOV — 1

Semigroups of Maps into Groups, Operator Doubles, and Complex Cobordisms
V. M. BUCHSTABER — 9

Nonlocal Hamiltonian Operators of Hydrodynamic Type: Differential Geometry and Applications
E. V. FERAPONTOV — 33

Nonselfintersecting Magnetic Orbits on the Plane. Proof of the Overthrowing of Cycles Principle
P. G. GRINEVICH and S. P. NOVIKOV — 59

Spin Generalization of the Calogero–Moser System and the Matrix KP Equation
I. KRICHEVER, O. BABELON, E. BILLEY, and M. TALON — 83

Symplectic and Poisson Geometry on Loop Spaces of Manifolds and Nonlinear Equations
OLEG MOKHOV — 121

Real Nonsingular Finite Zone Solutions of Soliton Equations
S. M. NATANZON — 153

Representations of Krichever–Novikov Algebras
O. K. SHEINMAN — 185

Huygens' Principle and Algebraic Schrödinger Operators
A. P. VESELOV — 199

The S. P. Novikov Seminar

V. M. BUCHSTABER AND S. P. NOVIKOV

Our seminar came into being in the mid-sixties. S. P. Novikov's works on the topology of foliations and on the topological invariance of rational Pontryagin classes were concluded in 1965. A new stage of research was about to begin, devoted to extraordinary (generalized) cohomology theories: complex cobordisms and K-theory. Novikov's older pupils (e.g. V. L. Golo) had moved on to other fields.

By that time K-theory had become extremely popular in wide circles of the mathematical community in connection with the work of Atiyah and Singer on the index of elliptic operators; besides, Adams discovered brilliant applications of K-theory in topology itself: the number of linearly independent vector fields on spheres was found, important subgroups in the stable homotopy groups of spheres were computed (the image of Whitehead's J-homomorphism). The possibility of constructing regular methods, based on K-theory, for the computation of homotopy groups (in particular those of spheres), more effective than the classical methods of Cartan-Serre-Adams, was being discussed (the Adams program).

Among the participants of the seminar during the first two-three years, the names of V. Buchstaber, A. Mishchenko, I. Bernstein, I. Volodin, S. Smirnov, S. Vishik, and F. Bogomolov should be singled out.

In 1966–67 S. P. Novikov [49] carried out his own program, creating methods for the regular computation of stable homotopy groups on the basis of complex cobordisms. It turned out that the Adams program for creating such methods on the basis of K-theory could not be realized in principle. In contrast, the new techniques of complex cobodisms were extremely rich in ideas and applications to topology and algebra, namely:

• formal groups, together with their applications in homotopy theory as well as in the study of fixed points of finite and compact transformation groups of smooth manifolds, including the remarkable properties of elliptic genera ([50, 11, 5, 31, 34, 59, 38]) discovered much later;

• the theory of multivalued formal groups, together with its applications to topology, algebra, analysis, including the relationship with generalized shift operator theory ([12, 6, 7, 8]);

• the algebra of operations from complex cobordism theory with its numerous topological applications and beautiful intrinsic algebraic structure that later led to the "operator (Heisenberg) double" of Hopf algebras, which is the quantum analog of the algebra of differential operators on a Lie group ([49, 13, 58, 9]).

These ideas are developed in the article by V. Buchstaber which also contains new results in this direction.

By the end of the sixties, the participants of the seminar, as well as its subject matter, changed noticeably. The research interests of several participants, e.g. I. Bernstein and S. Vishik, shifted to other fields of mathematics. A whole new generation of extremely talented young researchers appeared in the seminar: G. Kasparov, O. Bogoyavlenskii, S. Gusein-Zade, S. Brakhman, I. Krichever, V. Dubrovin, M. Bruk, V. Krasnov, M. Brodskii, S. Tankeev, F. Zak, R. Nadiradze, A. Peresetskii, A. Shokurov, N. Panov, V. Vedenyapin, and others. For about two years, B. G. Moishezon became co-director of the seminar. During this period its participants studied Kahler and algebraic geometry. Later several members of the seminar began working on the problems of algebraic geometry itself.

Among the participants were several others, who, beginning in 1974, greatly contributed to the application of methods coming from algebraic geometry to modern mathematical physics (more specifically, to the periodicity problems in soliton theory and in integrable systems). The work of these participants, carried out in the seventies and eighties, became widely known ([24, 52, 35, 41, 17, 23].

In this volume these ideas are developed in the papers by I. Krichever, S. Natanzon, A. Veselov. A striking example of feedback, namely, the application of the theory of nonlinear equations to algebraic geometry, was Novikov's conjecture on the characterization of the Jacobians of Riemann surfaces (the Riemann-Schottki problem). The first notable advance in this problem was the work of B. Dubrovin [18]. A complete proof was given by Shiota [61].

During the second half of the sixties, several participants of the seminar also studied the topology of non-simply-connected manifolds, developing the ideas that had arisen in foliation theory, as well as those appearing in the proof of the topological invariance of the rational Pontryagin classes. At the time, the final formulation of S. P. Novikov's conjecture on higher signatures [51] for manifolds with arbitrary fundamental group was put forward and was established for abelian fundamental groups π_1. In this period the seminar, besides its traditional contacts with the seminars of the leading Moscow mathematical schools, was in constant very intense interaction with the V. A. Rokhlin seminar in Leningrad.

This period also witnessed the appearance of ideas on the Hermitian analogs of algebraic K-theory, based on the language and basic concepts of the Hamiltonian formalism, an algebraic version of sorts of symplectic geometry [51].

The K-theory of infinite-dimensional complexes was then constructed, leading up to complete answers in many important cases, including computations for the classifying spaces of compact Lie groups and Eilenberg-MacLane complexes [10]. The correct higher analogs of algebraic K-theory K^0, K^1 (at the same time as Quillen, but on the basis of a different idea) were obtained [66].

Soon after that, the development of Fredholm representations was undertaken, both for the construction of a topological K-homology theory and for the higher signature problem ([32, 47]).

The S. P. Novikov conjecture on higher signatures eventually became widely known in mathematics. A huge number of papers is devoted to this conjecture. It gave the impetus for finding deep relationships between topology, algebra, and functional analysis ([48, 33, 16, 14, 15]). A conference was held on this topic in

Oberwolfach in September 1993. Two volumes of the proceedings of the conference [70] will include both research and survey papers illustrating the current status of the conjecture.

At the beginning of the seventies, the research interests of the participants of the seminar diverged: new seminars were organized, where the branches of topology and algebra mentioned above (cobordisms, formal groups, problems of nonsimply-connected manifolds, including Hermitian K-theory, problems of higher signatures and functional analysis techniques) were still studied, e.g. at the V. Buchstaber seminar or the A. Mishchenko seminar.

In the second half of the seventies, only one pure topologist began research at the S. P. Novikov seminar itself on the then very new subjectmatter related to Sullivan's ring approach to rational homotopy type: this was I. Babenko ([2, 3]).

Around 1970–71, the seminar concentrated on the study of special and generalized relativity (at the time one of the co-directors of the seminar was V. P. Myasnikov).

At the beginning of 1971, S. P. Novikov began working at the Landau Institute of Theoretical Physics and the interests of the seminar shifted more and more towards the mathematical problems of modern theoretical physics. Jointly with several of his pupils (O. Bogoyavlenskii, B. Dubrovin, I. Krichever), S. P. Novikov originated the qualitative theory of homogeneous cosmological models, solved the periodic problems of soliton theory, developed the theory of one-dimentional and two-dimensional Schrödinger operators in periodic electric and magnetic fields ([4, 24, 22, 53]). Later these ideas were developed and led to the creation of the theory of two-dimensional periodic and rapidly decreasing operators ([57, 29, 30, 37, 39]).

In the late seventies and in the eighties, new participants of the seminar joined in these directions of research, in particular A. Veselov, I. Taimanov, P. Grinevich, O. Mokhov, S. Tsarev, A. Lyskova, R. Novikov [26].

In the process of solving problems of physical nature, S. P. Novikov returned to topology: he found curious topological characteristics of the typical two-dimensional Schrödinger operator in a magnetic field and in a periodic lattice, that was later to play a crucial role in explaining the quantum Hall effect, initiated the notion of multivalued variational calculus in theoretical physics and mathematics, and constructed the analog of Morse theory for multivalued functions and functionals ([53, 54, 55, 56, 62]). A. Lyskova and I. Taimanov took part in these studies. The topic was later developed by quite a few researchers.

These topics were those where a whole new generation of the seminar's participants, working in topology, began their research: F. Voronov, A. Zorich, A. Lazarev, D. Millionshchikov, A. Alaniya, Le Tu Thang, I. Dynnikov, and others ([68, 67, 69, 43, 45, 46, 1, 44, 28]). In this volume these ideas are developed in the joint article by P. Grinevich and S. P. Novikov and the one by I. Dynnikov.

A curious cycle of new ideas in Euclidean geometry arose in the eighties as the result of the interaction with young theoretical physicists from the Landau Institute (Levitov, Kitaev, Kalugin): the beautiful concept of quasi-crystallic subgroup of the isometry groups of Euclidean space in the sense of Novikov–Veselov and other aspects of the geometry of quasi-crystals were successfully exploited by S. Piunikhin, V. Sadov, and Le Tu Thang [60]; the asymptotic problems of soliton theory led to the construction of a Hamiltonian theory of systems of hydrodynamic type (i.e., quasilinear systems of the first order), which had not appeared in the entire hundred-year history of this field. The Hamiltonian formalism of "hydrodynamic type" was discovered by S. P. Novikov and B. Dubrovin in 1983, giving rise to a new branch of Riemannian geometry ([27, 25, 26, 19]).

In the framework of this geometry, S. Tsarev found a method for integrating such systems ([63, 64, 65]). As the result of a cycle of further studies of several participants of the seminar (V. Avilov, S. P. Novikov, I. Krichever, G. Potemin), the complete analytic solution for the dispersive analog of the wave equation, undertaken in the early seventies by leading members of the Landau school (A. B. Gurevich and L. P. Pitaevskiĭ), was finally obtained. The algebraic geometry realization of the Flaschke–MacLaughlin mean of the (Whitham) soliton equations was developed by Krichever [36, 40].

These geometric ideas play an important role in the construction of the now very popular two-dimensional topological quantum field theory [20], [21]. In the present volume this cycle of ideas appears in the articles by O. Mokhov, E. Feropontov, and in the subsequent one in those of B. Dubrovin, M. Pavlov, V. Alekseev.

In the eighties, several participants of the seminar made significant contributions to the development of the geometry and topology of supermanifolds. In particular, F. Voronov and A. Zorich constructed the correct superanalog of the de Rham complex [67]. The ideas of supersymmetery are developed in this volume in the article by F. Voronov.

In connection with the operator quantization of boson strings and two-dimensional conformal field theory, S. P. Novikov and I. Krichever defined the correct analogs of the Laurent and Fourier series for the expansion of meromorphic tensor fields of any weight on Riemann surfaces. The operator quantization program for strings was actively developed by physicists in 1970–74 (Veneciano, Virasoro, Mandelstam, and others) for surfaces of genus 0. The absence of analogs of Laurent–Fourier series on surfaces of positive genus stopped further development of this topic in the mid-seventies.

Using a different approach (the continual integral), A. Polyakov succeeded in quantizing the string for Riemann surfaces (diagrams) of any genus.

The implementation of the program of operator or algebraic quantization of strings became possible as the result of the work of S. P. Novikov and I. Krichever at the end of the eighties. The "Krichever–Novikov bases" and the "Krichever–Novikov algebra" that they constructed are the Riemann analogs of the Laurent–Fourier bases and of the Virasoro algebra [42]. This topic is that of numerous papers by many authors. In this volume it is represented by O. Sheinman's article.

References

1. L. A. Alaniya, *On manifolds of the Alexander type*, Uspekhi Mat. Nauk **46** (1991), no. 1, 203–204; English transl. in Russian Math. Surveys **46** (1991).
2. I. K. Babenko, *On real homotopy properties of complete intersections*, Izv. Akad. Nauk SSSR **43** (1979), no. 5, 1004–1024; English transl. in Math. USSR-Izv. **15** (1980).
3. _____, *Growth and rationality problems in algebra and topology*, Uspekhi Mat. Nauk **41** (1986), 95–142; English transl. in Russian Math. Surveys **41** (1986).
4. O. I. Bogoyavlenskiĭ and S. P. Novikov, *Qualitative theory of homogeneous cosmological models*, Trudy Sem. Petovsk. **1** (1975), 7–44; English transl. in Selecta Math. Soviet. **2** (1982).
5. V. M. Bukhshtaber, *The Chern-Dold character in cobordisms.* I, Mat. Sb. **83** (1970), 575–595; English transl. in Math. USSR-Sb. **12** (1970).

6. _____, *Two-valued formal groups. Algebraic theory and applications to cobordism.* I, Izv. Akad. Nauk SSSR **39** (1975), no. 5, 1044–1064; English transl. in Math. USSR-Izv. **9** (1975).
7. _____, *Two-valued formal groups. Algebraic theory and applications to cobordism.* II, Izv. Akad. Nauk SSSR **40** (1976), no. 2, 289–325; English transl. in Math. USSR-Izv. **10** (1976).
8. _____, *Topological applications of the theory of two-valued formal groups*, Izv. Akad. Nauk SSSR **42** (1978), no. 1, 130–184; English transl. in Math. USSR-Izv. **12** (1978).
9. _____, *Operator doubles and semigroups of maps into groups*, Dokl. Akad. Nauk SSSR **341** (1995); English transl. in Soviet Math. Dokl. (to appear).
10. V. M. Bukhshtaber and A. S. Mishchenko, *K-theory on the category of infinite cell complexes*, Izv. Akad. Nauk SSSR **32** (1968), 560–604; English transl. in Math. USSR-Izv. **2** (1968).
11. V. M. Bukhshtaber, A. S. Mishchenko, and S. P. Novikov, *Formal groups and their role in the apparatus of algebraic topology*, Uspekhi Mat. Nauk **26** (1971), no. 2, 131–154; English transl. in Russian Math. Surveys **26** (1971).
12. V. M. Bukhshtaber and S. P. Novikov, *Formal groups, power systems and Adams operators*, Mat. Sb. **84** (1971), 81–118; English transl. in Math. USSR-Sb. **13** (1971).
13. V. M. Bukhshtaber and A. B. Shokurov, *The Landweber–Novikov algebra and formal vector fields on the line*, Funktsional. Anal. i Prilozhen. **12** (1978), no. 3, 1–11; English transl. in Functional Anal. Appl. **12** (1978).
14. A. Connes, M. Gromov, and H. Moscovici, *Conjecture de Novikov et fibrés presque plats*, C. R. Acad. Sci. Paris Sér. I Math. **310** (1990), 273–277.
15. _____, *Group cohomology with Lipschitz control and higher signatures*, Geom. Funct. Anal. **3** (1993), 1–78.
16. A. Connes and H. Moscovici, *Cyclic cohomology, the Novikov conjecture and hyperbolic group*, Topology **29** (1990), 1345–1388.
17. B. A. Dubrovin, *Theta-functions and nonlinear equations*, Uspekhi Mat. Nauk **36** (1981), no. 2, 11–80; English transl. in Russian Math. Surveys **36** (1981).
18. _____, *The Kadomtsev-Petviashvili equation and the relation between the periods of holomorphic differentials on Riemann surfaces*, Izv. Akad. Nauk SSSR **45** (1981), no. 5, 1015–1028; English transl. in Math. USSR-Izv. (1982), 285–296.
19. _____, *On the differential geometry of strongly integrable systems of hydrodynamical type*, Funktsional. Anal. i Prilozhen. **24** (1990), no. 4, 25–30; English transl. in Functional Anal. Appl. **24** (1990).
20. _____, *Integrable systems and classification of 2 dimensional topological field theories*, Integrable Systems, the Verdier Memorial Conference, Progress in Math., vol. 115, Boston, Birkhäuser, 1989, pp. 313–359.
21. _____, *Differential geometry of the space of orbits of a Coxeter group*, Preprint SISSA-89/94/FM, Trieste.
22. B. A. Dubrovin, I. M. Krichever, and S. P. Novikov, *Schrödinger equation in a periodic field and Riemann surfaces*, Dokl. Akad. Nauk SSSR **229** (1976), no. 1, 15–18; English transl. in Soviet Math. Dokl. **17** (1976).
23. _____, *Integrable systems.* I, Itogi Nauki i Tekhniki: Sovremennye Problemy Mat.: Fundamental'nye Napravleniya, vol. 4, VINITI, Moscow, 1985, pp. 179-284; English transl. in Encyclopedia of Math. Sci., vol. 4, (Dynamical Systems), IV, Springer-Verlag, Berlin, 1990.
24. B. A. Dubrovin, V. B. Matveev, and S. P. Novikov, *Non-linear equations of the Korteweg-de Vries type, finite-band linear operators and Abelian varieties*, Uspekhi Mat. Nauk **31** (1976), no. 1, 55–136; English transl. in Russian Math. Surveys **31** (1976).
25. B. A. Dubrovin and S. P. Novikov, *Hydrodynamics of weakly deformed soliton lattices. Differential geometry and Hamiltonian theory*, Uspekhi Mat. Nauk **44** (1989), no. 6, 29–98; English transl. in Russian Math. Surveys **44** (1989), 35–124.
26. _____, *Hydrodynamics of soliton lattices*, Soviet Sci. Rev. Sect. C: Math. Phys. Rev., vol. 9, Part 4, Harwood, Chur, 1993.
27. _____, *On Poisson brackets of hydrodynamic type*, Dokl. Akad. Nauk SSSR **279** (1984), no. 2, 294–297; English transl. in Soviet Math. Dokl. **30** (1985).
28. I. A. Dynnikov, *S. P. Novikov's problem on the semiclassical motion of an electron*, Mat. Zametki **48** (1993), 179–180; English transl. in Math. Notes **48** (1993).

29. P. G. Grinevich and R. G. Novikov, *Analogues of multisoliton potentials for the two-dimensional Schrödinger operator*, Funktsional. Anal. i Prilozhen. **19** (1985), no. 4, 32–42; English transl. in Functional Anal. Appl. **19** (1985).
30. P. G. Grinevich and S. P. Novikov, *A two-dimensional "inverse scattering problem" for negative energies, and generalized-analytic functions. I. Energies lower than the ground state*, Funktsional. Anal. i Prilozhen. **22** (1988), no. 1, 23–33; English transl. in Functional Anal. Appl. **22** (1988).
31. S. M. Gusein-Zade, *U-actions of a circle and fixed points*, Izv. Akad. Nauk SSSR **35** (1971), 1120–1136; English transl. in Math. USSR-Izv. **5** (1971).
32. G. G. Kasparov, *Topological invariants of elliptic operators. I. K-homology*, Izv. Akad. Nauk SSSR **39** (1975), no. 4, 796–838; English transl. in Math. USSR-Izv. **9** (1975).
33. _____, *Equivariant KK-theory and the Novikov conjecture*, Invent. Math. **91** (1988), 147–201.
34. I. M. Krichever, *Formal groups and the Atiyah–Hirzebruch formula*, Izv. Akad. Nauk SSSR **38** (1974), 1289–1304; English transl. in Math. USSR-Izv. **8** (1974).
35. _____, *Methods of algebraic geometry in the theory of nonlinear equations*, Uspekhi Mat. Nauk **32** (1977), no. 6, 183–108; English transl. in Russian Math. Surveys **32** (1977).
36. _____, *The method of averaging for two-dimensional "integrable" equations*, Funktsional. Anal. i Prilozhen. **22** (1988), no. 3, 37–52; English transl. in Functional Anal. Appl. **22** (1988).
37. _____, *Spectral theory of two-dimensional periodic operators and its applications*, Uspekhi Mat. Nauk **44** (1989), no. 2, 121–184; English transl. in Russian Math. Surveys **44** (1989).
38. _____, *Generalized elliptic genera and Baker-Akhiezer functions*, Mat. Zametki **47** (1990), no. 2, 35–45; English transl. in Math. Notes **47** (1990).
39. _____, *Perturbation theory in periodic problems for two-dimensional integrable systems*, Soviet Sci. Rev. Sect.: C Math. Phys. Rev., vol. 9, Harwood, Chur, 1991.
40. _____, *The τ-function of the universal Whitham hierarchy, matrix models and topological field theories*, Comm. Pure Appl. Math. **47** (1994), 437–475.
41. I. M. Krichever and S. P. Novikov, *Holomorphic bundles over algebraic curves, and nonlinear equations*, Uspekhi Mat. Nauk **35** (1980), no. 6, 47–68; English transl. in Russian Math. Surveys **35** (1980).
42. _____, *Algebras of Virasoro-type, Riemann surfaces and the structures soliton theory*, Funktsional. Anal. i Prilozhen. **21** (1987), no. 2, 46–63; English transl. in Functional Anal. Appl. **21** (1987).
43. A. Yu. Lazarev, *Novikov homology in knot theory*, Mat. Zametki **51** (1992), 53–57; English transl. in Math. Notes **51** (1992).
44. T. T. Le, *Structure of level surfaces of the Morse form*, Mat. Zametki **44** (1988), no. 1, 124–133; English transl. in Math. Notes **44** (1988).
45. D. V. Millionshchikov, *Imbeddings of minimal model of k-homotopy type into the algebras of smooth forms $\Lambda^*(M)$*, Uspekhi Mat. Nauk **43** (1988), no. 2, 147–148; English transl. in Russian Math. Surveys **43** (1988).
46. _____, *Some spectral sequences in analytical homotopy theory*, Mat. Zametki **47** (1990), no. 5, 52–61; English transl. in Math. Notes **47** (1990).
47. A. S. Mishchenko, *Hermitian K-theory. The theory of characteristic classes and methods of functional analysis*, Uspekhi Mat. Nauk **31** (1976), no. 2, 69–134; English transl. in Russian Math. Surveys **31** (1976).
48. _____, *Infinite dimensional representations of discrete groups, and higher signatures*, Izv. Akad. Nauk SSSR **38** (1974), 81–106; English transl. in Math. USSR-Izv. **8** (1974).
49. S. P. Novikov, *Methods of algebraic topology from the point of view of cobordism theory*, Izv. Akad. Nauk SSSR **31** (1967), 885–951; English transl. in Math. USSR-Izv. **1** (1967).
50. _____, *Adams operators and fixed points*, Izv. Akad. Nauk SSSR **32** (1968), 1245–1263; English transl. in Math. USSR-Izv. **2** (1968).
51. _____, *Algebraic construction and properties of Hermitian analogs of K-theory over rings with involution from the viewpoint of Hamiltonian formalism. Applications to differential topology and the theory of characteristic classes. I, II*, Izv. Akad. Nauk SSSR **34** (1970), 253–288; 475–500; English transls. in Math. USSR-Izv. **4** (1970).
52. _____, *A periodic problem for the Korteweg de Vries equation. I*, Funktsional. Anal. i Prilozhen. **8** (1974), no. 3, 54–66; English transl. in Functional Anal. Appl. **8** (1974).
53. _____, *Magnetic Bloch functions and vector bundles. Typical dispersion laws and their quantum numbers*, Dokl. Akad. Nauk SSSR **257** (1981), no. 3, 538–543; English transl. in Soviet Math. Dokl. **23** (1981).

54. _____, *The Hamiltonian formalism and a multivalued analogue of Morse theory*, Uspekhi Mat. Nauk **37** (1982), no. 5, 3–49; English transl. in Russian Math. Surveys **37** (1982).
55. _____, *Two–dimensional Schrödinger operators in periodic fields*, Itogi Nauki i Tekhniki: Sovremennye Problemy Mat., vol. 23, VINITI, Moscow, 1983, pp. 3–23; English transl. in J. Soviet Math. **28** (1985).
56. S. P. Novikov and I. A. Taimanov, *Periodic extremals of many-valued or not everywhere-positive functionals*, Dokl. Akad. Nauk SSSR **274** (1984), no. 1, 26–28; English transl. in Soviet Math. Dokl. **29** (1984).
57. S. P. Novikov and A. P. Veselov, *Two-dimensional Schrödinger operator: inverse scattering and evolutional equations*, Phys. D **18** (1986), 267–273.
58. S. P. Novikov, *Various doubles of Hopf algebras. Operator algebras on quantum groups, complex cobordisms*, Uspekhi Mat. Nauk **47** (1992), no. 5, 189–190; English transl. in Russian Math. Surveys **47** (1992).
59. S. Oshanine, *Sur les genres multiplicatifs définis par des intégrales elliptiques*, Topology **26** (1987), 143–151.
60. S. A. Piunikhin, T. T. Le, and V. A. Sadov, *The geometry of quasicrystals*, Uspekhi Mat. Nauk **48** (1993), no. 1, 41–102; English transl. in Russian Math. Surveys **48** (1993).
61. T. Shiota, *Characterization of Jacobian varieties in terms of soliton equations*, Invent. Math. **83** (1986), 333–382.
62. I. A. Taimanov, *Closed extremals on two-dimensional manifolds*, Uspekhi Mat. Nauk **47** (1992), no. 2, 143–185; English transl. in Russian Math. Surveys **47** (1992).
63. S. P. Tsarev, *Liouville Poisson brackets and one-dimensional Hamiltonian systems of hydrodynamic type, arising in the Bogolyubov-Whitman theory of averaging*, Uspekhi Mat. Nauk **39** (1984), no. 6, 209–210; English transl. in Russian Math. Surveys **39** (1984).
64. _____, *Poisson brackets and one-dimensional Hamiltonian systems of the hydrodynamic type*, Dokl. Akad. Nauk SSSR **282** (1985), no. 3, 534–537; English transl. in Soviet Math. Dokl. **31** (1985).
65. _____, *The geometry of Hamiltonian systems of hydrodynamic type. The generalized hodograph method*, Izv. Akad. Nauk SSSR **54** (1990), 1048–1068; English transl. in Math. USSR-Izv. **37** (1991).
66. I. A. Volodin, *Algebraic K-theory as an extraordinary homology theory on the category of associative rings with a unit*, Izv. Akad. Nauk SSSR **35** (1971), 844–873; English transl. in Math. USSR-Izv. **5** (1971).
67. T. Voronov, *Geometric integration theory on supermanifolds*, Soviet Sci. Rev. Sect.: C Math Phys. Rev., vol. 9, Part 1, Harwood, Chur, 1991.
68. F. F. Voronov and A. V. Zorich, *Complexe of forms on a supermanifold*, Funktsional. Anal. i Prilozhen. **20** (1986), no. 2, 58–59; English transl. in Functional Anal. Appl. **20** (1986).
69. A. V. Zorich, *Quasiperiodic structure of level surfaces of a Morse 1-form, close to a rational one–a problem of S. P. Novikov*, Izv. Akad. Nauk SSSR **51** (1987), no. 6, 1322–1344; English transl. in Math. USSR-Izv. **31** (1988).
70. S. Ferry, A. Ranicki, and J. Rosenberg (eds.), *Novikov conjectures, index theorems and rigidity*, Proc. Conf. Oberwolfach 1993, Vols. I,II., Cambridge Univ. Press, Cambridge, 1995.

Translated by A. B. SOSSINSKY

Semigroups of Maps into Groups, Operator Doubles, and Complex Cobordisms

V. M. BUCHSTABER

In this paper we develop the theory of Lie groups and Hopf algebras defined in terms of the space G^V of maps taking the space V to the group G. The results are based on a construction of a semigroup G_α^V associated to each action α of the group G in V. This semigroup is realized as the semigroup of multiplicative operators in the corresponding Novikov operator double. Important examples of semigroups G_α^V have geometric realizations in terms of complex cobordism theory: the semigroup of multiplicative operations in complex cobordisms coincides with the semigroup of maps into itself of the group of diffeomorphisms of the line, the quantum double of the Landweber–Novikov algebra is a subalgebra in the algebra of operations of the doubled theory of complex cobordisms. Part of the main results of this paper appeared in [2].

§1. Semigroups of maps into groups

Suppose G is a group with multiplication $m(g_2, g_1) = g_2 g_1$ and G' is the group with the opposite multiplication $m'(g_2, g_1) = g_1 g_2$. For a space V with a right action α of the group G, let us introduce the multiplications m_α and m'_α in the space of maps G^V:

(1) $\qquad m_\alpha : \varphi_2 * \varphi_1(v) = \varphi_2(v)\varphi_1(v\varphi_2(v))$,

(2) $\qquad m'_\alpha : \varphi_2 * \varphi_1(v) = \varphi_1(v\varphi_2(v)^{-1})\varphi_2(v)$,

where $\varphi_k \in G^V$, $k = 1, 2$, $vg = \alpha(v, g)$.

In the case when α acts trivially, the corresponding multiplication coincides with the pointwise multiplication in G^V and $(G')^V$.

Denote by G_α^V and $(G_\alpha^V)'$ the spaces with multiplication by m_α and m'_α, respectively. Let $i : G \to G^V$ be the map that takes the point g to the constant map to this point.

LEMMA 1.1. 1) *The spaces G_α^V and $(G_\alpha^V)'$ are semigroups with two-sided units;*
2) *the inclusion i induces homomorphisms $G \to G_\alpha^V$ and $G \to (G_\alpha^V)'$.*

PROOF. Let us check associativity for the multiplication m_α:

$$(\varphi_3 * (\varphi_2 * \varphi_1))(v) = \varphi_3(v)(\varphi_2 * \varphi_1)(v_1) = \varphi_3(v)(\varphi_2(v_1)\varphi_1(v_2)),$$

1991 *Mathematics Subject Classification.* Primary 16W30, 57R77.

©1995, American Mathematical Society

where $v_1 = v\varphi_3(v)$ and $v_2 = v_1\varphi_2(v_1)$. On the other hand,

$$((\varphi_3 * \varphi_2) * \varphi_1)(v) = (\varphi_3 * \varphi_2)(v)\varphi_1(v)(\varphi_3 * \varphi_2)(v) = (\varphi_3(v)\varphi_2(v_1))\varphi_1(v_2).$$

The other statements of the lemma can be verified just as easily. □

The semigroups G_α^V possess the following obvious functorial properties.

LEMMA 1.2. 1) *On the space V with action α of the group G_2, any homomorphism $\lambda\colon G_1 \to G_2$ induces the action $\lambda^*\alpha$ of the group G_1 and determines a homomorphism $\lambda_*^V\colon G_{1,\lambda^*\alpha}^V \to G_{2,\alpha}^V$.*

2) Let V_1 and V_2 be spaces with actions α_1 and α_2 of the group G, respectively. Then a map $\gamma\colon V_1 \to V_2$ that commutes with the actions α_1 and α_2 defines a homomorphism $\gamma^*\colon G_{\alpha_2}^{V_2} \to G_{\alpha_1}^{V_1}$.

For example, the isomorphism $\lambda\colon G' \to G$, $g \mapsto g^{-1}$ induces the isomorphism $\lambda^*\colon G_\alpha^V \to (G_\alpha^V)'$ since we have $(G_{\alpha_2}^V)' \cong (G')_{\lambda^*\alpha}^V$ according to (2).

In the case when the space V is supplied with a multiplication

$$\mu\colon V \times V \to V, \quad \mu(v_1, v_2) = v_1 v_2,$$

the following diagonal map is defined:

(3) $\quad \Delta\colon G^V \to (G \times G)^{V \times V}, \quad (\Delta\varphi)(v_1, v_2) = (\varphi(v_1 v_2), \varphi(v_1 v_2)).$

COROLLARY 1.1. *Suppose α, β, γ are actions of the group G on V compatible with the multiplication μ in the sense that*

(4) $\qquad\qquad (v_1 \cdot v_2)\alpha(g) = (v_1\beta(g))(v_2\gamma(g)).$

Then the diagonal defines the homomorphism

(5) $\qquad\qquad \Delta_{\alpha,\beta_\gamma}\colon G_\alpha^V \to (G \times G)_{\beta \times \gamma}^{V \times V}.$

PROOF. The map Δ splits into the composition

$$G^V \xrightarrow{\mu_*} G^{V \times V} \xrightarrow{\delta_*} (G \times G)^{V \times V},$$

where $\delta\colon G \times G \to G$ is the usual diagonal. Then, according to Lemma 1.2, the map δ_* induces the homomorphism $G_{\delta_*(\beta\times\gamma)}^{V\times V} \to (G \times G)_{\beta\times\gamma}^{V\times V}$, while the map μ_* induces $G_\alpha^V \to G_{\delta_*(\beta\times\gamma)}^{V\times V}$. □

For each group G there are three canonical actions of the group on itself: the right translation $r(g, g_1) = gg_1$, the left translation $l(g, g_1) = g_1^{-1}g$, and the inner conjugation $\mathrm{ad}(g, g_1) = g_1^{-1}gg_1$. The semigroups G_r^G, G_l^G, and G_{ad}^G corresponding to these actions will be the focus of our attention in the sequel.

An important particular case of Corollary 1.1 is the following

COROLLARY 1.2. *The group multiplication $G \times G \to G$ defines the diagonal homomorphisms*

$$\Delta_r \colon G_r^G \to (G \times G)_{r \times \mathrm{ad}}^{G \times G}, \quad \Delta_l \colon G_l^G \to (G \times G)_{\mathrm{ad} \times l}^{G \times G}, \quad \Delta_{\mathrm{ad}} \colon G_{\mathrm{ad}}^G \to (G \times G)_{\mathrm{ad} \times \mathrm{ad}}^{G \times G}$$

that satisfy the coassociativity conditions

$$(\Delta_r \times 1)\Delta_r = (1 \times \Delta_{\mathrm{ad}})\Delta_r;$$
$$(\Delta_{\mathrm{ad}} \times 1)\Delta_l = (1 \times \Delta_l)\Delta_l;$$
$$(\Delta_{\mathrm{ad}} \times 1)\Delta_{\mathrm{ad}} = (1 \times \Delta_{\mathrm{ad}})\Delta_{\mathrm{ad}}.$$

COROLLARY 1.3. *For any space V with a right action α of the group G, the diagonal homomorphism $\Delta \colon G \to G \times G$ can be extended to the homomorphism*

$$\Delta^V \colon G_\alpha^V \to (G \times G)_{\alpha \times \mathrm{ad}}^{V \times G}.$$

PROOF. To the map $f \colon V \to G$ we assign the map

$$\Delta(f) \colon V \times G \xrightarrow{\alpha} V \xrightarrow{f} G \xrightarrow{\Delta} G \times G.$$

On $V \times G$ the homomorphism Δ induces (from the action $\alpha \times \mathrm{ad}$ on $G \times G$) an action of the group G. The map α is equivariant with respect to the action of G. Therefore we can apply Lemma 1.2. □

Another important example is given by the canonical action of the Lie group G on its Lie algebra $\mathfrak{g}$. In this case we obtain the semigroup $G_{\mathrm{ad}}^{\mathfrak{g}}$ with the homomorphism $E^* \colon G_{\mathrm{ad}}^G \to G_{\mathrm{ad}}^{\mathfrak{g}}$ induced by the exponential map $E \colon \mathfrak{g} \to G$.

Consider the semigroup of maps V^V of the space V into itself, in which multiplication is the composition of maps. The action α defines a map $j_\alpha \colon G^V \to V^V$ that takes the map $\varphi \colon V \to G$ to the map $j_\alpha(\varphi) \colon V \to V \times G \to V$ given by $j_\alpha(\varphi)(v) = v\varphi(v)$. It is clear that the map j_α is interesting only in the case of nontrivial actions α. A direct verification establishes the following

LEMMA 1.3. *The map $j_\alpha \colon G^V \to V^V$ is a homomorphism.*

To familiarize ourselves with the group G_α^V, let us consider a few examples.

EXAMPLE 1.1. Suppose V is the finite set $\{1, \ldots, n\}$ and the action of the group G on V is given by the representation $\rho \colon G \to S_n$ of the group G into the symmetric group S_n. Then $G^V \cong G^n$, and the multiplication is given by the formula

$$(6) \qquad (g_1, \ldots, g_n) * (f_1, \ldots, f_n) = (g_1 f_{i_1}, \ldots, g_n f_{i_n}),$$

where $i_k = \rho(g_k)(k)$, $k = 1, \ldots, n$. In particular, when $V = \{\pm\}$ and $G = O(n)$ is the group of orthogonal matrices, while the action of G on V is given by multiplication by the determinant of the matrix, we obtain an unusual semigroup structure on $O(n) \times O(n)$ with the following multiplication defined in accordance to (6). Namely, if $\varphi, \psi \colon \{\pm 1\} \to O(n)$ and $\varphi(k) = A_k$, $\psi(k) = B_k$, $A_k, B_k \in O(n)$, $k = \pm 1$, then $(\varphi * \psi)(k) = A_k B_{1_k}$, where $1_k = k \cdot |A_k|$ and where $|A|$ is the determinant of the matrix A.

EXAMPLE 1.2. Suppose $V = G = \mathbb{R}^n$ and the Euclidean space $\mathbb{R}^n$ as the group of n-dimensional vectors acts on itself by ordinary translations. Consider the algebra of linear operators $L(\mathbb{R}^n)$ on $\mathbb{R}^n$ and let us regard $\mathbb{R}^n \times L(\mathbb{R}^n)$ as the set of maps of $\mathbb{R}^n$ into itself of the form $\varphi(v) = w + Av$, where $v, w \in \mathbb{R}^n$, $A \in L(\mathbb{R}^n)$. According to (1), we have

$$
\begin{aligned}
\varphi_2 * \varphi_1(v) &= \varphi_2(v) + \varphi_1(v + \varphi_2(v)) \\
&= (w_1 + w_2 + A_1 w_2) + (A_1 + A_2 + A_1 A_2)v.
\end{aligned}
\tag{7}
$$

Thus $\mathbb{R}^n \times L(\mathbb{R}^n) \cong \mathbb{R}^{n+n^2}$ is closed with respect to the multiplication (1) and one can see directly that (7) defines the structure of a local group in $\mathbb{R}^{n+n^2}$. In the particular case $n = 1$, let us consider the family of actions $\{\alpha_h: \alpha_h(x)(y) = y + hx\}$ of translations along the line $\mathbb{R}^1$. Then, according to (7), for the plane $\mathbb{R}^2$ we obtain:

$$(x_2, y_2) * (x_1, y_1) = (x_1 + x_2 + hy_1 x_2, y_1 + y_2 + hy_1 y_2),$$

i.e., the deformation of the ordinary addition ($h = 0$) into the noncommutative one ($h > 0$).

The following construction is related to a Lie group G and its Lie algebra $\mathfrak{g}$. The exponential map $E: \mathfrak{g} \to G$ defines the inclusion

$$\tau: G \times \mathfrak{g}^V \to G^V, \quad \tau(g, \xi)(v) = g \exp \xi(v), \tag{8}$$

where $v \in V$ and $\xi: V \to \mathfrak{g}$. Using the Campbell–Hausdorff formula, we see that the semigroup structure of G_α^V induces (by means of τ) a semigroup structure in $G \times \mathfrak{g}^V$.

LEMMA 1.4. *For a smooth action α of Lie group G on the manifold V and for the space of smooth maps G^V, the inclusion τ (see (8)) defines a group structure in $G \times \mathfrak{g}^V$ with the multiplication $(g_2, \xi_2) \circ (g_1, \xi_1) = (g_2 g_1, \xi)$, where*

$$\xi(v) = g_1^{-1} \xi_2(v) g_1 + \xi_1(v g_2). \tag{9}$$

PROOF. Consider smooth maps $\varphi_1, \varphi_2 \in G^V$ such that $\varphi_k(v) = g_k \exp h \xi_k(v)$. For their product in G_α^V, according to (1), we have the formula

$$
\begin{aligned}
\varphi_2 * \varphi_1(v) &= (g_2 \exp h \xi_2(v_1)) g_1 \exp h \xi_1(v \cdot \varphi_2(v)) \\
&\approx g_2 g_1 \exp h [g_1^{-1} \xi_2(v) g_1 + \xi_1(v g_2) + o(h)].
\end{aligned}
$$

This formula implies the statement of the lemma. $\square$

Suppose V is a linear space and the action α is given by the representation $\rho: G \to L(V)$ of the group G in the algebra of linear operators on V. Denote by $L(V, \mathfrak{g})$ the space of linear operators from V to $\mathfrak{g}$. Lemma 1.4 directly implies the following

COROLLARY 1.3. *The inclusion $G \times L(V, \mathfrak{g}) \to G \times \mathfrak{g}^V$ induces on $G \times L(V, \mathfrak{g})$ a group structure with the multiplication*

$$(g_2, A_2) \circ (g_1, A_1)(v) = (g_2 g_1, g_1^{-1}(A_2 v) g_1 + A_1(v \cdot \rho(g_2))) \quad A_k \in L(V, \mathfrak{g}).$$

To conclude this section, let us present one more general statement about the semigroups G_α^V.

LEMMA 1.5. *Any linear representation $\rho\colon G \to GL(W)$ of the group G in the linear space W (we can then regard W as a left G-module) can be extended to a linear representation $\rho_\alpha\colon G_\alpha^V \to GL(W^V)$ of the semigroup G_α^V in the linear space W^V in the following way. Suppose $\varphi\colon V \to G$ and $f\colon V \to W$; then*

(10) $$\rho_\alpha(\varphi)(f)(v) = \rho_\alpha(\varphi(v))f(v\varphi(v)).$$

In the case of trivial action α, this lemma is the basis for constructing important representations of the group G^V (see [9]). For nontrivial actions α another particular case is also important, namely the case when W is a trivial G-module.

§2. Operator doubles of Hopf algebras

Consider a Hopf algebra X over k with unit $1 \in k$, antipode $\gamma\colon X \to X$, multiplication $m\colon X \otimes X \to X$, comultiplication $\Delta\colon X \to X \otimes X$, and augmentation $\varepsilon\colon X \to k$. Here k is a field, but most of the constructions below work when k is a commutative ring. All tensor products, unless otherwise specified, are over k.

DEFINITION 2.1 ([8]). A *Milnor module* M over the Hopf algebra X is an algebra with unit $1 \in k$ which is also a module over X satisfying

(11) $\quad x(uv) = \sum x'_n(u)x''_n(v), \qquad x \in X, \quad u,v \in M, \quad \Delta x = \sum x'_n \otimes x''_n.$

DEFINITION 2.2 ([8]). The *operator double* MX (*O-double*) of a Milnor module M over the Hopf algebra X is an algebra such that

(1) the algebras M and X are embedded in MX as subalgebras and the k-linear map $M \otimes X \to MX$, $u \otimes x \to ux$ induced by these embeddings is an additive isomorphism;

(2) the following commutation rule holds in MX:

(12) $\quad xu = \sum x'_n(u)x''_n, \qquad x \in X, \quad u \in M, \quad \Delta x = \sum x'_n \otimes x''_n.$

When the action of X on M is trivial, we obviously have $MX \cong M \otimes X$.

In the cases when it will be necessary to stress the dependence of the O-double MX on the action of X on M, we shall use the notation $M_a X$.

Consider the Hopf algebra $X^* = \mathrm{Hom}_k(X,k)$ dual to the Hopf algebra X. Denote by r, l, and ad the actions of X on X^* defined by the formulas

$$\langle r(x)\sigma, y\rangle = \langle \sigma, yx\rangle;$$
$$\langle l(x)\sigma, y\rangle = \langle \sigma, \gamma(x)y\rangle;$$
$$\langle \mathrm{ad}(x)\sigma, y\rangle = \langle \sigma, \sum \gamma(x'_n)yx''_n\rangle,$$

where $x, y \in X$, $\sigma \in X^*$ and $\langle \cdot, \cdot \rangle$ is the canonical pairing $X^* \otimes X \to k$.

LEMMA 2.1. *With respect to the actions r, l, and ad, the algebra X^* is a Milnor module over X and therefore the O-doubles X_r^*X, X_l^*X, and X_{ad}^*X are defined.*

Denote by j_x the ring homomorphism $X \to L(M)$ induced by the action $a\colon M \otimes X \to M$ of the Hopf algebra X on M. The algebra $M_a X$ described in Definition 2.2 was called the operator double by S. P. Novikov, because Definition 2.1 immediately implies the following statement.

LEMMA 2.2. *The map $j_a\colon M_a X \to L(M)$, $j_a(ux) = j_M(u)j_X(x)$ is a homomorphism of rings.*

DEFINITION 2.3 ([8]). The *semigroup of multiplicative elements* $m(M_a X)$ of the operator double $M_a X$ is the multiplicative semigroup of elements $y \in M_a X$ such that the maps $j_a(y)\colon M \to M$ are ring homomorphisms.

The paper [8] gives some motivation, using the example of cohomology operations in the theory of complex cobordisms, of the importance of describing such semigroups. The results presented below solve the corresponding problem in the case of O-doubles defined by an action α of the group G on the space V.

Suppose X^* is the Hopf algebra of functions on G with values in k, while X is the dual Hopf algebra. As usual in the construction of doubles ([4], [8], [10]), we assume that a basis $\{e_k\}$ has been chosen in X^* and $\{e^k\}$ is the dual basis in X. Then we can write the identity operator from X^* to X^* in the form $\sum e_k e^k \in X^* X$ and expand any function $f \in X^*$ as $f(g) = \sum e_k(g) e^k(f(g))$.

In the role of M, let us take the ring of α-regular functions on the space V, i.e., functions such that, given any action α and any function $p(v) \in M$, the function $p(vg)$ belongs to M for a fixed g and belongs to X^* for a fixed v. Then the following action is defined $a\colon M \otimes X \to M$, $a(p(v), x) = \langle x, p(vg) \rangle \in M$.

LEMMA 2.3. *The given action a defines a Milnor module structure on M over the Hopf algebra X and therefore the O-double $M_a X$ is defined.*

PROOF. Suppose $p_1(v), p_2(v) \in M$. For a fixed v we have the product of two functions $p_1(vg)p_2(vg)$ on the group G and therefore

$$\langle x, p_1(vg)p_2(vg) \rangle = \langle \Delta x, p_1(vg) \otimes p_2(vg) \rangle,$$

since the comultiplication Δ on X and the multiplication in X^* are adjoint with respect to the scalar product. $\square$

A direct consequence of the previous results is the following

THEOREM 2.1. *The map $j_\alpha^0\colon G_\alpha^V \to M_a X$, $j_\alpha(\varphi) = \sum e_k(\varphi) e^k$ induces the homomorphism $G_\alpha^V \to m(M_a X)$ defined as follows: for each $\varphi \in G_\alpha^V$ the element $j_a(j_\alpha^0(\varphi))$ is the ring homomorphism of the ring of functions M induced by the map $j_\alpha(\varphi)\colon V \to V$.*

Let us return again to the general case of O-doubles. A straightforward verification establishes the following

LEMMA 2.4. 1) *Let $M_a X$ be some O-double. Then the homomorphism of Hopf algebras $\lambda\colon Y \to X$ induces a Milnor action Y on M and determines a ring homomorphism of the O-doubles $\lambda_*\colon M_{\lambda^* a} Y \to M_a X$.*

2) *Let M and N be Milnor modules over the Hopf algebra X with actions a and b on X respectively. Then any ring homomorphism $h\colon M \to N$ equivariant with respect to these actions can be extended to a ring homomorphism of the O-doubles $h_*\colon M_a X \to N_b X$.*

The restriction of the homomorphism j_α^0 to the subgroup $G \subset G_\alpha^V$ coincides with the canonical embedding $j\colon G \to X$. Lemmas 1.2 and 2.4 show that well-known universality properties of the homomorphism j can be carried over under certain conditions to the homomorphism j_α^0.

§3. Quantum doubles of Hopf algebras

Suppose that as above X and X^* are dual Hopf algebras over k. Fix an additive basis $\{e^k\}$ in X and its dual basis $\{e_k\}$ in X^*. Using these bases, let us express the operations in the Hopf algebras:

$$m(e^i \otimes e^j) = e^i e^j = \sum m_k^{ij} e^k \, ; \quad m^*(e_i \otimes e_j) = e_i e_j = \sum \lambda_{ij}^k e_k \, ;$$

$$\Delta e^k = \sum \lambda_{ij}^k e^i \otimes e^j \, ; \quad \Delta^* e_k = \sum m_k^{ij} e_i \otimes e_j \, ;$$

$$\gamma(e^k) = \sum \gamma_l^k e^l \, ; \quad \gamma^*(e_l) = \sum \gamma_l^k e_k \, .$$

Denote by X^0 the Hopf algebra X with the opposite comultiplication, i.e., $\Delta^0 e^k = \sum \lambda_{ij}^k e^j \otimes e^i$.

DEFINITION 3.1 ([4],[10]). The *quantum double* of the Hopf algebra X is the Hopf algebra $\mathcal{D}(X)$ such that

(1) the algebras X^* and X^0 are embedded in $\mathcal{D}(X)$ as Hopf subalgebras and the k-linear map $X^* \otimes X^0 \to \mathcal{D}(X)$, $u \otimes x \to ux$, induced by these embeddings is a coalgebra isomorphism (here $X^* \otimes X^0$ is regarded as the tensor product of coalgebras over k);

(2) the commutation rule of elements $u \in X^*$ and $x \in X^0$ in $\mathcal{D}(X)$ is described by the equation

(13) $$\mathfrak{R}\Delta(a) = (\sigma \circ \Delta)(a)\mathfrak{R},$$

where Δ is the comultiplication in $\mathcal{D}$, the operator σ is the transposition in $\mathcal{D}(X) \otimes \mathcal{D}(X)$, i.e., $\sigma(a \otimes b) = b \otimes a$, while $\mathfrak{R}$ is the image of the canonical element $\sum e_k \otimes e^k \in X^* \otimes X^0$ under the embedding $X^* \otimes X^0 \subset \mathcal{D}(X) \otimes \mathcal{D}(X)$.

In order to avoid cumbersome formulas, we shall identify the elements from X^* and X^0 with their images in $\mathcal{D}(X)$.

Now in $\mathcal{D}(X) \otimes \mathcal{D}(X) \otimes \mathcal{D}(X)$ let us put

$$\mathfrak{R}_{13} = \sum e_s \otimes 1 \otimes e^s, \quad \mathfrak{R}_{12} = \sum e_s \otimes e^s \otimes 1, \quad \mathfrak{R}_{23} = \sum 1 \otimes e_s \otimes e^s.$$

An important consequence of the construction of the quantum double $\mathcal{D}(X)$ is the following

THEOREM 3.1 ([4]). *The element $\mathfrak{R}$ is invertible in the algebra $\mathcal{D}(X) \otimes \mathcal{D}(X)$ and gives a solution to the quantum Yang–Baxter equation (QYBE):*

(14) $$\mathfrak{R}_{12}\mathfrak{R}_{13}\mathfrak{R}_{23} = \mathfrak{R}_{23}\mathfrak{R}_{13}\mathfrak{R}_{12}.$$

PROOF. The theorem follows from the next relations (see [4], [10]):

$$(\gamma^* \otimes \mathrm{id})(\mathfrak{R}) \cdot \mathfrak{R} = 1 \, ; \quad (\mathfrak{R} \otimes 1)(\Delta \otimes \mathrm{id})(\mathfrak{R}) = (\sigma\Delta \otimes 1)(\mathfrak{R} \otimes 1),$$

$$\mathfrak{R} \otimes 1 = \mathfrak{R}_{12} \, , \quad (\Delta \otimes \mathrm{id})\mathfrak{R} = \mathfrak{R}_{13}\mathfrak{R}_{23} \, ; \quad (\sigma\Delta \otimes 1)(\mathfrak{R}) = \mathfrak{R}_{23}\mathfrak{R}_{13}.$$

A direct verification shows that in order to write out the commutation rule in the canonical basis $\{e_i e^j\} \in \mathcal{D}(X)$, it suffices to use only the equation $\mathfrak{R}\Delta e^k = (\sigma\Delta)(e^k)\mathfrak{R}$, i.e., the relation

$$\sum \lambda_{ij}^k m_t^{li} e_l e^j = \sum \lambda_{nq}^k m_t^{qp} e^n e_p.$$

THEOREM 3.2. ([8]) *Suppose X is a cocommutative Hopf algebra. Then the quantum double $\mathcal{D}(X)$ coincides with the O-double $X^*_{\mathrm{ad}}X$.*

PROOF. By assumption $X^0 = X$ and X^* is a commutative algebra. Therefore the operators $\mathfrak{R}_{12}$ and $\mathfrak{R}_{13}$ commute. Hence, by Theorem 3.1, we have the relation

$$\mathfrak{R}_{12}\mathfrak{R}_{23} = \mathfrak{R}_{13}^{-1}\mathfrak{R}_{23}\mathfrak{R}_{13}\mathfrak{R}_{12}. \tag{15}$$

Let us express the left- and right-hand side of equation (15) in the canonical basis of the algebra $\mathcal{D}(X) \otimes \mathcal{D}(X) \otimes \mathcal{D}(X)$:

$$\mathfrak{R}_{12}\mathfrak{R}_{23} = \sum e_i \otimes e^i e_j \otimes e^j,$$
$$\mathfrak{R}_{13}^{-1}\mathfrak{R}_{23}\mathfrak{R}_{13}\mathfrak{R}_{12} = \sum \gamma^*(e_l) e_s e_q \otimes e_n e^q \otimes e^l e^n e^s$$
$$= \sum e_k e_s e_q \otimes e_n e^q \otimes \gamma(e^k) e^n e^s$$
$$= \sum e_i \otimes a^{in}_{jq} e_n e^q \otimes e^j$$

where

$$a^{in}_{jq} = \left\langle \sum \lambda^i_{ksq} \gamma(e^k) e^n e^s, \; e^j \right\rangle.$$

On the other hand, the commutation rule in $X^*_{\mathrm{ad}}X$ is of the form (see [8])

$$e^i e_j = \sum \lambda^i_{sqk} p_{1s} p_{2k}(e_j) e^q,$$
$$p_{1s} p_{2k}(e_j) = \sum b^n_{skj} e_n,$$
$$b^n_{skj} = \langle e^n, \; p_{1s} p_{2k}(e_j) \rangle = \langle \gamma(e^k) e^n e^s, \; e_j \rangle.$$

Since $\lambda^i_{ksq} = \lambda^i_{sqk}$, the two commutation rules coincide. □

§4. The Hopf comodule structure of O-doubles

We begin with a natural dualization of the Milnor module over a Hopf algebra.

DEFINITION 4.1. By a *Hopf comodule* (or *Milnor comodule*) over a Hopf algebra X we mean a k-algebra M with unit provided M is a comodule over X with coaction $b\colon M \to X \otimes M$ such that $b(uv) = b(u)b(v)$, i.e., such that b is a homomorphism of rings.

Suppose M is a Milnor module over the Hopf algebra X and $a\colon X \otimes M \to M$, $a(x, u) = x(u)$ is the homomorphism that determines this structure. Then the following homomorphism is defined

$$b\colon M \to X^* \otimes M, \quad b(u) = \sum e_k \otimes e^k(u). \tag{16}$$

Clearly, (16) is the composition of the homomorphism $a^*\colon M \to \mathrm{Hom}(X, M)$ with the canonical isomorphism $\mathrm{Hom}(X, M) \cong X^* \otimes M$.

A direct verification establishes the following

LEMMA 4.1. *The Milnor module over the Hopf algebra X is a Milnor comodule with the coaction* (16) *over the dual Hopf algebra X^**.

COROLLARY 4.1. *For any Milnor module M over a cocommutative Hopf algebra X with left action a on M, the diagonal homomorphism $\Delta \colon X \to X \otimes X$ extends to a ring homomorphism $\Delta \colon M_a X \to X^*_{\mathrm{ad}} X \otimes M_a X$.*

PROOF. The algebra $X^* \otimes M$ is a Milnor module over the Hopf algebra $X \otimes X$ with action $\mathrm{ad} \otimes a$. For a cocommutative algebra X, the diagonal homomorphism $\Delta \colon X \to X \otimes X$ is a homomorphism of Hopf algebras and therefore, according to Lemma 2.4, the following ring homomorphism is defined:

$$(X^* \otimes M)_{\Delta^* a} X \to (X^* \otimes M)_{(\mathrm{ad} \otimes a)}(X \otimes X) \cong X^*_{\mathrm{ad}} X \otimes M_a X.$$

On the other hand, the homomorphism $b \colon M \to X^* \otimes M$, being dual to the action a, is a homomorphism of Milnor X-modules. Therefore the homomorphism $M_a X \to (X^* M)_{\Delta^*} X$ is defined. $\square$

Now using Theorem 3.2 and Corollary 4.1, we obtain

THEOREM 4.1. *The operator double $M_a X$ of a cocommutative Hopf algebra X is a Hopf comodule over the quantum double $\mathcal{D}(X)$.*

§5. Some information from the theory of complex cobordisms

A systematic exposition of the foundations of complex cobordism theory can be found, for example, in [11].

Let $U(n)$ be the unitary group of linear transformations of complex linear space $\mathbb{C}^n$. Denote by i_n the standard embedding of unitary groups $i_n \colon U(n) \subset U(n+1)$.

Consider the universal n-dimensional complex vector bundle $\xi \to BU(n) = \lim_{k \to \infty} G_{n, n+k}$, where $G_{n, n+k}$ is the Grassman manifold of complex n-dimensional linear subspaces of $\mathbb{C}^{n+k}$. The embedding i_n induces the map $Bi_n \colon BU(n) \to BU(n+1)$ such that $(Bi_n)^* \xi_{n+1} = \xi_n \oplus 1$, where 1 denotes the complex trivial linear bundle.

Denote by $M_n = MU(n)$ the Thom space $T(\xi_n) = D(\xi_n)/\partial D(\xi_n)$, where $D(\xi_n)$ and $\partial D(\xi_n)$ are the fiber spaces associated to ξ_n with fiber the unit disk D^{2n} and the sphere S^{2n-1} in $\mathbb{C}^n$, respectively. The point from M_n corresponding to $\partial D(\xi_n)$ is chosen for the base point. An important role is played by the following multiplicativity property of Thom spaces.

Suppose $\xi_k \to V_k$, $k = 1, 2$, are complex vector bundles. Then for the bundle $\xi_1 \times \xi_2 \to V_1 \times V_2$ we have the homeomorphism $T(\xi_1 \times \xi_2) \cong T(\xi_1) \wedge T(\xi_2)$. Here and below $\wedge$ denotes the smash product of pointed spaces, i.e., $X \wedge Y = X \times Y / X \vee Y$, where $X \vee Y$ is the wedge of pointed spaces. In particular, we have $T(\xi_n \otimes 1) \cong S^2 M_n$, where the sphere S^2 is regarded as the Thom space of the trivial complex one-dimensional bundle over the point. Therefore the map Bi_n induces the map $Mi_n \colon S^2 M_n \to M_{n+1}$.

DEFINITION 5.1. *The Thom spectrum in the theory of complex cobordisms is the sequence of Thom spaces of universal bundles with the maps that join them:*

$$MU = \{M_n = MU(n), Mi_n\}.$$

The Thom spectrum MU for pairs of CW-complexes $V \subset W$ is used to define the cobordism groups

$$U^q(W, V) = \lim_{n \to \infty} [S^{2n-q}(W/V), MU(n)]$$

and the bordism groups

$$U_q(W, V) = \lim_{n \to \infty} [S^{2n+q}, (W/V) \wedge MU(n)].$$

Here $[\cdot, \cdot]$ denotes the set of homotopy classes of base point preserving maps.

The point in W/V corresponding to V is chosen to be the base point. We set $U^q(W) = U^q(W, \emptyset)$ and $U_q(W) = U_q(W, \emptyset)$, where $\emptyset$ is the empty set. Recall that by definition $W/\emptyset = W_+$ is the disjoint union of W with the point $*$ regarded as the base point.

One can give a purely geometric description of the groups $U^q(\cdot)$ and $U_q(\cdot)$ by using the notion of bordism of a map of smooth manifolds in the stable tangent (or normal) bundle, provided a complex bundle structure is fixed.

Let us introduce the graded groups

$$U^*(W) = \sum U^q(W), \quad U_*(W) = \sum U_q(W).$$

The standard embedding $U(n) \times U(m) \subset U(n+m)$ induces a map of Thom spaces $\mu_{n,m} \colon M_n \wedge M_m \cong M(\xi_n \times \xi_m) \to M_{n+m}$ and thereby a map of Thom spectra

$$\mu \colon MU \wedge MU \to MU.$$

This map allows us to introduce the structure of a graded commutative ring in $U^*(W)$.

According to the Milnor–Novikov theorem, we have the isomorphism

$$U^*(\text{point}) = \Omega_U = \mathbb{Z}[a_1, \ldots, a_n, \ldots], \quad \deg a_n = -2n.$$

DEFINITION 5.2. (1) A *Thom U-class* of a complex bundle $\zeta \to V$ is the class of cobordisms $u(\zeta) \in U^{2n}(T(\zeta))$ whose representative is the map of Thom spaces $T\zeta \to MU(n)$ corresponding to the classifying map $f \colon V \to BU(n)$, $f^*\xi_n = \zeta$.

(2) The *Euler class* of a complex n-dimensional bundle $\zeta \to V$ is the class of bordisms $\chi(\zeta) = \mathcal{S}_0^* u(\zeta) \in U^{2n}(V)$, where $\mathcal{S}_0 \colon V \to T\zeta$ is the embedding corresponding to the zero section of the bundle ζ.

Denote by $u(n) \in U^{2n}(MU(n))$ the Thom class of the universal bundle $\xi_n \to BU(n)$. It is clear that the identity map $MU(n) \to MU(n)$ is a representative of the class $u(n)$.

According to Conner and Floyd [11], there exists a complete characteristic Chern class

$$C(\zeta) = 1 + C_1(\zeta) + \cdots = C_k(\zeta) + \ldots, \quad C_k(\zeta) \in U^{2k}(V)$$

of the complex vector bundle $\zeta \to V$ which is uniquely determined by the following properties:

1) $C_k(\zeta) = 0$, if $k > \dim_{\mathbb{C}} \zeta$;
2) $C_k(\zeta) = \chi(\zeta)$, if $k > \dim_{\mathbb{C}} \zeta$;
3) $C(\zeta_1 + \zeta_2) = C(\zeta_1)C(\zeta_2)$.

This result directly implies the following facts:

(1) $U^*(\prod \mathbb{C}P(\infty)) = \Omega_U[[u_1, \ldots, u_n]]$, where $u_m = C_1(\eta)$ is the first Chern class of the universal one-dimensional bundle $\xi \to \mathbb{C}P(\infty)$ over the infinite-dimensional projective space, regarded here as the mth factor of the product space $\prod \mathbb{C}P(\infty)$.

(2) $U^*(BU(n)) = \Omega_U[[C_1, \ldots, C_n]]$, where $C_k = C_k(\xi_n)$.

(3) The map $h_n \colon \prod \mathbb{C}P(\infty) \to BU(n)$ determined by the embedding of the maximal torus $T_n \subset U(n)$ induces a monomorphism

$$U^*(BU(n)) \to U^*\left(\prod \mathbb{C}P(\infty)\right)$$

under which C_k is taken to the kth elementary symmetric polynomial in the variables $u_1, \ldots, u_n$.

(4) The embedding $\mathcal{S}_0 \colon BU(n) \to MU(n)$ defined by the zero section of the universal bundle ξ_n induces the monomorphism $h_n^* \mathcal{S}_0^* \colon U^*(MU(n)) \to U^*(\prod \mathbb{C}P(\infty))$ whose image coincides with the ring of all symmetric formal series in $u_1, \ldots, u_n$ divisible by the monomial $u_1 \ldots u_n = h_n^* \mathcal{S}_0^* u(n)$.

Using the construction of complex cobordism theory in terms of the Thom spectrum, we directly obtain the existence of a multiplicative transformation

$$\varepsilon \colon U^*(W) \to H^*(W; Z),$$

which is entirely determined by the property that for the Thom classes $u(\zeta) \in U^{2n}(T\zeta)$, the classes $\varepsilon(u(\zeta)) \in H^{2n}(T\zeta)$ coincide with the ordinary Thom classes in cohomology.

Under the transformation ε, the Chern–Conner–Floyd class $C_k(\zeta)$ of the complex vector bundle $\zeta \to W$ is taken to the ordinary Chern cohomology classes $C_k^H(\zeta)$, i.e., $\varepsilon C_k(\zeta) = C_k^H(\zeta)$.

Denote by $\mathbb{Z}_+^n$ the semigroup of all n-dimensional vectors with nonnegative integer coordinates. Put $\mathbb{Z}_+^\infty = \varinjlim \mathbb{Z}_+^n$. Suppose $(q) \in \mathbb{Z}_+^\infty$ is an infinite-dimensional vector all of whose coordinates except the qth is zero, while the qth one is equal to 1. Then the semigroup $\mathbb{Z}_+^\infty$ coincides with the semigroup of all finite linear combinations $\sum n_q(q)$, where n_q are nonnegative integers. For each element $w = \sum n_q(q)$ let us put $\|\omega\| = \sum n_q \cdot q$ and $|\omega| = \sum n_q$. For

$$\omega = \sum_{q=1}^{l} n_q(q)$$

denote by $\sigma_{(\omega)}$ the smallest symmetric polynomial in $u_1, \ldots, u_n$, $n \geq |\omega|$, containing the monomial

$$(u_1 \ldots u_{n_1})(u_{n_1+1}^2 \ldots u_{n_1+n_2}^2) \ldots (u_{|\omega|-n_l+1}^l \ldots u_{|\omega|+n_l}^l)$$

and by $C_\omega(\xi_n)$ the unique element of the ring $U^{2\|\omega\|}(BU(n))$, $n > |\omega|$, that satisfies $h_n^* C_\omega(\xi_n) = \sigma_\omega$. For example, the polynomial $\sigma_{(q)}$ is the Newton polynomial $\sum u_i^q$, the polynomial $\sigma_{(k(1))}$ is the elementary symmetric polynomial $\sum u_1 \ldots u_k$, while the class $C_{k(1)}(\xi_n)$ is C_k.

For each $\omega \in \mathbb{Z}_+^\infty$, let us put

$$s_{\omega,n} = s_\omega u(n) = u(n) C_\omega(\xi_n) \in U^{2(n+\|\omega\|)}(MU(n)).$$

It follows from the functoriality of Thom classes and of characteristic classes that the sequence of elements $\{s_{\omega,n} | n = 1, 2, \ldots\}$ defines a stable cohomology operation s_ω of degree $2\|\omega\|$ in the theory $U^*(\cdot)$ which acts as follows.

Suppose $a \in U^q(W)$. Choose a representative $f: S^{2n-q} W_+ \to MU(n)$ of the class a and put

$$s_\omega a = f^* s_{\omega,n} \in U^{2(n+\|\omega\|)}(S^{2n-q} W_+) \cong U^{q+2\|\omega\|}(W).$$

Denote by $S = \sum_{n \geq 0} S^n$ the free topological commutative group with topological basis s_ω, $\omega \in \mathbb{Z}_+^\infty$. Let

$$A_U = \sum_{n > -\infty}^\infty A_U^n$$

be the group of all stable cohomology operations in the theory of complex cobordisms. It is clear that we have the isomorphism

$$A_U = U^*(MU), \qquad U^k(MU) = \varprojlim_n U^{2n+k}(MU(n)),$$

the inverse limit being induced by the maps Mi_n. With respect to composition, A_U is a graded ring. The operator of multiplication by a scalar $\sigma \in \Omega_U$ is a stable cohomology operation. Therefore we have an embedding of rings $\Omega_U \subset A_U$ and A_U is a module (that we regard as a left module) over Ω_U.

Using the description of the rings $U^*(MU(n))$ given above, we see that $A_U \cong \Omega_U S$, i.e., elements of A_U may be uniquely expressed in the form $\sum_\omega a_\omega s_\omega$. The following results are due to Novikov [7] and Landweber [6]:

(1) $S = \sum S^n$ is a subalgebra in A_U.

(2) The algebra S is a cocommutative Hopf algebra with diagonal

$$\Delta: S \to S \otimes S, \quad \Delta s_\omega = \sum_{\omega=\omega_1+\omega_2} s_{\omega_1} \otimes s_{\omega_2}.$$

Here the indices ω_1 and ω_2 are added as elements of the semigroup $\mathbb{Z}_+^\infty$.

(3) The multiplication in S can be uniquely determined from the following properties:

(a) for any $n \geq 1$, the ring $\mathbb{Z}[[u_1, \ldots, u_n]]$ is a module over the Hopf algebra S such that $s_{(q)} u_m = u_m^{q+1}$ and $s_\omega u_m = 0, \omega \neq (q)$ for all $m = 1, \ldots, n$ and

$$s_\omega(ab) = \sum_{\omega=\omega_1+\omega_2} s_{\omega_1}(a) s_{\omega_2}(b) \quad \text{for all } a, b \in \mathbb{Z}[[u_1, \ldots, u_n]];$$

(b) The element $s \in S^n$ vanishes if and only if we have $s(a) = 0$ for all $a \in \mathbb{Z}[[u_1, \ldots, u_n]]$.

DEFINITION 5.3. The algebra S is called a *Landweber–Novikov algebra*.

It follows directly from the description of the multiplication in the $U^*(\cdot)$ theory and the description of the algebra S that for any CW-complex W the ring $U^*(W)$ is a Milnor module over the Hopf algebra S (see Definition 2.1). The complete description of all the operations in A_U is concluded by the following result due to Novikov [7].

The commutation rule in the algebra A_U is

$$s_\omega \sigma = \sum_{\omega=\omega_1+\omega_2} s_{\omega_1}(\sigma) s_{\omega_2}.$$

The operation $\varphi \in A_U$ is called *multiplicative* if it determines a ring homomorphism $\varphi: U^*(V) \to U^*(V)$ for any space V. As any other operation, φ is determined by the sequence of its values $\{\varphi u(n), \ n = 1, \dots\}$ on the universal elements $u(n) \in U^n(MU(n))$, but in view of the monomorphism

$$h_n^* S_0^* : U^*(MU(n)) \to U^*\left(\prod \mathbb{C}P(\infty)\right), \quad h_n^* S_0^* u(n) = u_1 \dots u_n$$

each multiplicative operation is uniquely determined by its value on the class $u = c_1(\xi_1) \in U^2(\mathbb{C}P(\infty))$, i.e., by the series

$$(17) \qquad \varphi u = \varphi(u) = u + \sum_{k \geq 0} \varphi_k u^{k+1}, \quad \varphi_k \in \Omega_U.$$

Note that the condition $\varphi(u) = u + (u^2)$ follows from the stability of the operation.

Denote by mA_U the set of all multiplicative operations. As we noted above, mA_U may be identified with the set of series of the form (17) from the ring $\Omega_U[[u]]$. It is easy to show that the multiplicative operation φ belongs to the Landweber–Novikov algebra $S \subset S^U$ if and only if all the coefficients of the series $\varphi(u)$ belong to the ring of integers $\mathbb{Z} = \Omega_U^0 \subset \Omega_U$.

Denote the set of such operations by mS. With respect to the composition of operations, mA_U is a semigroup, while mS is a group. The semigroup mA_U and the group mS play the central role in subsequent sections.

Consider the Hopf algebra $S^* = \text{Hom}(S, \mathbb{Z})$ dual to the Landweber–Novikov algebra over the ring of integers $\mathbb{Z}$. Fix an additive basis $\{s_\omega, \omega \in \mathbb{Z}_+^\infty\}$ in S and denote by s^ω the dual basis in S^*. It follows immediately from the description of the diagonal Δ in S that the elements $s_{(q)}$ are primitive in S (i.e., $\Delta s_{(q)} = s_{(q)} \otimes 1 + 1 \otimes s_{(q)}$, $q = 1, 2, \dots$) and

$$S^* = \mathbb{Z}[s^{(1)}, \dots, s^{(q)}, \dots].$$

DEFINITION 5.4. A *Chern–Dold character* in the theory of complex cobordisms is a multiplictive transformation of cohomology theory

$$\text{ch}_U : U^*(W) \to H^*(W; \Omega \otimes \mathbb{Q})$$

uniquely determined by the fact that for $X = $ (point) the homomorphism ch_U coincides with the canonical embedding $\Omega_U \subset \Omega_U \otimes \mathbb{Q}$. Here $\mathbb{Q}$ is the field of rational numbers.

The theory of the operator ch_U was constructed by the author in [1]. It was called the Chern–Dold character because it may be obtained by the Dold construction, which generalizes the well-known Chern character from K-theory to the case of arbitrary cohomology theories.

The action of the algebra A_U on the ring $\Omega_U \otimes \mathbb{Q}$ allows us to regard the rings $H^*(W; \Omega_U \otimes \mathbb{Q})$ as A_U-modules. A most important property of the Chern–Dold character ch_U is that it determines an A_U-module homomorphism.

To the operation $s_\omega \in S$, let us assign the transformation
$$s_\omega^H : U^*(W) \to H^*(W; \mathbb{Z}), \quad s_\omega^H(a) = \varepsilon s_\omega(a),$$
where $\varepsilon : U^*(W) \to H^*(W; \mathbb{Z})$.

It follows at once from the definitions that for $W = \text{(point)}$ we obtain a homomorphism $s_\omega^H : \Omega_U \to \mathbb{Z}$ calculating the normal cohomological characteristic numbers, i.e., if the smooth manifold M^{2n} with complex normal bundle ν represents the class $\sigma \in \Omega_U^{-2n}$, then
$$s_\omega^H(\sigma) = \langle C_\omega^H(\nu), \langle M^{2n} \rangle \rangle,$$
where $C_\omega^H = \varepsilon C_\omega$ is the Chern class and $\langle M^{2n} \rangle$ is the fundamental cycle of the manifold M^{2n} in homology. Let us introduce the ring $\Omega_U(\mathbb{Z})$ by setting
$$\Omega_U(\mathbb{Z}) = \sum_{n \geq 0} \Omega_U^{-2n}(\mathbb{Z}),$$
$$\Omega_U^{-2n}(\mathbb{Z}) = \{ \sigma \in \Omega_U^{-2n}(\mathbb{Z}) \otimes \mathbb{Q}, \, s_\omega^H(\sigma) \in \mathbb{Z} \text{ for all } \omega \}.$$

It is clear that the ring $\Omega_U^{-2n}(\mathbb{Z})$ is closed with respect to the action of the algebra A^U. Now consider the transformation
$$\mathrm{ch}_U : U^*(\mathbb{C}P(\infty)) = \Omega_U[[u]] \to H^*(\mathbb{C}P(\infty); \Omega_U \otimes \mathbb{Q}) = \Omega_U \otimes \mathbb{Q}[[t]],$$
where $u \in U^2(\mathbb{C}P(\infty))$ and $t \in H^2(\mathbb{C}P(\infty); \mathbb{Z})$ are the first Chern classes in complex cobordisms and in the cohomology of the universal bundle $\xi_1 \to \mathbb{C}P(\infty)$. We have
$$\mathrm{ch}_U = t + \sum_{k \geq 1} \alpha_k t^{k+1}, \quad \alpha_k \in \Omega_U^{-2n} \otimes \mathbb{Q}.$$

Therefore, for the class $u(n) \in U^{2n}(MU(n))$, we obtain
$$\mathrm{ch}\, u(n) = t(n) + \sum \alpha^\omega s_\omega^H \in H^{2n}(MU(n), \Omega_U \otimes \mathbb{Q}).$$
Here $t(n) = \varepsilon u(n)$, $s_\omega^H = s_\omega^H(u(n))$ and $\alpha^\omega = \prod \alpha_q^{n_q}$ for $\omega = \sum n_q(q)$. Thus we arrive at the following result ([1]):

(1) $\Omega_U(\mathbb{Z}) = \mathbb{Z}[\alpha_1, \dots \alpha_q, \dots]$.

(2) The Chern–Dold character ch_U can be factored into a composition with the transformation
$$\mathrm{ch}_U^H : U^*(W) \to H^*(W; \Omega_U(\mathbb{Z}))$$
defined by the following explicit formula
$$\mathrm{ch}_U^H(a) = \sum \alpha^\omega s_\omega^H(a).$$

In particular, for $W = \text{(point)}$, the transformation ch_U^H defines the canonical representation of the cobordism class $\sigma \in \Omega_U^{-2n}$ in the form of a homogeneous polynomial in the graded variables $\alpha_1, \dots \alpha_q, \dots$.

COROLLARY 5.1. *The pairing*

$$\Omega_U(\mathbb{Z}) \otimes S \to \mathbb{Z}, \quad (\sigma, s_\omega) \mapsto s_\omega^H(\sigma)$$

induces an isomorphism of the ring $\Omega_U(\mathbb{Z})$ *with the Hopf algebra* S^*.

§6. The Landweber–Novikov algebra is the enveloping algebra of the Lie algebra of formal vector fields on the line

The result stated in the title of this section was obtained jointly by A. V. Shokurov and the author in 1978 ([3]). Here we give an exposition of this result convenient for the subsequent sections. Details about the Lie algebras of formal vector fields may be found in [5].

Consider the group $G = \text{Diff}_1(\mathbb{Z})$ of formal diffeomorphisms of the line of the form

$$x(t) = t + x_1 t^2 + \cdots + x_k t^{k+1} + \ldots, \quad x_k \in \mathbb{Z},$$

in which multiplication is, of course, the composition of diffeomorphisms. By the *coordinates* of a diffeomorphism we mean the set of its coefficients, i.e., we put $x = (x_1, \ldots x_k, \ldots)$. Thus if $x = (x_k)$ and $y = (y_k)$ are two diffeomorphisms, then the coordinates of the diffeomorphism $z = y \cdot x$ will be the coefficients of the series

$$(18) \qquad x(y(t)) = \left(t + \sum y_j t^{j+1}\right) + \sum x_k (t + y_j t^{j+1})^{k+1}.$$

For the ring of functions on the group G, let us take the polynomial ring P with integer coefficients in the coordinates $x \in G$. To each cobordism class $\sigma \in \Omega_U^{-2n}$ we assign a function on the group G. For any $x \in G$ consider the multiplicative operation $\varphi_x \in mS$ that takes the value $\varphi_x(u)$ on a canonical element $u \in U^2(\mathbb{C}P(\infty))$ equal to the series $x(u)$ and put $\sigma(x) = \varepsilon\varphi_x(\sigma)$. Thus we have defined the map $\lambda \colon \Omega_U \to P$.

THEOREM 6.1. (1) λ *is a ring homomorphism and can be extended to a ring homomorphism* $\lambda \colon \Omega_U(\mathbb{Z}) \to P$.

(2) *Suppose* $\{\alpha_k\}$ *are the coefficients of the Chern–Dold character. Then we have* $\lambda(\alpha_k)(x) = x_k$, *i.e., the coefficients* α_k *are taken to the coordinate polynomials on the group G and we have the isomorphism*

$$\Omega_U(\mathbb{Z}) = \mathbb{Z}[\alpha_1, \ldots, \alpha_k, \ldots] \cong P$$

under which to the action of the group of multiplicative operations on mS on the cobordism ring $\Omega_U(\mathbb{Z})$ *corresponds the action of the group G by right translations on its function ring P.*

PROOF. (1) Suppose $\sigma_1, \sigma_2 \in \Omega_U$. Then

$$\lambda(\sigma_1\sigma_2)(x) = \varepsilon\varphi_x(\sigma_1\sigma_2) = (\varepsilon\varphi_x(\sigma_1))(\varepsilon\varphi_x(\sigma_2)) = \lambda(\sigma_1)(x) \cdot \lambda(\sigma_2(x)).$$

Since $\varepsilon s_\omega(\sigma) \in \mathbb{Z}$ for any $\sigma \in \Omega_U(\mathbb{Z})$, it follows that λ can be extended to a homomorphism $\Omega_U(\mathbb{Z}) \to P$.

(2) We have $\mathrm{ch}_U u = t + \sum \alpha_k t^{k+1} = \alpha(t)$, therefore

$$\varphi_x \mathrm{ch}_U u = t + \sum \varphi_x(\alpha_k) t^{k+1}.$$

Using the fact that the Chern–Dold character $\mathrm{ch}_U u$ commutes with the action of the operations, we obtain

$$(19) \quad \varphi_x \mathrm{ch}_U u = \mathrm{ch}_U \varphi_x(u) = \mathrm{ch}_U \left(u + \sum x_n u^{n+1} \right) = \alpha(t) + \sum x_n \alpha(t)^{n+1},$$

i.e., the action of the operation φ_x on α_k is given by the formula

$$(20) \quad \varphi_x(\alpha_k) = \alpha_k + \sum x_n [\alpha(t)^{n+1}]_{k+1},$$

where $[\alpha(t)]_k$ is the coefficient at t^{k+1} in the series $a(t)$, and therefore

$$\lambda(\alpha_k) = \varepsilon \varphi_x(\alpha_k) = x_k.$$

Thus we have proved that λ determines an isomorphism.

Now consider the representation of the group G on its ring of functions induced by right translations. According to (18), for any $y \in G$ we have

$$x(\lambda(\alpha_k))(y) = \lambda(\alpha_k)(yx) = y_k + \sum x_n [y(t)^{n+1}]_{k+1}$$
$$= \lambda \left(\alpha_k + \sum x_n [\alpha(t)^{n+1}]_{k+1} \right)(y).$$

Using formula (20), we obtain $x(\lambda(\alpha_k)) = \lambda(\varphi_x(\alpha_k))$. □

The Lie algebra of the group $G = \mathrm{Diff}_1(\mathbb{Z})$ is the Lie algebra of formal vector fields on the line $L_1(1)$. The enveloping algebra $\mathfrak{g}$ of the Lie algebra $L_1(1)$ is the algebra of differential operators D_ω, $\omega \in \mathbb{Z}_+^\infty$ whose action on the ring of functions $a(x) \in P$ is given by the formula

$$(21) \quad a(x \cdot y) = \sum_\omega D_\omega a(x) y^\omega.$$

Therefore, $\mathfrak{g}$ is the algebra of all left-invariant differential operators on the group G.

Now Theorem 6.1 directly implies

COROLLARY 6.1. *The map $S \to \mathfrak{g}$, $s_\omega \mapsto D_\omega$ defines an isomorphism of the Landweber–Novikov algebra onto the enveloping algebra of the Lie algebra $L_1(1)$ that induces an isomorphism of the S-module $\Omega_U(\mathbb{Z})$ with the $\mathfrak{g}$-module P.*

Suppose $A^U(\mathbb{Z})$ is the algebra of operations of the form $\sum a_\omega s_\omega$, where the coefficients a_ω are in $\Omega_U(\mathbb{Z})$.

COROLLARY 6.2. *The algebra $\Omega_U(\mathbb{Z})$ is isomorphic to the algebra of all differential operators on the group $\mathrm{Diff}_1(\mathbb{Z})$.*

Novikov has proved that Theorem 6.1 implies

COROLLARY 6.3. *Under the isomorphism $\Omega_U(\mathbb{Z}) \cong S^*$, the embedding $\Omega_U \subset \Omega_U(\mathbb{Z})$ induces the embedding $A_U \subset S_r^* S$, i.e., A_U is the operator double of the S-module S^* with respect to the Milnor action of the Hopf algebra S dual to the action of S on itself by right translations.*

Further we shall need a result from [3] that describes the image of the ring Ω_U in P in terms of the group of diffeomorphisms.

DEFINITION 6.1. The *Hirzebruch genus* associated with the series

$$x(t) = t + \sum x_k t^{k+1}, \; x_k \in \mathbb{Q},$$

is the ring homomorphism $L_x: \Omega_U \to \mathbb{Q}$ that to each cobordism class $[M^{2n}] \in \Omega_U^{-2n}$ assigns the value given by the formula

$$(22) \qquad L_x[M^{2n}] = \left(\prod_{q=1}^n \frac{t_q}{x(t_q)}, \langle M^{2n} \rangle \right),$$

where M^{2n} is a smooth manifold whose stable tangent bundle $\tau(M^{2n})$ is a complex bundle with complete Chern class in cohomology

$$C^H(\tau) = 1 + C_1^H(\tau) + \cdots + C_n^H(\tau) = \prod_{q=1}^n (1 + t_q)$$

and $\langle M^{2n} \rangle$ is the fundamental cycle in homology.

DEFINITION 6.2. The *Todd genus* T is the Hirzebruch genus associated to the series $(1 - \exp(-t))$.

THEOREM 6.2. *Let $\mathrm{Diff}_1(\mathbb{Q})$ be the group of formal diffeomorphisms of the line over the field of rational numbers. Then the homomorphism $\mathrm{Diff}_1(\mathbb{Q}) \to \mathbb{Q}$ taking each polynomial $\lambda([M^{2n}]) \in P$ to its value on the diffeomorphism $x(t)$ coincides with the Hirzebruch genus $L_x[M^{2n}]$.*

PROOF. As shown above, the coefficients of the Chern–Dold character α_k are taken to the coordinate functions Diff_1 by the homomorphism λ. Therefore it suffices to prove that $L_x(\alpha_k) = x_k$. But this immediately follows if we compare the formulas for ch_U and L_x. □

Let us identify the ring P of integer polynomials on the group $\mathrm{Diff}_1(\mathbb{Q})$ with the graded ring $\mathbb{Z}[\alpha_1, \ldots, \alpha_k, \ldots]$, $\deg \alpha_k = -2n$.

Denote by $\mathcal{U}^{-2n} \subset P$ the group of all homogeneous (of degree $-2n$) integer polynomials which remain an integer after a left shift by the diffeomorphism $T(t) = 1 - \exp(-t)$ and denote by $\mathcal{U}$ the graded ring $\sum_{n \geq 0} \mathcal{U}^{-2n}$.

COROLLARY 6.4. *Under the identification of the ring $\Omega_U(\mathbb{Z})$ with the ring P, the ring of complex cobordisms Ω_U is identified with the ring $\mathcal{U}$.*

PROOF. Suppose $a(x) \in \mathcal{U}^{-2n}$. According to (21), we have

$$T(a(x)) = a(T \cdot x) = \sum (D_\omega a)(T) x^\omega.$$

Now using Corollary 6.1 from Theorem 6.2, we obtain

$$T(a(x)) = \sum Ts_\omega(a)x^\omega.$$

Therefore, if $T(a(x)) \in P$, then $Ts_\omega(a) \in \mathbb{Z}$ for all $\omega \in \mathbb{Z}_+^\infty$. According to the Stong–Hattori theorem ([11]), the element $a \in \Omega_U^{-2n}(\mathbb{Z})$ belongs to the ring Ω_U^{-2n} if and only if the Todd genus of the elements $s_\omega(a)$ is an integer for all $\omega \in \mathbb{Z}_+^\infty$. □

§7. The semigroup of multiplicative operations in complex cobordisms is the semigroup of maps into itself of the group of formal diffeomorphisms of the line

For the group $G = \text{Diff}_1(\mathbb{Z})$ with the ring of functions P, the map $f: G \to G$ is called *polynomial* if $a(f(x))$ is a polynomial for any polynomial $a(x) \in P$. It is clear that any such map is given by the following series

$$(23) \quad f(x;t) = t + \sum f_k(x)t^{k+1}, \quad f_k(x) \in P.$$

Denote by G^G the set of all polynomial maps of the group G into itself. Then the semigroup G_r^G (see §1) corresponding to the action of G on itself by right translations is defined. Let us describe the multiplication in G_r^G explicitly. Let $f_q \in G^G$, $q = 1, 2$, and $x \in G$. Then

$$f_q(x;t) = t + \sum f_{q,k}(x)t^{k+1},$$

$$f_{q,k}(x) \in P, \quad q = 1, 2, \quad x(t) = t + \sum x_k t^{k+1}, \quad x_k \in \mathbb{Z}.$$

In G_r^G, we have

$$(24) \quad \begin{aligned} (f_2 f_1)(x,t) &= f_2(x,t) f_1(x f_2(x,t); t) \\ &= f_2(x,t) + \sum f_{1,k}(x f_2(x,t)) f_2^{k+1}(x,t), \end{aligned}$$

where $x f_2(x,t) = x(t) + \sum f_{2,n}(x)x(t)^{n+1}$.

Now consider the ring $\mathcal{U}^* \subset P$ introduced in §6. Denote by $G^G(\mathcal{U})$ the subset of G^G consisting of the series

$$f(x,t) = t + \sum f_k(x)t^{k+1}, \quad f_k(x) \in \mathcal{U}.$$

A direct verification shows that $G^G(\mathcal{U})$ is closed with respect to the multiplication in G_r^G and, therefore, the semigroup $G_r^G(\mathcal{U})$ is defined.

THEOREM 7.1. *The semigroup of multiplicative operations mA_U is isomorphic to the semigroup of maps $G_r^G(\mathcal{U})$.*

PROOF. Suppose φ_1, φ_2 are two multiplicative operations. Then they are determined by their values on the class $u \in U^2(\mathbb{C}P(\infty))$:

$$\varphi_q u = \varphi_q(u) = u + \sum \varphi_{q,k} u^{k+1}, \quad \varphi_{q,k} \in \Omega_U, \quad q = 1, 2.$$

Under the embedding $\Omega_U \to \Omega_U(\mathbb{Z}) = \mathbb{Z}[\alpha_1, \ldots, \alpha_k, \ldots]$, we can identify $\varphi_{q,k}$ with the polynomials $\varphi_{q,k}(\alpha)$, where $\alpha = (\alpha_1, \ldots, \alpha_k)$.

By the definition of multiplication in mA_U, we have

$$(25) \qquad (\varphi_2 \varphi_1) u = \varphi_2 \varphi_1(u) = \varphi_2(u) + \sum \varphi_{q,k}(\varphi_2 \alpha) \varphi_2(u).$$

Thus by assigning to each operation $\varphi \in mA_U$ the series $\varphi(u) = \varphi(\alpha; u)$, we obtain, according to Corollary 6.4, the bijection $mA_U \cong G^G(\mathcal{U})$. Further,

$$\varphi_2 \operatorname{ch}_U(u) = t + \sum_{n \geq 0} (\varphi_2 \alpha_n) t^{n+1},$$

$$\operatorname{ch}_U \varphi_2(u) = \alpha(t) + \sum \varphi_{q,k} \alpha(t)^{k+1},$$

i.e., $\varphi_2 \alpha = \alpha \varphi_2(\alpha; u)$. Therefore, comparing the multiplication formulas, we see that the semigroups mA_U and $G^G(\mathcal{U})$ are isomorphic. $\square$

§8. The quantum double of the Landweber–Novikov algebra is a subalgebra in the algebra of operations of the doubled theory of complex cobordisms

The doubled theory of complex cobordisms (bordisms) is constructed for smooth manifolds whose stable tangent bundle τ possesses a fixed splitting into two complex bundles $\tau = \tau_l \otimes \tau_1$.

Suppose $U_{n,m} = U(n) \times U(m)$ is the product of unitary groups. The standard embedding $i_{n,m}^{k,q} : U_{n,m} \subset U_{n+k, m+q}$ induces a map of classifying spaces

$$Bi_{n,m}^{k,q} : BU_{n,m} \to BU_{n+k, m+q}, \quad (Bi_{n,m}^{k,q})^* (\xi_{n+k} \times \xi_{m+q}) = \xi_n \times \xi_m \oplus 1_{k+q}$$

and the corresponding map of Thom spaces

$$Mi_{n,m}^{k,q} : M_{n+k, m+q} \cong M_{n+k} \wedge M_{m+q} \to S^{2(k+q)} M_n \wedge M_m \cong S^{2(k+q)} M_{n,m}.$$

DEFINITION 8.1. The *Thom spectrum* in the theory of doubled complex cobordisms $DU^*(\cdot)$ is the following sequence of Thom spaces together with their joining maps:

$$DMU = \{M_{n,m} = M_n \wedge M_m : Mi_{n,m}^{k,q}\}.$$

Thus we have

$$DU^q(W, V) = \varinjlim_{n,m} [S^{2(n+m)-q}(W/V), M_{n,m}],$$

$$DU_q(W, V) = \varinjlim_{n,m} [S^{2(n+m)+q}, (W/V) \wedge M_{n,m}].$$

Let us put, as usual,

$$DU^q(W) = DU^q(W_+), \quad DU_q(W) = DU_q(W_+),$$

$$DU^*(W, V) = \sum_q DU^q(W, V), \quad DU_*(W, V) = \sum_q DU_q(W, V).$$

In the case $W = (\text{point})$, by using the Thom isomorphism from the theory of complex cobordisms $U_*(\cdot)$ [11], we obtain

$$\Omega_{DU}^{-2q} = DU^*(\text{point}) = \varinjlim_{n,m}[S^{2(n+m+q)}, M_n \wedge M_m]$$

$$= \varinjlim_n \varinjlim_m [S^{2(n+m+q)}, M_n \wedge M_m] = \varinjlim_n U_{2(n+q)}(M_n) \cong U_{2q}(BU).$$

Multiplication in the theory $DU^*(\cdot)$ is defined by means of the map of spectra $D_\mu \colon (DMU) \wedge (DMU) \to DMU$ induced by the sequence of canonical maps $M_{n,m} \wedge M_{k,q} \to M_{n+k,m+q}$. It is easy to verify that the isomorphism $\Omega_{DU}^{-2q} \cong U_{2q}(BU)$ can be extended to an isomorphism of rings $\Omega_{DU}^* \cong U_*(BU)$, where in $U_*(BU)$ one takes the multiplication in the bordisms of the H-space BU induced by the multiplication $BU \times BU \to BU$ (via the Whitney sum).

If we assign to each manifold M^{2n} with split stable complex tangent bundle $\tau = \tau_l \otimes \tau_r$ its complex cobordism class, we obtain a multiplicative transformation $\pi \colon DU^*(\cdot) \to U^*(\cdot)$. The assignments

$$\iota_l \colon (M^{2n}, \tau) \to (M^{2n}, \tau \oplus 0); \qquad \iota_r \colon (M^{2n}, \tau) \to (M^{2n}, 0 \oplus \tau)$$

induce the corresponding multiplicative transformations

$$\iota_l, \iota_r \colon U^*(\cdot) \to DU^*(\cdot).$$

A direct verification yields the following

LEMMA 8.1. *The multiplication μ in the theory of complex cobordims splits into the composition of homomorphisms*

$$\iota_l \otimes \iota_r \colon U^*(\cdot) \otimes U^*(\cdot) \to DU^*(\cdot) \otimes DU^*(\cdot),$$

$$DU^*(\cdot) \otimes DU^*(\cdot) \xrightarrow{D\mu} DU^*(\cdot) \xrightarrow{\pi} U^*(\cdot).$$

Now let us describe the algebra of stable cohomology operations A_{DU} in the theory $DU^*(\cdot)$. In the role of the universal Thom class $u_{n,m} \in DU^{2(n+m)}(M_{n,m})$, let us choose the class $(\iota_l u(n))(\iota_r u(m))$ and put $s_{\omega,n}^l = u_{n,m} \iota_l C_\omega(\xi_n)$, $s_{\omega,n}^r = u_{n,m} \iota_r C_\omega(\xi_m)$.

It can be verified in a standard way that the sequences $\{s_{\omega,n}^l\}$ and $\{s_{\omega,n}^r\}$ determine the stable cohomology operations $s_\omega^l, s_\omega^r \in A_{DU}$, $\omega \in \mathbb{Z}_+^\infty$. The operations $s_{\omega_1}^l$ and $s_{\omega_2}^r$ commute for all ω_1 and ω_2.

Denote by DS the free topological commutative group with topological basis $s_{\omega_1}^l, s_{\omega_2}^r$, $\omega_1, \omega_2 \in \mathbb{Z}_+^\infty$. We obtain

(1) DS is a subalgebra of A_{DU};
(2) the algebra DS is isomorphic to the tensor product of Hopf algebras $S^l \otimes S^r$, where $S^l \cong \tau_l S$ and $S^r \cong \tau_l S$.
(3) $A_{DU} \cong \Omega_{DU}(S^l \otimes S^r)$.

Thus it only remains to describe the representation of the algebra $S^l \otimes S^r$ in the ring Ω_{DU}.

Denote by mA_{DU} the semigroup of all multiplicative operations in the theory $DU^*(\cdot)$. Using, as usual, the identification of the ring $U^*(MU(n))$ with its image in $U^*(\prod \mathbb{C}P(\infty))$ under the monomorphism $h_n^* S_0^*$, we obtain

LEMMA 8.2. *Let u_l, u_r be the Thom classes $u_{1,0}$, $u_{0,1}$ from $DU^2(\mathbb{C}P(\infty))$. Each operation $\psi \in mA_{DU}$ is determined by its values on the classes $u_{1,0}$ and $u_{0,1}$, i.e., by the series*

$$\psi u_l = \psi_l(u_l) = u_l + \sum \psi_{l,k} u_l^{k+1}, \quad \psi u_r = \psi_r(u_r) = u_r + \sum \psi_{r,k} u_r^{k+1},$$

where $\psi_{l,k}$, $\psi_{r,k} \in \Omega_{DU}$.

Further we shall need the following fundamental result from the theory of complex cobordisms (see [7, Appendix 1]).

Consider the tensor product of one-dimensional complex universal bundles $\xi_1' \otimes \xi_1'' \to \mathbb{C}P(\infty) \times \mathbb{C}P(\infty)$. Then the following results hold.

(1) The Chern class

$$C_1(\xi_1' \otimes \xi_1'') \in U^2(\mathbb{C}P(\infty) \times \mathbb{C}P(\infty)) \cong \Omega_U[[u,v]],$$

where $u = C_1(\xi_1')$, $v = C_1(\xi_1'')$ is determined by the formal series

$$C_1(\xi_1' \otimes \xi_1'') = F(u,v) = u + v + \sum a_{k,q} u^k v^q,$$

which defines the formal group over Ω_U.

(2) The coefficients of the series $a_{k,q}$, $\deg a_{k,q} = -2(k+q-1)$ generate the ring $\Omega_U = \mathbb{Z}[a_1, \ldots, a_n, \ldots]$, $\deg a_n = -2n$.

Now let us present a similar description of the ring Ω_{DU}. The universal bundle $\xi_1 \to \mathbb{C}P(\infty)$ has two canonical first Chern classes in the theory $DU^*(\cdot)$:

$$u_l = C_{1,l}(\xi_1) = \iota_l C_1(\xi_1), \quad u_r = C_{1,r}(\xi_1) = \iota_r C_1(\xi_1).$$

LEMMA 8.3. *Over the ring Ω_{DU}, two formal groups are defined*

$$C_{1,l}(\xi_1' \otimes \xi_1'') = F_l(u_l, v_l) = u_l + v_l + \sum a_{k,q}^l u_l^k v_l^q, \quad a_{k,q}^l \in \operatorname{Im} \iota_l \Omega_U,$$

$$C_{1,r}(\xi_1' \otimes \xi_1'') = F_r(u_r, v_r) = u_r + v_r + \sum a_{k,q}^r u_r^k v_r^q, \quad a_{k,q}^r \in \operatorname{Im} \iota_r \Omega_U,$$

as well as the series

$$u_l = \varphi(u_r) = u_r + \sum \varphi_k u_r^{k+1},$$

that determines a strong isomorphism of these groups, where $\pi \varphi_k = 0$.

PROOF. Since $DU^*(\cdot)$ is a multiplicative cohomology theory, for any class $u \in DU^2(\mathbb{C}P(\infty))$ such that εu is a generator of the group $H^2(\mathbb{C}P, \mathbb{Z}) = \mathbb{Z}$, we have the isomorphism $DU^*(\mathbb{C}P(\infty)) \cong \Omega_{DU}[[u]]$. Therefore,

$$DU^*(\mathbb{C}P(\infty)) \cong \Omega_{DU}[[u_l]] \cong \Omega_{DU}[[u_r]]$$

and so $u_l = \varphi(u_r) = u_r + \sum \varphi_k u_r^{k+1}$.

Now consider the map $\gamma \colon \mathbb{C}P(\infty) \times \mathbb{C}P(\infty) \to \mathbb{C}P(\infty)$ given by $\gamma^*\xi_1 = \xi'_1 \otimes \xi''_1$. We have

$$F_l(\varphi(u_r), \varphi(v_r)) = C_{1,l}(\xi'_1 \otimes \xi''_1) = \gamma^* C_{1,l}(\xi_1) = \gamma^* \varphi(C_{1,r}(\xi_1))$$
$$= \varphi(C_{1,r}(\gamma^*\xi_1)) = \varphi(C_{1,r}(\xi'_1 \otimes \xi''_1)) = \varphi(F_r(u_r, v_r)). \quad \square$$

Now denote by $\Omega_{U,l} = \mathbb{Z}[a_n^l]$, $\Omega_{U,r} = \mathbb{Z}[a_n^r]$ the subrings of Ω_{DU} generated by the coefficients of the formal groups $F_l(u, v)$ and $F_r(u, v)$ and by $B = \mathbb{Z}[\varphi_n]$ the subring of Ω_{DU} generated by the coefficients of the series $\varphi(u)$. It follows from the previous computation that in the ring Ω_{DU}^*, up to decomposable elements, we have the relation

$$a_{k,q}^l - a_{k,q}^r \cong \binom{k+q}{k} \varphi_{k+q-1},$$

where $\binom{k+q}{k}$ is a the binomial coefficient. Hence the rings $\Omega_{U,l}$ and $\Omega_{U,r}$ together do not generate the ring Ω_{DU}.

The proof of the next result is left to the reader.

THEOREM 8.1. *We have the following isomorphisms*

$$\Omega_{DU}^* \cong \mathbb{Z}[a_n^l, \varphi_k] \cong \mathbb{Z}[\varphi_k, a_q^r].$$

LEMMA 8.4. *The subring $B = \mathbb{Z}[\varphi_n] \subset \Omega_{DU}$ is closed with respect to the action of the algebra $S^l \otimes S^r$ and is therefore a Milnor module over the Hopf algebra $S \otimes S$.*

PROOF. Consider the universal multiplicative cohomology operations

$$\mathcal{S}^l, \mathcal{S}^r \colon DU^*(V) \to DU^*(V)[[t_k, \tau_q]]$$

such that

$$\mathcal{S}^l u_l = u_l + \sum t_k u_l^{k+1} = \mathcal{S}^l(u_l), \quad \mathcal{S}^r u_l = u_l,$$
$$\mathcal{S}^l u_r = u_r, \quad \mathcal{S}^r u_r = u_r + \sum \tau_q u_r^{q+1} = \mathcal{S}^r(u_r).$$

Here t_k and τ_q are formal variables. It follows from the universality of the Thom class $u_{n,m}$ that

$$\mathcal{S}^l u_{n,m} = \sum s^l_{\omega,n} t^\omega, \quad \mathcal{S}^r u_{n,m} = \sum s^r_{\omega,m} \tau^\omega,$$

i.e., $\mathcal{S}^l$ and $\mathcal{S}^r$, as elements of the algebra $S^l \otimes S^r[[t_k, \tau_q]]$, are the generating series for all the operations in $S^l \otimes S^r$.

For $V = \mathbb{C}P(\infty)$ we have

$$\mathcal{S}^l(u_l) = \mathcal{S}^l u_l = \mathcal{S}^l u_r + \sum (\mathcal{S}^l \varphi_n)(\mathcal{S}^l u_r)^{n+1} = u_r + \sum (\mathcal{S}^l \varphi_n) u_r^{n+1}.$$

Therefore, setting $u_r = t$, we obtain

(26) $$t + \sum (\mathcal{S}^l \varphi_n) t^{n+1} = \mathcal{S}^l(\varphi(t)) = (\varphi \mathcal{S}^l)(t),$$

i.e., the operations s_ω^l act on φ_n by right translations. Further,

$$\mathcal{S}^r u_l = \mathcal{S}^r u_r + \sum (\mathcal{S}^r \varphi_n)(\mathcal{S}^r u_r)^{n+1}.$$

Therefore, setting $u_r = t$ again, we obtain

$$\varphi(t) = \mathcal{S}^r(t) + \sum (\mathcal{S}^r \varphi_n)(\mathcal{S}^r(t))^{n+1},$$

i.e.,

(27) $$t + \sum \mathcal{S}^r \varphi_n t^{n+1} = \varphi((\mathcal{S}^r)^{-1}(t)) = ((\mathcal{S}^r)^{-1}\varphi)(t),$$

so that the operations s_ω^r act on φ_n by left translations.

COROLLARY 8.1. *The operator double $B(S \otimes S)$ associated with the Milnor action of $S^l \otimes S^r$ on B is a subalgebra of A_{DU}.*

Now everything is ready to obtain the result appearing in the heading of the present section.

THEOREM 8.2. *Suppose $S^* = \mathbb{Z}[s^{(1)}, \ldots, s^{(q)}, \ldots]$ is the Hopf algebra dual to the Hopf algebra S. Then the isomorphism $S^* \to B$, $s^{(q)} \to \varphi_q$ together with the diagonal homomorphism $\Delta: S \to S \otimes S$ determine an embedding of the quantum double $D(S)$ in $B(S \otimes S) \subset A_{DU}$.*

PROOF. Formulas (26), (27) show that the diagonal homomorphism $\Delta: S \to S \otimes S$ induces the action ad of the Hopf algebra S on $B \cong S^*$. Therefore $BS \cong S^*_{\text{ad}} S$. According to Theorem 3.2, we have the isomorphism $D(S) \cong S^*_{\text{ad}} S$. Thus the embedding $BS \subset B(S \otimes S)$ defines an embedding of the quantum double $D(S)$ of the Landweber–Novikov algebra into A_{DU}. $\square$

References

1. V. M. Buchstaber, *The Chern–Dold character in cobordisms*, Mat. Sb. **83** (1970), 575–595; English transl. in Math. USSR-Sb. **12** (1970).
2. _____, *Operator doubles and semigroups of maps into groups*, Doklady Akad. Nauk **341** (1995), 1–3; English transl. in Soviet Math. Dokl. (to appear).
3. V. M. Buchstaber and A. B. Shokurov, *The Landweber–Novikov algebra and formal vector fields on the line*, Funktsional. Anal. i Prilozhen. **12** (1978), 1–11; English transl. in Functional Anal. Appl. **12** (1978).
4. V. G. Drinfeld, *Quantum groups*, Zap. Nauchn. Sem. Leningrad. Otdel. Mat. Inst. Steklov. (LOMI) **155** (1986), 18–49; English transl. in J. Soviet Math. **41** (1988), no. 2.
5. D. B. Fuchs, *Cohomology of infinite-dimensional Lie algebras*, Plenum Press, New York, 1984.
6. P. S. Landweber, *Cobordism operations and Hopf algebras*, Trans. Amer. Math. Soc. **129** (1967), 94–110.
7. S. P. Novikov, *Algebraic topology methods from the point of view of cobordisms*, Izv. Akad. Nauk SSSR Ser. Mat. **31** (1967), 855–951; English transl. in Math. USSR-Izv. **1** (1967).
8. _____, *Various doubles of Hopf algebras. Operator algebras on quantum groups, complex cobordisms*, Uspekhi Mat. Nauk **47** (1992), 189–190; English transl. in Russian Math. Surveys.
9. E. Pressley and G. Segal, *Loop groups*, Clarendon Press, Oxford, 1986.
10. L. D. Faddeev, N. Yu. Reshetikhin, and L. A. Takhtadjan, *Quantization of groups and Lie algebras*, Algebra i Analiz **1** (1989), 178–206; English transl. in Leningrad J. Math. **1** (1990).
11. R. Stong, *Notes on cobordism theory*, Princeton Univ. Press, Princeton, NJ, 1968.

Translated by A. B. SOSSINSKY

Nonlocal Hamiltonian Operators of Hydrodynamic Type: Differential Geometry and Applications

E. V. FERAPONTOV

ABSTRACT. In this survey we discuss nonlocal Hamiltonian operators of hydrodynamic type

$$A^{ij} = g^{ij}d - g^{is}\Gamma^j_{sk}u^k_x + \sum_\alpha w^i_{\alpha k}u^k_x d^{-1}w^j_{\alpha l}u^l_x, \qquad d = \frac{d}{dx},$$

for which the skew-symmetry condition and the Jacobi identity assume the form of Gauss–Peterson–Codazzi equations in the theory of submanifolds of flat spaces. Nonlocal Hamiltonian operators of hydrodynamic type arise naturally in applications, when one applies one of the following standard constructions:
- Dirac reduction,
- recursion scheme,
- reciprocal transformation

to the local Hamiltonian operator $\delta^{ij}d/dx$. Nonlocal Hamiltonian operators of hydrodynamic type play an important role as Hamiltonian structures for "semi-Hamiltonian" systems.

§1. Introduction

Let us consider an infinite-dimensional phase space consisting of vector functions $u = \{u^i(x), i = 1, \dots, n\}$, where the Poisson bracket of two functionals $F = \int f(u, u_x, \dots)\,dx$ and $G = \int g(u, u_x, \dots)\,dx$ is given by the formula

$$\{I, J\} = \int \frac{\delta F}{\delta u^i} A^{ij} \frac{\delta G}{\delta u^j}\,dx; \tag{1.1}$$

here A^{ij} is an operator of hydrodynamic type

$$A^{ij} = g^{ij}(u)d + b^i_{jk}(u)u^k_x, \qquad d = \frac{d}{dx}. \tag{1.2}$$

The theory of such brackets was developed by B. A. Dubrovin and S. P. Novikov in [1]. A fundamental observation was that this theory is essentially differential-geometric. Indeed, if we take $\det g^{ij} \neq 0$ (such Poisson brackets are called *nondegenerate*) and represent b^i_{jk} in the form $b^i_{jk} = -g^{is}\Gamma^j_{sk}$, it is not difficult to show that under point transformations $\widetilde{u}^i = \widetilde{u}^i(u^1, \dots, u^n)$ the coefficients g^{ij} transform as

1991 *Mathematics Subject Classification.* Primary 58F05.

©1995, American Mathematical Society

components of a type $(2, 0)$ tensor, while Γ^j_{sk} transform as Christoffel symbols of an affine connection. The condition for the operator (1.2) to be Hamiltonian (i.e., to define a bracket that is skew-symmetric and satisfies the Jacobi identity) imposes strict constraints on g^{ij} and Γ^j_{sk}.

THEOREM 1. 1. *The bracket defined by* (1.1) *and* (1.2) *is skew-symmetric if and only if the tensor g^{ij} is symmetric (i.e., defines a pseudo-Riemannian metric) and the connection Γ^j_{sk} is compatible with the metric:* $\nabla_k g^{ij} = 0$.

2. *The bracket defined by* (1.1) *and* (1.2) *satisfies the Jacobi identity if and only if the connection Γ^j_{sk} is symmetric and its curvature tensor vanishes.*

In other words, the metric $ds^2 = g_{ij}\, du^i\, du^j$ is flat (here $g_{ik} g^{kj} = \delta^j_i$), and Γ^j_{sk} are the coefficients of the corresponding Levi–Civita connection. It follows from this that for Hamiltonian operators of the form (1.2) we have an infinite-dimensional analog of the Darboux theorem: in the flat coordinates, $g^{ij} = \varepsilon^i \delta^{ij}$ ($\varepsilon^i = \pm 1$), $\Gamma^j_{sk} = 0$, and we obtain a particularly simple expression for A^{ij} with constant coefficients: $A^{ij} = \varepsilon^i \delta^{ij} d$.

If for the Hamiltonian we select the hydrodynamic functional $H = \int h(u)\, dx$ whose density does not explicitly depend on the derivatives $u_x, u_{xx}, \ldots$, we obtain a Hamiltonian system of hydrodynamic type

$$u^i_t = A^{ij} \frac{\delta H}{\delta u^j} = v^i_j(u) u^j_x,$$

where the matrix v^i_j is given by the formula $v^i_j = \nabla^i \nabla_j h$. One may consult the surveys [3, 4, 6] for the necessary information concerning differential geometry, integrability, and applications of Hamiltonian systems of hydrodynamic type.

Theorem 1 provides the basis for a whole series of generalizations:
- multidimensional Poisson brackets [2, 7];
- degenerate Poisson brackets [10];
- Poisson brackets with an explicit x-dependence [11];
- higher order homogeneous Poisson brackets [8, 9].
- Poisson brackets with an infinite number of field variables u^i [12];
- linear Poisson brackets [13].

It should be emphasized that in all the generalizations the operators A^{ij} remain local, i.e., they do not explicitly depend on d^{-1}. The first nonlocal generalization of Hamiltonian operators of hydrodynamic type was proposed by O. I. Mokhov and the author in [14], see also [15, 16, 32]:

$$(1.3) \qquad A^{ij} = g^{ij} d - g^{is} \Gamma^j_{sk} u^k_x + c u^i_x d^{-1} u^j_x, \qquad c = \mathrm{const}.$$

Formally, A^{ij} can be treated as a linear combination of the local operator (1.2) (henceforth, we assume $\det g^{ij} \neq 0$) and the nonlocal term $u^i_x d^{-1} u^j_x$.

REMARK. For a single field variable $u(x)$, the operator $u_x d^{-1} u_x$ first appeared in [20] as the Hamiltonian operator of the Krichever–Novikov equation:

$$u_t = u_{xxx} - \frac{3}{2} \frac{u^2_{xx}}{u_x} + \frac{h(u)}{u_x} = u_x d^{-1} u_x \frac{\delta H}{\delta u};$$

here

$$H = \int \left(\frac{1}{2} \frac{u_{xx}^2}{u_x^2} + \frac{1}{3} \frac{h(u)}{u_x^2} \right) dx,$$

and $h(u) = c_3 u^3 + c_2 u^2 + c_1 u + c_0$ is an arbitrary polynomial of the third order. It was noted that the expression $u_x d^{-1} u_x \, \delta H/\delta u$ is local for any local functional $H = \int h(u, u_x, \dots) \, dx$. A similar nonlocal term appears in the second Hamiltonian structure of MKdV equation [12, p. 232]:

$$u_t = u_{xxx} - 6u^2 u_x = A \frac{\delta H}{\delta u},$$

where

$$A = d^3 - 2(du^2 + u^2 d) + 4u_x d^{-1} u_x, \qquad H = \int \frac{u^2}{2} dx.$$

The conditions required for the operator (1.3) to be Hamiltonian depend on the constant c in a nontrivial way.

THEOREM 2 [14]. 1. *The bracket defined by (1.1) and (1.3) is skew-symmetric if and only if the tensor g^{ij} is symmetric (i.e., defines a pseudo-Riemannian metric) and the connection Γ_{sk}^j is compatible with the metric: $\nabla_k g^{ij} = 0$.*

2. The bracket defined by (1.1) and (1.3) satisfies the Jacobi identity if and only if the connection Γ_{sk}^j is symmetric and has constant curvature c (i.e., $R_{kl}^{ij} = c(\delta_k^i \delta_l^j - \delta_k^j \delta_l^i)$).

In other words, the metric $ds^2 = g_{ij} du^i du^j$ has constant curvature c (here $g_{ik} g^{kj} = \delta_i^j$), and Γ_{sk}^j are the coefficients of the corresponding Levi–Civita connection. For Hamiltonian operators (1.3), the Darboux theorem (in the form in which it holds for local operators (1.2)) is no longer valid, since a metric of nonzero curvature cannot be reduced to constant coefficients by a point transformation. Nevertheless, the nonlocal Hamiltonian formalism is in a sense equivalent to the local one, and in [16] the author constructed a Bäcklund-type transformation (the so-called reciprocal transformation), that eliminates the nonlocal "tail" from the operator (1.3). In this sense the Darboux theorem remains valid. The relevant construction is described in §4. Let us discuss several examples.

EXAMPLE 1.1. The metric of the unit sphere $(S^1)^2 + (S^2)^2 + (S^3)^2 = 1$ in the spherical-conical parametrization

$$S^1 = \sqrt{\frac{(R^1 - a^1)(R^2 - a^1)}{(a^2 - a^1)(a^3 - a^1)}}, \qquad S^2 = \sqrt{\frac{(R^1 - a^2)(R^2 - a^2)}{(a^1 - a^2)(a^3 - a^2)}},$$

$$S^3 = \sqrt{\frac{(R^1 - a^3)(R^2 - a^3)}{(a^1 - a^3)(a^2 - a^3)}}$$

(a^1, a^2, a^3 are constants), assumes the form

$$ds^2 = (dS^1)^2 + (dS^2)^2 + (dS^3)^2 = \frac{(R^2 - R^1)(dR^1)^2}{4P(R^1)} + \frac{(R^1 - R^2)(dR^2)^2}{4P(R^2)},$$

where $P(R) = (R - a^1)(R - a^2)(R - a^3)$. This metric has constant curvature $c = 1$. Inserting the corresponding expressions for g^{ij} and Γ^j_{sk} in (1.3), we arrive at the operator

$$A = 4 \begin{pmatrix} \dfrac{P(R^1)}{R^2 - R^1} & 0 \\ 0 & \dfrac{P(R^2)}{R^1 - R^2} \end{pmatrix} d$$

$$+ 2 \begin{pmatrix} \dfrac{P(R^1)}{R^2 - R^1}\left(\dfrac{R^1_x - R^2_x}{R^2 - R^1} + \dfrac{P'(R^1)R^1_x}{P(R^1)}\right) & \dfrac{P(R^1)R^2_x - P(R^2)R^1_x}{(R^1 - R^2)^2} \\ \dfrac{P(R^2)R^1_x - P(R^1)R^2_x}{(R^1 - R^2)^2} & \dfrac{P(R^2)}{R^1 - R^2}\left(\dfrac{R^2_x - R^1_x}{R^1 - R^2} + \dfrac{P'(R^2)R^2_x}{P(R^2)}\right) \end{pmatrix}$$

$$+ \begin{pmatrix} R^1_x d^{-1} R^1_x & R^1_x d^{-1} R^2_x \\ R^2_x d^{-1} R^1_x & R^2_x d^{-1} R^2_x \end{pmatrix}.$$

If for the Hamiltonian we take $H = -\int (R^1 + R^2)\,dx$, the resulting Hamiltonian system will be of hydrodynamic type:

$$R^1_t = (3R^1 + R^2 - a)R^1_x, \qquad R^2_t = (3R^2 + R^1 - a)R^2_x.$$

Up to the unessential constant $a = 2(a^1 + a^2 + a^3)$, this system coincides with the Riemann invariant form of the shallow water equations (gas dynamics with $\gamma = 2$). Note that besides the nonlocal operator A, the system of shallow water equations (as well as gas dynamics with arbitrary γ) possesses three local Hamiltonian structures of type (1.2) [21].

In the n-dimensional case the metric of the unit sphere S^n in the spherical-conical coordinates assumes the form

$$ds^2 = \sum g_{ii}(dR^i)^2, \qquad g_{ii} = \dfrac{\prod_{k \neq i}(R^k - R^i)}{4P(R^i)},$$

where $P(R) = \prod_{s=1}^{n+1}(R - a^s)$. The corresponding operator (1.3) together with the Hamiltonian $H = -\int \sum R^i\,dx$ generate the system of hydrodynamic type

$$(1.4) \qquad R^i_t = \left(\sum R^k + 2R^i - a\right)R^i_x,$$

which up to the unessential constant $a = 2\sum a^s$ coincides with the dispersionless limit of the first nontrivial flow in the Coupled KdV hierarchy (see [40, 18]). Equations (1.4) possess also $n + 1$ local Hamiltonian structures, see Example 2.1 in §2.

Nonlocal operators of type (1.3) naturally arise in applications as the "second" Hamiltonian structures of higher order evolution equations, integrable via the inverse scattering transform.

EXAMPLE 1.2. The metric of the unit sphere $(S^1)^2 + (S^2)^2 + (S^3)^2 = 1$ in the coordinates u^1, u^2 of the stereographic projection

$$S^1 = u^1/P, \quad S^2 = u^2/P, \quad S^3 = (P-1)/P$$

(here $P = ((u^1)^2 + (u^2)^2 + 1)/2$) assumes the form

$$ds^2 = (dS^1)^2 + (dS^2)^2 + (dS^3)^2 = ((dR^1)^2 + (dR^2)^2)/P^2.$$

This metric generates the Hamiltonian operator (1.3):

$$A = P^2 \begin{pmatrix} 1 & 0 \\ 0 & 1 \end{pmatrix} d + P \begin{pmatrix} u^1 u_x^1 + u^2 u_x^2 & u^1 u_x^2 - u^2 u_x^1 \\ u^2 u_x^1 - u^1 u_x^2 & u^1 u_x^1 + u^2 u_x^2 \end{pmatrix} + \begin{pmatrix} u_x^1 d^{-1} u_x^1 & u_x^1 d^{-1} u_x^2 \\ u_x^2 d^{-1} u_x^1 & u_x^2 d^{-1} u_x^2 \end{pmatrix}.$$

If for the Hamiltonian we take

$$H = \int \frac{u^2 u_x^1 - u^1 u_x^2}{(2P-1)P} dx,$$

then the resulting Hamiltonian system will be of the form

(1.5)
$$u_t^1 = u_{xx}^2 + \frac{u^2 u_x^{1\,2} - 2u^1 u_x^1 u_x^2 - u^2 u_x^{2\,2}}{P},$$

$$u_t^2 = -u_{xx}^1 - \frac{u^1 u_x^{2\,2} - 2u^2 u_x^1 u_x^2 - u^1 u_x^{1\,2}}{P},$$

which identically coincides with the equations of Heisenberg magnet

(1.6) $$S_t = S \times S_{xx}, \qquad S^2 = 1,$$

rewritten in the coordinates u^1, u^2. In equations (1.5), let us introduce the new independent variable y by the formula

(1.7) $$dy = \frac{1}{P} dx + \frac{u^2 u_x^1 - u^1 u_x^2}{P^2} dt$$

(transformations of this type are known as "reciprocal" [36]). In the new variables y, t, the operator A assumes the canonical form

$$A = \begin{pmatrix} 1 & 0 \\ 0 & 1 \end{pmatrix} \frac{d}{dy}$$

(thus establishing the validity of the Darboux theorem, see §4 for the relevant calculations), while the Hamiltonian H, being a linear homogeneous expression in the first derivatives, remains unchanged:

$$H = \int \frac{u^2 u_y^1 - u^1 u_y^2}{(2P-1)P} dy.$$

In the variables y, t, equations (1.5) assume the form

$$\begin{pmatrix} u^1 \\ u^2 \end{pmatrix}_t = \begin{pmatrix} 1 & 0 \\ 0 & 1 \end{pmatrix} \frac{d}{dy} \begin{pmatrix} \delta H/\delta u^1 \\ \delta H/\delta u^2 \end{pmatrix},$$

or, explicitly,

(1.8) $$u_t^1 = \left(\frac{u_y^2}{P^2}\right)_y, \qquad u_t^2 = -\left(\frac{u_y^1}{P^2}\right)_y.$$

EXAMPLE 1.3. The metric of constant curvature $c = -1$ of the Lobachevskiĭ half-plane has the form
$$ds^2 = ((du^1)^2 + (du^2)^2)/(u^1)^2$$
and generates the Hamiltonian operator (1.3):
$$A = (u^1)^2 \begin{pmatrix} 1 & 0 \\ 0 & 1 \end{pmatrix} d + u^1 \begin{pmatrix} u_x^1 & u_x^2 \\ -u_x^2 & u_x^1 \end{pmatrix} - \begin{pmatrix} u_x^1 d^{-1} u_x^1 & u_x^1 d^{-1} u_x^2 \\ u_x^2 d^{-1} u_x^1 & u_x^2 d^{-1} u_x^2 \end{pmatrix}.$$

If for the Hamiltonian we take
$$H = \int \frac{u^2 u_x^1 - u^1 u_x^2}{(2P-1)P} dx, \qquad P = \frac{1}{2}((u^1)^2 + (u^2)^2 + 1),$$
then the resulting Hamiltonian system will be of the form

(1.9) $\qquad u_t^1 = (u^1)^2 \left(\dfrac{u_x^2}{P^2} \right)_x, \qquad u_t^2 = -(u^1)^2 \left(\dfrac{u_x^1}{P^2} \right)_x - u^1 \dfrac{{u_x^1}^2 + {u_x^2}^2}{P^2}.$

Introducing the complex variable $q = u^2/u^1 + i(P-1)/u^1$, one can rewrite (1.9) in the form

(1.10) $\qquad iq_t + \left(\dfrac{q}{\sqrt{1+q\bar{q}}} \right)_{xx} = 0.$

This equation was obtained in [30] within the framework of the so-called WKI spectral problem. In equations (1.9), let us introduce a new independent variable y by the formula

(1.11) $\qquad dy = \dfrac{1}{u^1} dx - \dfrac{u_x^2}{P^2} dt.$

In the variables y, t, the operator A assumes the canonical form
$$A = \begin{pmatrix} 1 & 0 \\ 0 & 1 \end{pmatrix} \frac{d}{dy}$$
(thus establishing the validity of the Darboux theorem, see §4 for the relevant calculations), while the Hamiltonian H, being a linear homogeneous expression in the first derivatives, remains unchanged:
$$H = \int \frac{u^2 u_y^1 - u^1 u_y^2}{(2P-1)P} dy.$$

In the variables y, t, equations (1.9) assume the form
$$\begin{pmatrix} u^1 \\ u^2 \end{pmatrix}_t = \begin{pmatrix} 1 & 0 \\ 0 & 1 \end{pmatrix} \frac{d}{dy} \begin{pmatrix} \delta H/\delta u^1 \\ \delta H/\delta u^2 \end{pmatrix},$$
which identically coincides with that from example (1.2). Hence equations (1.6) and (1.10) are reciprocally related. (The equivalence of (1.6) and (1.10) was demonstrated earlier in [30, 31].)

Further generalizations of nonlocal Hamiltonian operators (1.3) lean in the direction of modifying the nonlocal "tail":

(1.12) $\qquad A^{ij} = g^{ij} d - g^{is} \Gamma^j_{sk} u_x^k + w_k^i u_x^k d^{-1} w_l^j u_x^l.$

The conditions required for the operator (1.12) to be Hamiltonian impose certain relationships on the metric $g^{ij}(u)$, the connection $\Gamma^j_{sk}(u)$, and the affinor $w^i_j(u)$.

THEOREM 3 [17]. 1. *The bracket defined by* (1.1) *and* (1.12) *is skew-symmetric if and only if the tensor* g^{ij} *is symmetric* (*i.e., defines a pseudo-Riemannian metric*) *and the connection* Γ^j_{sk} *is compatible with the metric:* $\nabla_k g^{ij} = 0$.

2. *The bracket defined by* (1.1) *and* (1.12) *satisfies Jacobi identity if and only if the connection* Γ^j_{sk} *is symmetric, the metric* g_{ij} (*with lower indices*) *and the affinor* w^i_j *satisfy Gauss–Peterson–Codazzi equations*:

$$g_{ik} w^k_j = g_{jk} w^k_i, \qquad \nabla_k w^i_j = \nabla_j w^i_k,$$
$$R^{ij}_{kl} = w^i_k w^j_l - w^j_k w^i_l \qquad (R^{ij}_{kl} \equiv g^{is} R^j_{skl}).$$

In other words, the classical Gauss–Peterson–Codazzi equations of the theory of hypersurfaces M^n in a pseudo-Euclidean space E^{n+1} are nothing but the Jacobi identity for the Poisson bracket (1.1), (1.12)! Here the metric g_{ij} plays the role of the first quadratic form of M^n, and the affinor w^i_j, the role of the Weingarten operator (shape-operator). If M^n is a hyperplane in E^{n+1}, its Weingarten operator vanishes and we obtain the Hamiltonian operator (1.2). If M^n is a unit hypersphere, then its Weingarten operator $w^i_j = \delta^i_j$ yields the operator (1.3) with $c = 1$. It should be emphasized that Hamiltonian operators (1.3) are distinguished in the class of all nonlocal operators (1.12) by the following important property: it is only for them that the expression $A^{ij} \delta H / \delta u^j$ is local for any local functional H. For A^{ij} of the general form (1.12), the class of functionals generating local Hamiltonian flows is sharply restricted. The peculiar correspondence between nonlocal operators (1.12) and hypersurfaces $M^n \subset E^{n+1}$ will be explained in §3, where it will be demonstrated that the operator (1.12) arises as the result of a Dirac reduction of the flat operator $\delta^{IJ} d/dx$ defined in the ambient space E^{n+1} (here $I, J = 1, \ldots, n+1$), to the hypersurface M^n.

Further generalizations involve "lengthening" the nonlocal tail of the Hamiltonian operator:

$$(1.13) \qquad A^{ij} = g^{ij} d - g^{is} \Gamma^j_{sk} u^k_x + \sum_{\alpha=1}^{N} w^i_{\alpha k} u^k_x d^{-1} w^j_{\alpha l} u^l_x.$$

THEOREM 4 [17]. 1. *The bracket defined by* (1.1) *and* (1.13) *is skew-symmetric if and only if the tensor* g^{ij} *is symmetric* (*i.e., defines a pseudo-Riemannian metric*) *and the connection* Γ^j_{sk} *is compatible with the metric:* $\nabla_k g^{ij} = 0$.

2. *The bracket defined by* (1.1) *and* (1.13) *satisfies the Jacobi identity if and only if the connection* Γ^j_{sk} *is symmetric, the metric* g_{ij} (*with lower indices*), *and the set of affinors* $w^i_{\alpha j}$ *satisfy the equations*:

$$g_{ik} w^k_{\alpha j} = g_{jk} w^k_{\alpha i}, \qquad \nabla_k w^i_{\alpha j} = \nabla_j w^i_{\alpha k},$$
$$R^{ij}_{kl} = \sum_{\alpha=1}^{N} w^i_{\alpha k} w^j_{\alpha l} - w^j_{\alpha k} w^i_{\alpha l}.$$

Moreover, the set of affinors is commutative: $[w_\alpha ; w_\beta] = 0$.

The equations we have written are also well known in differential geometry—they are nothing but the Gauss–Peterson–Codazzi equations for submanifolds M^n

with flat normal connection in pseudo-Euclidean space E^{n+N}. Here g_{ij} is the first quadratic form of M^n, and w_α are the Weingarten operators corresponding to the field of pairwise orthogonal unit normals $\vec{n}_\alpha$. (By definition, $M^n \subset E^{n+N}$ is said to be a submanifold with *flat normal connection* if the family of Weingarten operators is commutative).

REMARK. Strictly speaking, operators of the form (1.13) that appear in applications may have a slightly more general nonlocal tail:

$$\sum_{\alpha=1}^N \varepsilon_\alpha w^i_{\alpha k} u^k_x d^{-1} w^j_{\alpha l} u^l_x, \qquad \varepsilon_\alpha = \pm 1.$$

From the complex viewpoint, the ε_α are eliminated by the substitution $w_\alpha \to iw_\alpha$, but in the real situation they must be taken into account. Geometrically, $\varepsilon_\alpha = -1$ means that the corresponding normal $\vec{n}_\alpha$ is space-like.

Nonlocal Hamiltonian operators are very natural objects not only from the point of view of their differential geometry. There exist several more or less standard constructions in the theory of integrable systems (in particular, hydrodynamic type systems), that naturally lead to nonlocalities introduced above:
- recursion scheme (§2);
- Dirac reduction (§3);
- reciprocal transformations (§4);
- Hamiltonian formalism for "semi-Hamiltonian systems" (§5);
- averaging theory (§6).

§2. Recursion scheme

Let us consider two local Hamiltonian operators of hydrodynamic type

$$A^{ij} = g^{ij} d - g^{is} \Gamma^j_{sk} u^k_x, \qquad \widetilde{A}^{ij} = \widetilde{g}^{ij} d - \widetilde{g}^{is} \widetilde{\Gamma}^j_{sk} u^k_x$$

generated by flat metrics g^{ij} and $\widetilde{g}^{ij}$ (assumed nondegenerate). To give the necessary and sufficient conditions of the compatibility of A and $\widetilde{A}$ (recall that two Hamiltonian operators are called *compatible*, if their linear combinations $\lambda A + \widetilde{\lambda} \widetilde{A}$ are Hamiltonian as well), we introduce the affinor $r^i_j = \widetilde{g}^{ik} g_{kj}$.

THEOREM 5. *Two Hamiltonian operators of hydrodynamic type are compatible if and only if the Nijenhuis tensor of the affinor r^i_j vanishes, i.e.,*

$$N^i_{jk} = r^p_j \partial_p r^i_k - r^p_k \partial_p r^i_j - r^i_p (\partial_j r^p_k - \partial_k r^p_j) = 0.$$

In a somewhat different form, the necessary and sufficient conditions of compatibility were formulated in [26].

REMARK 1. The compatibility criterion of Hamiltonian operators of hydrodynamic type resembles that of finite-dimensional Poisson bivectors: two skewsymmetric Poisson bivectors ω^{ij} and $\widetilde{\omega}^{ij}$ are compatible if and only if the Nijenhuis tensor of the corresponding recursion operator $r^i_j = \widetilde{\omega}^{ik} \omega_{kj}$ vanishes. Note that in our situation the operator r^i_j does not coincide with the recursion operator corresponding to A and $\widetilde{A}$.

REMARK 2. If the spectrum of r^i_j is simple, the vanishing of the Nijenhuis tensor implies the existence of coordinates $R^1, \ldots, R^n$ for which all objects r^i_j, g^{ij}, $\widetilde{g}^{ij}$ become diagonal. Moreover, the ith eigenvalue of r^i_j depends only on the coordinate R^i. In the case when all eigenvalues are nonconstant, one can introduce them as new coordinates. In these coordinates $r^i_j = \mathrm{diag}(R^1, \ldots, R^n)$, $g^{ij} = \mathrm{diag}(g^{ii})$, $\widetilde{g}^{ij} = \mathrm{diag}(g^{ii} R^i)$. This remark generalizes a similar observation from [26] in the particular case of compatible Hamiltonian pairs associated with Coxeter groups.

Examples of compatible Hamiltonian pairs naturally arise in the theory of multi-Hamiltonian systems of hydrodynamic type, see, e.g., [21–25, 27, 41]. An interesting class of compatible Poisson brackets can be associated with the space of orbits of Coxeter groups [26]. Compatible Poisson brackets of hydrodynamic type can be obtained also as the result of averaging local compatible Hamiltonian structures of integrable equations [1, 3, 4, 6]. Despite a broad variety of examples, the general classification of compatible Hamiltonian operators of hydrodynamic type is still lacking and remains one of the most interesting unsolved problem of the theory. Some particular results in this direction can be found in [16, 18, 21].

Given two local compatible operators A and $\widetilde{A}$, we construct the recursion operator $R = \widetilde{A}A^{-1}$. The operator R defines an infinite sequence of compatible Hamiltonian operators A, $\widetilde{A} = RA$, $\widetilde{\widetilde{A}} = R\widetilde{A}$, and so on. Although starting from $\widetilde{\widetilde{A}}$, all these operators will be in general nonlocal, their nonlocality always has the special form (1.3), (1.12), (1.13), or (1.14).

EXAMPLE 2.1. The system of hydrodynamic type

$$R^i_t = \left(\sum_1^n R^k + 2R^i \right) R^i_x$$

coincides with the dispersionless limit of the first nontrivial flow in the coupled KdV hierarchy [18, 40]. This system is $(n+1)$-Hamiltonian and possesses $n+1$ compatible local Hamiltonian structures A^α, generated by $n+1$ diagonal metrics

$$ds^{\alpha 2} = \sum_1^n g^\alpha_{ii} \, dR^{i2}, \qquad g^\alpha_{ii} = \frac{\prod_{k \neq i}(R^k - R^i)}{(R^i)^\alpha},$$

$\alpha = 0, \ldots, n$. All these metrics are flat and can be obtained as degenerations of the well-known expression for the Euclidean metric in elliptic coordinates:

$$ds^2 = \sum_{i=1}^n \frac{\prod_{k \neq i}(R^k - R^i)}{4P(R^i)} dR^{i2},$$

where $P(R)$ is an arbitrary polynomial of degree n. Thus the system under study can be represented in the $(n+1)$-Hamiltonian form

$$R^i_t = (A^0)^{ij} \frac{\delta H^0}{\delta R^j} = \cdots = (A^n)^{ij} \frac{\delta H^n}{\delta R^j}$$

(the densities of the corresponding Hamiltonians $H^0, \ldots, H^n$ are certain homogeneous polynomials in the variables R^i; the explicit form of these polynomials is not essential now). Let us introduce the recursion operator R by the formula $A^1 = RA^0$. A straightforward calculation gives

$$R = \begin{pmatrix} R^1 & \cdots & 0 \\ \vdots & \ddots & \vdots \\ 0 & \cdots & R^n \end{pmatrix} + \frac{1}{2} \begin{pmatrix} R^1_x & \cdots & R^1_x \\ \vdots & \ddots & \vdots \\ R^n_x & \cdots & R^n_x \end{pmatrix} d^{-1}.$$

One can verify directly that

$$A^2 = RA^1, \quad \ldots, \quad A^n = RA^{n-1}.$$

The operator $A^{n+1} = RA^n$ is already of the nonlocal form (1.3):

$$(A^{n+1})^{ij} = (g^{n+1})^{ii} \delta^{ij} d - (g^{n+1})^{ii} \Gamma^j_{ik} R^k_x + \tfrac{1}{4} R^i_x d^{-1} R^j_x.$$

It is generated by the diagonal metric

$$ds^{n+1\,2} = \sum_1^n g^{n+1}_{ii} dR^{i\,2}, \qquad g^{n+1}_{ii} = \frac{\prod_{k \neq i}(R^k - R^i)}{(R^i)^{n+1}}$$

of constant curvature $1/4$. Further applications of the recursion operator R lead to Hamiltonian operators with ever lengthening nonlocal tails.

§3. Dirac reduction

We consider Euclidean space $E^{n+N}(u^1, \ldots, u^{n+N})$ with the corresponding Hamiltonian operator

(3.1) $$A^{IJ} = \delta^{IJ} d, \qquad I, J = 1, \ldots, n+N.$$

Let us introduce a set of phase constraints

(3.2) $$\varphi^1(u) = \cdots = \varphi^N(u) = 0,$$

(φ^i are assumed independent of $u_x, u_{xx}, \ldots$). The constraints (3.2) define an n-dimensional submanifold $M^n \subset E^{n+N}$. The well-known Dirac construction [29] allows one to reduce the Hamiltonian operator $\delta^{IJ} d$ to a submanifold of the phase space. Our aim is to write out the explicit form of this reduction in a suitable curvilinear coordinate system $R^1, \ldots, R^n$ on the submanifold M^n. As one can expect, the form of this reduction is completely determined by the differential geometry of the submanifold M^n. It turns out that a "good" expression for the reduced operator can be obtained only for some special submanifolds M^n, the so-called submanifolds with flat normal connection.

We recall the necessary definitions. Let $g_{ij} dR^i dR^j$ be the first quadratic form (metric) of submanifold M^n, and $h_{\alpha ij} dR^i dR^j$ be the second quadratic forms corresponding to the field of pairwise orthogonal unit normals $\vec{n}_\alpha$ (here $i, j = 1, \ldots, n$, $\alpha = n+1, \ldots, n+N$). Let us introduce the Weingarten operators w_α:

$$w^i_{\alpha j} = g^{ik} h_{\alpha kj}.$$

Submanifolds with flat normal connection are characterized by the commutativity of the Weingarten operators: $[w_\alpha, w_\beta] = 0$. For these submanifolds the Gauss–Peterson–Codazzi equations assume the form

$$g_{ik}w_{\alpha j}^k = g_{jk}w_{\alpha i}^k, \quad \nabla_k w_{\alpha j}^i = \nabla_j w_{\alpha k}^i \quad \text{(Peterson-Codazzi)},$$

$$R_{kl}^{ij} = \sum_{\alpha=1}^N w_{\alpha k}^i w_{\alpha l}^j - w_{\alpha k}^j w_{\alpha l}^i \quad \text{(Gauss)}.$$

Here ∇ is the Levi–Civita connection of the metric g_{ij}, and $R_{kl}^{ij} = g^{is} R_{skl}^j$. Among the most important examples of submanifolds with flat normal connection are all hypersurfaces, coadjoint orbits of simple Lie groups (more generally, orbits of s-representation of symmetric spaces), and submanifolds with a holonomic net of lines of curvature.

THEOREM 6 [19]. *The Dirac reduction of the Hamiltonian operator $\delta^{IJ}d$ to a submanifold M^n with flat normal connection assumes the form* (1.13):

$$A^{ij} = g^{ij}d - g^{is}\Gamma_{sk}^j R_x^k + \sum_{\alpha=n+1}^{n+N} w_{\alpha k}^i R_x^k d^{-1} w_{\alpha l}^j R_x^l.$$

The formal proof of Theorem 6 will be given at the end of this section.

COROLLARY. *The Dirac reduction of the Hamiltonian operator $\delta^{IJ}d$ to a unit hypersphere S^n assumes the form* (1.3):

$$A^{ij} = g^{ij}d - g^{is}\Gamma_{sk}^j R_x^k + R_x^i d^{-1} R_x^j.$$

Indeed, S^n has only one Weingarten operator $w_j^i = \delta_j^i$. The Dirac reduction to a unit hypersphere was discussed earlier in [28], however, without writing out the reduced operator explicitly.

REMARK. Given an arbitrary submanifold $M^n \subset E^{n+N}$ with the field of pairwise orthogonal unit normals $\vec{n}_\alpha$, one can introduce the expansions

$$d\vec{n}_\alpha = \omega_\alpha^\beta \vec{n}_\beta \mod TM^n.$$

The 1-forms ω_α^β ($\omega_\alpha^\beta + \omega_\beta^\alpha = 0$) are said to be the forms of *normal connection*. Submanifolds with flat normal connection can be characterized equivalently by the requirement $\omega_\alpha^\beta \equiv 0$, or $d\vec{n}_\alpha \in TM^n$ for any $\alpha = n+1, \ldots, n+N$. If one applies the procedure of Dirac reduction to an arbitrary submanifold M^n, the local part of the reduced operator will remain the same, while the nonlocal tail

$$\sum_{\alpha=n+1}^{n+N} w_{\alpha k}^i R_x^k d^{-1} w_{\alpha l}^j R_x^l,$$

becomes

$$\sum_{\alpha=n+1}^{n+N} w_{\alpha k}^i R_x^k (\nabla^\perp)^{-1} w_{\alpha l}^j R_x^l.$$

Here $\nabla^\perp$ denotes parallel transport in the normal connection, i.e., the expression $(\nabla^\perp)^{-1}\Phi_\alpha$ is to be understood as the solution ϕ_α of the linear system

$$\nabla^\perp \phi_\alpha \equiv \frac{d}{dx}\phi_\alpha + \omega_\alpha^\beta \phi_\beta = \Phi_\alpha.$$

If the normal connection is flat, then $\omega_\alpha^\beta \equiv 0$ and $\nabla^\perp = d$.

We illustrate the procedure of Dirac reduction for two particular examples.

Example 3.1. Hydrodynamic analog of the Newmann problem. Let us consider Euclidean space $E^{n+1}(u^1, \ldots, u^{n+1})$ with the Hamiltonian system

$$u_t^I = \delta^{IJ} d \frac{\delta H}{\delta u^J}, \qquad I, J = 1, \ldots, n+1,$$

defined by the quadratic Hamiltonian $H = \int \sum_1^{n+1} a^I (u^I)^2 dx$. This system is linear and describes $n+1$ noninteracting waves $u_t^I = 2a^I u_x^I$, $a^I = \text{const}$. In order to reduce this system to a unit hypersphere $S^n \subset E^{n+1}$, we must proceed as follows:

- introduce any suitable curvilinear coordinates R^i on the sphere S^n;
- reduce the operator $\delta^{IJ} d$ to S^n; the result will be the nonlocal operator A^{ij} of the form (1.3);
- restrict the Hamiltonian H to S^n;
- write out the reduced equations $R_t^i = A^{ij} \delta H/\delta R^j$.

In our case it is convenient to work in the spherical-conical coordinates

$$u^1 = \sqrt{\frac{\prod(R^i - a^1)}{\prod_{K \neq 1}(a^K - a^1)}}, \quad \ldots, \quad u^{n+1} = \sqrt{\frac{\prod(R^i - a^{n+1})}{\prod_{K \neq n+1}(a^K - a^{n+1})}}.$$

In this parametrization, the metric of S^n assumes the form

$$ds^2 = \sum_{i=1}^n g_{ii} (dR^i)^2, \qquad g_{ii} = \frac{\prod_{k \neq i}(R^k - R^i)}{4P(R^i)},$$

where $P(R) = \prod_1^{n+1}(R - a^K)$. In the same coordinates R^i, the Hamiltonian H can be rewritten as follows:

$$H = \int \sum_1^{n+1} a^I (u^I)^2 dx = -\int \left(\sum_1^n R^i - c\right) dx, \qquad c = \sum a^K.$$

The reduced system has the form $R_t^i = A^{ij} \delta H/\delta R^j$, where A^{ij} is the nonlocal operator (1.3) generated by the metric g_{ii}. These equations are of the form

(3.3)
$$R_t^i = \left(\sum R^k + 2R^i\right) R_x^i$$

(see Example 1.1). In some sense the system obtained can be called the "hydrodynamic analog" of the Newmann problem. Indeed, just like the Newmann problem in classical mechanics, this system is obtained as the result of Dirac reduction of the linear system with quadratic Hamiltonian to the unit hypersphere.

Example 3.2. Hydrodynamic analog of geodesics on a quadric. Let us consider Euclidean space $E^{n+1}(u^1, \ldots, u^{n+1})$ with the Hamiltonian system

$$u_t^I = \delta^{IJ} d \frac{\delta H}{\delta u^J} = 0, \qquad I, J = 1, \ldots, n+1,$$

defined by the trivial Hamiltonian with zero density. Let us reduce this system to a hypersurface $M^n \subset E^{n+1}$. The reduced Hamiltonian operator is of the form

$$A^{ij} = g^{ij} d - g^{is} \Gamma^j_{sk} R^k_x + w^i_k R^k_x d^{-1} w^j_l R^l_x,$$

where g^{ij} and w^i_j are the metric and the Weingarten operator of M^n, respectively. The reduced system will be of the form

$$R_t^i = A^{ij} \frac{\delta H}{\delta R^j} = A^{ij}(0) = w^i_k R^k_x d^{-1}(0) = w^i_k R^k_x$$

(we take $1 = d^{-1}(0)$). The matrix w^i_k of the reduced system is nothing but the Weingarten operator of the hypersurface M^n. The equations

$$(3.4) \qquad R_t^i = w^i_k R^k_x$$

can be called the "hydrodynamic analog" of the equations of geodesics. Indeed, in classical mechanics, equations of geodesics also can be obtained as the result of Dirac reduction of the system with zero Hamiltonian (i.e., of free motion) to a hypersurface M^n.

Let us consider, for example, a quadric

$$a^1 (u^1)^2 + \cdots + a^{n+1} (u^{n+1})^2 = \left(\prod a^K \right)^{-1}, \qquad a^K = \text{const}.$$

In the coordinates $R^1, \ldots, R^n$ of the lines of curvature, we have

$$u^1 = \frac{1}{a^1 \sqrt{\prod R^i}} \sqrt{\frac{\prod (R^i - a^1)}{\prod_{K \neq 1} (a^K - a^1)}}, \ldots,$$

$$u^{n+1} = \frac{1}{a^{n+1} \sqrt{\prod R^i}} \sqrt{\frac{\prod (R^i - a^{n+1})}{\prod_{K \neq n+1} (a^K - a^{n+1})}},$$

and the Weingarten operator becomes diagonal [33]: $w^i_j = w^i \delta^i_j$. Here the eigenvalues w^i, called the *principal curvatures* of the hypersurface M^n, are given by $w^i = R^i \sqrt{\prod R^k}$. Hence system (3.4) assumes the diagonal form

$$(3.5) \qquad R_t^i = R^i \sqrt{\prod R^k} \, R^i_x.$$

Equations (3.5) can be called the "hydrodynamic analog" of geodesics on a quadric. It is interesting to note that equations (3.5) are nothing but the dispersionless limit of coupled Harry–Dym equations [42].

Thus applying the procedure of Dirac reduction, we arrive at two remarkable systems of hydrodynamic type:
- the hydrodynamic analog of the Newmann problem (3.3):

$$R^i_t = \left(\sum R^k + 2R^i\right) R^i_x;$$

- the hydrodynamic analog of geodesics on a quadric (3.5):

$$R^i_t = R^i \sqrt{\prod R^k}\, R^i_x.$$

According to [43], in the classical situation equations of geodesics on a quadric can be transformed into that of the Newmann problem through a suitable reparametrization of trajectories. The "hydrodynamic analog" of this construction reads as follows: equations $R^i_t = R^i \sqrt{\prod R^k}\, R^i_x$, after being rewritten in the new independent variables y, t, where

$$dy = \frac{2}{\sqrt{\prod R^k}}\, dx - \sum R^k\, dt$$

assume the form $R^i_t = (\sum R^k + 2R^i)\, R^i_y$.

There exist three important classes of submanifolds with flat normal connection:

1. Hypersurfaces $M^n \subset E^{n+1}$.

2. Orbits of coadjoint representation of simple Lie groups. We shall discuss this construction in more detail. Let $\mathfrak{g}^*$ be the dual space to a simple Lie algebra $\mathfrak{g}$. Then $\mathfrak{g}^*$ is naturally endowed with two compatible Hamiltonian operators $A^{IJ} = \eta^{IJ} d$ and $\omega^{IJ} = c^{IJ}_K u^K$, where η^{IJ} and c^{IJ}_K (here I, J, $K = 1, \ldots,$ dim $\mathfrak{g}$) are the Killing form and the structure constants of $\mathfrak{g}$, respectively. Let $M^n \subset \mathfrak{g}^*$ be an orbit of the coadjoint representation. Reducing A to M^n, we obtain

$$A^{ij} = g^{ij} d - g^{is} \Gamma^j_{sk} R^k_x + \sum_{\alpha=n+1}^{\dim \mathfrak{g}} w^i_{\alpha k} R^k_x d^{-1} w^j_{\alpha l} R^l_x.$$

The corresponding Weingarten operators w_α have the following important properties:

- The eigenvalues of w_α are constants (depending on α). We recall that submanifolds with all Weingarten operators having constant eigenvalues are called *isoparametric*.
- The net of lines of curvature of M^n is not holonomic (in general), i.e., it cannot be chosen for the coordinate system on M^n. We recall that the lines of curvature of a submanifold with flat normal connection are integral lines of the common eigendistributions of the commuting family w_α.

Restricting ω^{IJ} to M^n, we obtain the nondegenerate bivector ω^{ij}. After the reduction, the operators A^{ij} and ω^{ij} remain compatible. This construction is ultimately related to the theory of reductions of integrable systems associated with the spectral problem

$$\psi_x = M\psi, \qquad M \in \mathfrak{g}.$$

3. Submanifolds with a holonomic net of lines of curvature. For these submanifolds the metric and the Weingarten operators in the coordinates R^i of the lines of curvature take the diagonal form $\sum g_{ii} dR^{i^2}$ and $w^i_{\alpha j} = w^i_\alpha \delta^i_j$, respectively. The Gauss–Peterson–Codazzi equations can be rewritten as follows:

$$\partial_j \ln \sqrt{g_{ii}} = \frac{\partial_j w^i_\alpha}{w^j_\alpha - w^i_\alpha} \quad \text{for any } \alpha, i \neq j \quad \text{(Peterson–Codazzi)},$$

$$R^{ij}_{ij} = \sum_{\alpha=n+1}^{n+N} w^i_\alpha w^j_\alpha \quad \text{(Gauss)}$$

(the remaining components of the curvature tensor vanish). In the coordinates R^i, the Hamiltonian operator (1.13) takes the form

$$A^{ij} = g^{ii} \delta^{ij} d - g^{ii} \Gamma^j_{ik} R^k_x + \sum_{\alpha=n+1}^{n+N} w^i_\alpha R^i_x d^{-1} w^j_\alpha R^j_x.$$

The Dirac reduction to a submanifold with a holonomic net of lines of curvature naturally arises, e.g., in the following situation. In the ambient Euclidean space E^{n+N}, let us consider the Hamiltonian system $u^I_t = \delta^{IJ} d \, \delta H/\delta u^J$ which is diagonalizable, i.e., can be transformed into the Riemann invariant form

(3.6) $$R^I_t = \lambda^I(R) R^I_x, \quad I = 1, \ldots, n+N.$$

Here R^I is a certain orthogonal curvilinear coordinate system in the Euclidean space E^{n+N}. Let us impose the reduction

(3.7) $$R^{n+1} = \text{const}, \ldots, R^{n+N} = \text{const}.$$

The remaining equations

(3.8) $$R^i_t = \lambda^i(R) R^i_x, \quad i = 1, \ldots, n,$$

describe the so-called quasisimple waves in system (3.6). In order to put equations (3.8) in Hamiltonian form, we must reduce the operator $\delta^{IJ} d$ to the submanifold (3.7). This submanifold (as, in fact, any coordinate submanifold of any orthogonal system in Euclidean space) has a flat normal connection and a holonomic net of lines of curvature.

Now let us turn to the proof of Theorem 6. For the sake of simplicity, we shall restrict ourselves to the case when the submanifold $M^n \subset E^{n+N}$ has a holonomic net of lines of curvature. In this case we can include M^n in a suitable orthogonal curvilinear coordinate system R^I, $I = 1, \ldots, n+N$, i.e., introduce orthogonal coordinates R^I in E^{n+N} such that the equations of M^n assume the simplest form

$$R^\alpha = 0, \quad \alpha = n+1, \ldots, n+N.$$

Although such an inclusion is not unique, the final result will not depend on the particular choice of the coordinate system R^I. In what follows $I, J, K =$

$1, \ldots, n+N$; $\alpha, \beta, \gamma = n+1, \ldots, n+N$; $i, j, k = 1, \ldots, n$ (the first n coordinates R^i can be viewed as the coordinates on M^n). In the variables R^I, the Hamiltonian operator $\delta^{IJ} d$ assumes the form

$$A^{IJ} = g^{II} \delta^{IJ} d - g^{II} \Gamma^J_{IK} R^K_x.$$

It defines the following Poisson brackets

$$\{R^I(x), R^J(y)\} = g^{II}(R(x)) \delta^{IJ} \delta'(x-y) - g^{II}(R(x)) \Gamma^J_{IK} R^K_x \delta(x-y).$$

In E^{n+N}, let us consider the Hamiltonian system

$$R^I_t = A^{IJ} \frac{\delta H}{\delta R^J},$$

with an arbitrary Hamiltonian H. According to the Dirac procedure [29], the reduction of this system to M^n has the form

(3.9) $$R^i_t = A^{iJ} \frac{\delta H}{\delta R^J} + \int \{R^i(x), R^\beta(y)\} m_\beta(y) \, dy,$$

where the multipliers $m_\beta(y)$ are specified by

(3.10) $$0 = R^\alpha_t = A^{\alpha J} \frac{\delta H}{\delta R^J} + \int \{R^\alpha(x), R^\beta(y)\} m_\beta(y) \, dy.$$

Let us first rewrite equations (3.9), putting $R^\alpha = 0$ after all the Poisson brackets are calculated:

$$R^i_t = A^{ij} \frac{\delta H}{\delta R^j} + A^{i\beta} \frac{\delta H}{\delta R^\beta}$$
$$+ \int (g^{ii}(R(x)) \delta^{i\beta} \delta'(x-y) - g^{ii}(R(x)) \Gamma^\beta_{ik} R^k_x \delta(x-y)) m_\beta(y) \, dy$$
$$= A^{ij} \frac{\delta H}{\delta R^j} - g^{ii} \Gamma^\beta_{ik} R^k_x \left(\frac{\delta H}{\delta R^\beta} + m_\beta(x) \right).$$

As to the diagonal metrics $g_{II} \, dR^{I^2}$, the only nonzero Christoffel symbols are

$$\Gamma^I_{IJ} = \frac{\partial_J g_{II}}{2 g_{II}}, \qquad \Gamma^J_{II} = -\frac{\partial_J g_{II}}{2 g_{JJ}} \quad (I \neq J),$$

one can represent R^i_t in the form

(3.11) $$R^i_t = A^{ij} \frac{\delta H}{\delta R^j} + \frac{\partial_\beta \sqrt{g_{ii}}}{\sqrt{g_{ii}} \sqrt{g_{\beta\beta}}} R^i_x \left(\frac{\delta H}{\delta R^\beta} + m_\beta(x) \right).$$

Here $A^{ij} = g^{ij} d - g^{is} \Gamma^j_{sk} R^k_x$. Now let us find $m_\beta(x)$ from equations (3.10), putting again $R^\alpha = 0$ after all the Poisson brackets are calculated:

$$0 = A^{\alpha j} \frac{\delta H}{\delta R^j} + A^{\alpha \beta} \frac{\delta H}{\delta R^\beta}$$
$$+ \int (g^{\alpha\alpha}(R(x)) \delta^{\alpha\beta} \delta'(x-y) - g^{\alpha\alpha}(R(x)) \Gamma^\beta_{\alpha k} R^k_x \delta(x-y)) m_\beta(y) \, dy$$
$$= g^{\alpha\alpha} \left(\frac{d}{dx} \left(\frac{\delta H}{\delta R^\alpha} + m_\alpha(x) \right) - \Gamma^\beta_{\alpha k} R^k_x \left(\frac{\delta H}{\delta R^\beta} + m_\beta(x) \right) - \Gamma^j_{\alpha k} R^k_x \frac{\delta H}{\delta R^j} \right).$$

Hence

$$\frac{d}{dx}\left(\frac{\delta H}{\delta R^\alpha}+m_\alpha(x)\right)-\Gamma^\beta_{\alpha k}R^k_x\left(\frac{\delta H}{\delta R^\beta}+m_\beta(x)\right)=\Gamma^j_{\alpha k}R^k_x\frac{\delta H}{\delta R^j}.$$

Using the expressions for the Christoffel symbols given above, these equations can be rewritten in the form

$$\frac{d}{dx}\left(\frac{\delta H/\delta R^\alpha+m_\alpha(x)}{\sqrt{g_{\alpha\alpha}}}\right)=\frac{\partial_\alpha\sqrt{g_{jj}}}{\sqrt{g_{jj}}\sqrt{g_{\alpha\alpha}}}R^j_x\frac{\delta H}{\delta R^j}.$$

Thus

$$\frac{\delta H}{\delta R^\alpha}+m_\alpha(x)=\sqrt{g_{\alpha\alpha}}\,d^{-1}\left(\frac{\partial_\alpha\sqrt{g_{jj}}}{\sqrt{g_{jj}}\sqrt{g_{\alpha\alpha}}}R^j_x\frac{\delta H}{\delta R^j}\right).$$

Inserting these expressions in (3.11), we arrive at the reduced equations expressed in Hamiltonian form

$$R^i_t=\left(A^{ij}+\sum_{\alpha=n+1}^{n+N}\frac{\partial_\alpha\sqrt{g_{ii}}}{\sqrt{g_{ii}}\sqrt{g_{\alpha\alpha}}}R^i_x d^{-1}\frac{\partial_\alpha\sqrt{g_{jj}}}{\sqrt{g_{jj}}\sqrt{g_{\alpha\alpha}}}R^j_x\right)\frac{\delta H}{\delta R^j}.$$

It remains to note that the quantity

$$w^i_\alpha=\frac{\partial_\alpha\sqrt{g_{ii}}}{\sqrt{g_{ii}}\sqrt{g_{\alpha\alpha}}}$$

is nothing but the eigenvalue of the Weingarten operator w_α corresponding to the unit normal $(g_{\alpha\alpha})^{-1/2}\partial/\partial R^\alpha$ of the submanifold M^n. This completes the proof.

§4. Reciprocal transformations

It was pointed out in [5] that an arbitrary Hamiltonian system of hydrodynamic type

$$(4.1)\qquad u^i_t=\delta^{ij}\frac{d}{dx}\frac{\partial h(u)}{\partial u^j}=v^i_j(u)u^j_x\qquad\left(v^i_j=\frac{\partial^2 h}{\partial u^i\partial u^j}\right)$$

remains Hamiltonian (in the local sense) after a linear change of the independent variables

$$(4.2)\qquad X=bx+at,\qquad T=\widetilde{b}x+\widetilde{a}t$$

(b, a, $\widetilde{b}$, $\widetilde{a}$ are constants). It turns out, however, that Hamiltonian systems (4.1) are actually preserved by a much broader class of transformations

$$(4.3)\qquad dX=B(u)\,dx+A(u)\,dt,\qquad dT=\widetilde{B}(u)\,dx+\widetilde{A}(u)\,dt,$$

where $B(u)\,dx+A(u)\,dt$ and $\widetilde{B}(u)\,dx+\widetilde{A}(u)\,dt$ are two integrals of system (4.1). If B, A, $\widetilde{B}$, $\widetilde{A}$ are constants, we return to linear transformations (4.2). Transformations of the type (4.3) were first introduced in gas dynamics and are known as *reciprocal* [36]. Reciprocal transformations of hydrodynamic type systems were extensively investigated by the author in [33–35]. Applying reciprocal transformation (4.3) to system (4.1), we arrive at the new system

(4.4) $$u_T^i = V_j^i(u) u_X^j,$$

where the transformed matrix $V_j^i(u)$ is given by

$$V_j^i = (v_j^i B - \delta_j^i A)(\delta_j^i \widetilde{A} - v_j^i \widetilde{B})^{-1}.$$

Although in general system (4.4) is no longer Hamiltonian in the local sense (4.1), it is always possible to put it into Hamiltonian form for an appropriate nonlocal Hamiltonian operator (1.13) (or (1.14)). However, the proof of this statement is beyond the scope of this survey.

There exist a special subclass of transformations of reciprocal type that are of particular interest. To introduce them we first recall that an arbitrary Hamiltonian system (4.1) possesses $n+2$ canonical integrals:

$$\text{Hamiltonian}: \quad H = h\, dx + \left(\sum h_s^2/2 + 1/2\right) dt;$$
$$\text{Impulse}: \quad P = \left(\sum u^{s^2}/2 + 1/2\right) dx + \left(\sum h_s u^s - h\right) dt;$$
$$\text{Casimirs}: \quad U^i = u^i\, dx + h_i\, dt, \qquad i = 1, \ldots, n$$

(the constant $1/2$ in the expressions for H and P is essential in what follows). Let us consider those reciprocal transformations (4.3) for which both integrals $B(u)\,dx + A(u)\,dt$ and $\widetilde{B}(u)\,dx + \widetilde{A}(u)\,dt$ are linear combinations of the canonical ones:

(4.5)
$$dX = B(u)\,dx + A(u)\,dt = \alpha H + \beta P + c_i U^i + b\,dx + a\,dt,$$
$$dT = \widetilde{B}(u)\,dx + \widetilde{A}(u)\,dt = \widetilde{\alpha} H + \widetilde{\beta} P + \widetilde{c}_i U^i + \widetilde{b}\,dx + \widetilde{a}\,dt.$$

Here α, β, c_i, b, a and $\widetilde{\alpha}$, $\widetilde{\beta}$, $\widetilde{c}_i$, $\widetilde{b}$, $\widetilde{a}$ are arbitrary constants.

THEOREM 7 [33]. *System* (4.4), *obtained from* (4.1) *by applying reciprocal transformation* (4.5), *will be Hamiltonian in the nonlocal sense* (1.3) (*i.e., with respect to the nonlocal operator corresponding to a metric of constant curvature* c), *if and only if the constants satisfy the constraints*:

(4.6)
$$(\alpha + a)^2 + (\beta + b)^2 - \sum c_i^2 - a^2 - b^2 = c,$$
$$(\alpha + a)(\widetilde{\alpha} + \widetilde{a}) + (\beta + b)(\widetilde{\beta} + \widetilde{b}) - \sum c_i \widetilde{c}_i - a\widetilde{a} - b\widetilde{b} = 0,$$
$$(\widetilde{\alpha} + \widetilde{a})^2 + (\widetilde{\beta} + \widetilde{b})^2 - \sum \widetilde{c}_i^2 - \widetilde{a}^2 - \widetilde{b}^2 = 0.$$

The metric g_{ij} *of the corresponding Hamiltonian operator* (1.3) *is given by*

(4.7) $$ds^2 = g_{ij}\,du^i du^j = \frac{1}{(B\widetilde{A} - A\widetilde{B})^2} \sum_s (\delta_i^s \widetilde{A} - v_i^s \widetilde{B})(\delta_j^s \widetilde{A} - v_j^s \widetilde{B}).$$

Transformations (4.5) are closely related to Lie spherical transformations [33].

EXAMPLE 4.1. Let the only nonzero constants be b, a, $\tilde{b}$, $\tilde{a}$. Then the transformation (4.5) reduces to (4.2) and the transformed system is Hamiltonian in the local sense (i.e., $c = 0$).

EXAMPLE 4.2. Let the only nonzero constants be $\beta = 1$, $\tilde{a} = 1$. Then the transformation (4.5) assumes the form

$$(4.8) \quad dX = P = \left(\sum u^{s^2}/2 + 1/2\right)dx + \left(\sum h_s u^s - h\right)dt, \quad dT = dt,$$

and the transformed system will be Hamiltonian in the nonlocal sense (1.3) with the Hamiltonian operator A^{ij}

$$A^{ij} = g^{ij}\frac{d}{dX} - g^{is}\Gamma^{j}_{sk}u^k_X + u^i_X \left(\frac{d}{dX}\right)^{-1} u^j_X$$

generated by the metric of constant curvature $c = 1$:

$$(4.9) \quad ds^2 = \frac{du^{1^2} + \cdots + du^{n^2}}{P^2}, \quad P = \sum \frac{u^{s^2}}{2} + \frac{1}{2}$$

(apply formula (4.7) with $\tilde{A} = 1$, $\tilde{B} = 0$, $B = P$). In the 2-component case, the corresponding operator A^{ij} was written out explicitly in Example 1.2, see the Introduction. The x-part of the transformation (4.8):

$$(4.10) \quad dX = \left(\sum u^{s^2}/2 + 1/2\right)dx = P\,dx$$

can be viewed as a reparametrization of the phase space. Let us take two hydrodynamic functionals

$$(4.11) \quad F = \int f(u)\,dX, \quad G = \int g(u)\,dX,$$

in the X-parametrization. Applying transformation (4.10), we can rewrite them in the x-parametrization as follows:

$$(4.12) \quad F = \int \tilde{f}(u)\,dx, \quad G = \int \tilde{g}(u)\,dx,$$

where $\tilde{f} = fP$, $\tilde{g} = gP$. It is now a straightforward exercise to verify the identity [17]:

$$\int \frac{\partial f}{\partial u^i} A^{ij} \frac{\partial g}{\partial u^j} dX = \int \frac{\partial \tilde{f}}{\partial u^i} \delta^{ij} \frac{d}{dx} \frac{\partial \tilde{g}}{\partial u^j} dx.$$

This means that the Poisson bracket of any two functionals (4.11) calculated with respect to the Hamiltonian operator A^{ij} in the X-parametrization is equal to the Poisson bracket of the same functionals (4.12) with respect to the Hamiltonian operator $\delta^{ij}\,d/dx$ in the x-parametrization. Hence the reparametrization (4.10) establishes an explicit link between the nonlocal operator A^{ij} and the local operator $\delta^{ij}\,d/dx$, thus showing the validity of the Darboux theorem for nonlocal operators of type (1.3) with $c = 1$.

EXAMPLE 4.3. Let the only nonzero constants be $c_1 = 1$, $\widetilde{a} = 1$. Then transformation (4.5) assumes the form

(4.13) $$dX = U^1 = u^1\, dx + h_1\, dt, \qquad dT = dt,$$

and the transformed system will be Hamiltonian in the nonlocal sense (1.3) with the following Hamiltonian operator A^{ij}

$$A^{ij} = g^{ij}\frac{d}{dX} - g^{is}\Gamma^j_{sk}u^k_X - u^i_X\left(\frac{d}{dX}\right)^{-1}u^j_X$$

generated by the metric of constant curvature $c = -1$:

(4.14) $$ds^2 = ((du^1)^2 + \cdots + (du^n)^2)/(u^1)^2$$

(apply formula (4.7) with $\widetilde{A} = 1$, $\widetilde{B} = 0$, $B = u^1$). In the 2-component case the corresponding operator A^{ij} was written explicitly in Example 1.3, see the Introduction. The x-part of transformation (4.13),

(4.15) $$dX = u^1\, dx,$$

can be viewed as a reparametrization of the phase space. Let us take two hydrodynamic functionals

(4.16) $$F = \int f(u)\, dX, \qquad G = \int g(u)\, dX,$$

in the X-parametrization. Applying transformation (4.15), we can rewrite them in the x-parametrization as follows:

(4.17) $$F = \int \widetilde{f}(u)\, dx, \qquad G = \int \widetilde{g}(u)\, dx,$$

where $\widetilde{f} = fu^1$, $\widetilde{g} = gu^1$. It is now a straightforward exercise to verify the identity:

$$\int \frac{\partial f}{\partial u^i} A^{ij} \frac{\partial g}{\partial u^j}\, dX = \int \frac{\partial \widetilde{f}}{\partial u^i} \delta^{ij} \frac{d}{dx} \frac{\partial \widetilde{g}}{\partial u^j}\, dx.$$

This means that the Poisson bracket of any two functionals (4.16) calculated with respect to the Hamiltonian operator A^{ij} in the X-parametrization is equal to the Poisson bracket of the same functionals (4.17) with respect to the Hamiltonian operator $\delta^{ij}\, d/dx$ in the x-parametrization. Hence reparametrization (4.15) establishes an explicit link between the nonlocal operator A^{ij} and the local operator $\delta^{ij}\, d/dx$, thus showing the validity of the Darboux theorem for nonlocal operators of type (1.3) with $c = -1$.

§5. Hamiltonian formalism for semi-Hamiltonian systems

According to the results of S. P. Tsarev [5], a system of hydrodynamic type in Riemann invariants

$$(5.1) \qquad R^i_t = v^i(R) R^i_x, \qquad i = 1, \ldots, n,$$

is Hamiltonian in the local sense if and only if there exist a diagonal flat metric $\sum g_{ii} dR^{i^2}$, satisfying the relations

$$(5.2) \qquad \partial_j \ln \sqrt{g_{ii}} = \frac{\partial_j v^i}{v^j - v^i} \quad \text{for any } i \neq j.$$

Cross-differentiating (5.2), one arrives at the identities

$$(5.3) \qquad \partial_k \frac{\partial_j v^i}{v^j - v^i} = \partial_j \frac{\partial_k v^i}{v^k - v^i} \quad \text{for any } i \neq j \neq k \neq i.$$

Identities (5.3) are thus automatically satisfied by the eigenvalues v^i of any Hamiltonian system. However, the class of systems (5.1) satisfying identities (5.3) is wider than that of locally Hamiltonian systems. This observation justifies the following

DEFINITION [5]. System (5.1) is called *semi-Hamiltonian* if there exist a diagonal metric $\sum g_{ii} dR^{i^2}$ (not necessarily flat) satisfying relations (5.2), or, equivalently, if the eigenvalues v^i satisfy identities (5.3).

An interesting example of semi-Hamiltonian system, which is not Hamiltonian in the local sense, arises in chemical kinetics (see [44, 45] and the example below).

Semi-Hamiltonian systems possess an infinite number of commuting flows

$$R^i_\tau = w^i(R) R^i_x,$$

where the functions w^i solve the linear system

$$(5.4) \qquad \frac{\partial_j w^i}{w^j - w^i} = \frac{\partial_j v^i}{v^j - v^i}, \qquad i \neq j.$$

All semi-Hamiltonian systems can be integrated by the generalized hodograph transform [5].

In [17] we have presented a series of considerations showing that any semi-Hamiltonian system is in fact Hamiltonian if the nonlocal operator (1.13) (or (1.14)) is appropriately selected. Let us recall the main steps of our construction.

1. Given the semi-Hamiltonian system (5.1), first determine coefficients g_{ii} of the corresponding diagonal metric from equations (5.2). Note that each g_{ii} is determined up to transformations of the form $g_{ii} \to g_{ii}/\varphi^i(R^i)$, where φ^i is an arbitrary function of the variable R^i. One can easily show that the condition of being semi-Hamiltonian is equivalent to the components R^i_{kkj} of the curvature tensor of this metric being zero when $i \neq j \neq k \neq i$.

2. Find the expansions of the remaining components $R^{ij}_{ij} = g^{ii} R^{j}_{iij}$ of the curvature tensor over the "squares" of the solutions w^i_α of the linear system (5.4):

$$R^{ij}_{ij} = \sum_\alpha \varepsilon_\alpha w^i_\alpha w^j_\alpha, \qquad \varepsilon_\alpha = \pm 1. \tag{5.5}$$

It was conjectured in [17] that such an expansion is always possible (a rigorous proof of this conjecture is still lacking). Note that in general the sum in (5.5) will be infinite.

3. Write out the Hamiltonian operator

$$A^{ij} = g^{ij} d - g^{is} \Gamma^j_{sk} R^k_x + \sum_\alpha \varepsilon_\alpha w^i_\alpha R^i_x d^{-1} w^j_\alpha R^j_x \tag{5.6}$$

(equations (5.2), (5.4), and (5.5) ensure the Jacobi identity, see §3). The system under study is automatically Hamiltonian with respect to the nonlocal operator A^{ij} (see [17]). Since the metric g_{ii} is determined up to n arbitrary functions $\varphi^i(R^i)$, our system can actually be expressed in Hamiltonian form in infinitely many different ways. All these Hamiltonian operators are mutually compatible.

EXAMPLE 5.1. The system of chromatography equations in Riemann invariants is of the form [17]:

$$R^i_t = \left(R^i \prod_1^n R^k \right)^{-1} R^i_x, \qquad i = 1, \ldots, n. \tag{5.7}$$

It was proved in [45] that for $n \geq 3$ it does not possess local Hamiltonian structures. The corresponding equations (5.2) assume the form

$$\partial_j \ln \sqrt{g_{ii}} = 1/(R^j - R^i), \qquad i \neq j. \tag{5.8}$$

Hence

$$g_{ii} = \frac{\prod_{k \neq i}(R^k - R^i)^2}{\varphi^i(R^i)}, \tag{5.9}$$

where $\varphi^i(R^i)$ are n arbitrary functions of one variable. As noted in [45], for the chromatography equations, the linear system (5.4) can be explicitly integrated:

$$w^i = \partial_i \left(\sum_{k=1}^n \frac{f^k(R^k)}{\prod_{l \neq k}(R^l - R^k)} \right),$$

where $f^k(R^k)$ are again n arbitrary functions of one variable. Let us take n particular solutions w^i_α ($\alpha = 1, \ldots, n$) as follows:

$$w^i_1 = \partial_i \left(\frac{\sqrt{\varphi^1(R^1)}}{\prod_{l \neq 1}(R^l - R^1)} \right), \ldots, w^i_n = \partial_i \left(\frac{\sqrt{\varphi^n(R^n)}}{\prod_{l \neq n}(R^l - R^n)} \right),$$

where $\varphi^i(R^i)$ are the same as in (5.9). A direct computation performed by M. V. Pavlov (see also [17]) shows that when the functions w^i_α are chosen in this way, we have

$$R^{ij}_{ij} = -\sum_1^n w^i_\alpha w^j_\alpha.$$

Thus the chromatography equations possess infinitely many nonlocal Hamiltonian structures. It is remarkable that the nonlocal tails of all of them are of finite length n.

§6. Averaging theory

The system of averaged KdV equations has the form

(6.1) $$R^i_t = v^i(R) R^i_x, \qquad i = 1, 2, 3,$$

where the functions $v^i(R)$ are given by

$$v^1 = \frac{R^1 + R^2 + R^3}{3} - \frac{2}{3} \frac{(R^2 - R^1)K}{K - E},$$

$$v^2 = \frac{R^1 + R^2 + R^3}{3} - \frac{2}{3} \frac{(R^2 - R^1)(1 - s^2)K}{E - (1 - s^2)K},$$

$$v^3 = \frac{R^1 + R^2 + R^3}{3} + \frac{2}{3} \frac{(R^3 - R^1)(1 - s^2)K}{E}.$$

Here $s^2 = (R^2 - R^1)/(R^3 - R^1)$ and $E(s)$ and $K(s)$ are complete elliptic integrals:

$$E' = \frac{E - K}{s}, \qquad K' = \frac{E - (1 - s^2)K}{(1 - s^2) s}.$$

System (6.1) possesses two local Hamiltonian structures A_0 and A_1 (see [1, 6]) generated by two diagonal flat metrics g_0^{ii} and g_1^{ii}:

$$g_0^{11} = -\frac{1}{4} \frac{s^2 K^2}{(K - E)^2}, \quad g_0^{22} = \frac{1}{4} \frac{s^2(1 - s^2)K^2}{(E - (1 - s^2)K)^2}, \quad g_0^{33} = -\frac{1}{4} \frac{(1 - s^2)K^2}{E^2},$$

and

$$g_1^{11} = \frac{s^2 K^2}{(K - E)^2} R^1, \quad g_1^{22} = -\frac{s^2(1 - s^2)K^2}{(E - (1 - s^2)K)^2} R^2, \quad g_1^{33} = \frac{(1 - s^2)K^2}{E^2} R^3.$$

These Hamiltonian structures are known to be the result of averaging two local compatible Hamiltonian operators B_0 and B_1 of the KdV equation:

$$B_0 = d, \qquad B_1 = d^3 + ud + du.$$

It is remarkable that system (6.1) also possesses a nonlocal Hamiltonian operator A_2 of type (1.3) generated by diagonal metric g_2^{ii} of constant curvature $c = -1$ [38]:

$$g_2^{11} = -4 \frac{s^2 K^2}{(K - E)^2} R^{1^2}, \qquad g_2^{22} = 4 \frac{s^2(1 - s^2)K^2}{(E - (1 - s^2)K)^2} R^{2^2},$$

$$g_2^{33} = -4 \frac{(1 - s^2)K^2}{E^2} R^{3^2}.$$

The Hamiltonian structure A_2 can be interpreted as the result of "averaging" the third nonlocal Hamiltonian operator B_2 of the KdV equation, which has a specific "constant curvature" nonlocal term:

$$B_2 = RB_1 = d^5 + 2(d^3 u + ud^3) + u_x d^2 - d^2 u_x + 4udu - u_x d^{-1} u_x.$$

Here R is the recursion operator:

$$R = B_1 B_0^{-1} = d^2 + 2u + u_x d^{-1}.$$

Similar nonlocal Hamiltonian structures appear in the case of averaged Sine-Gordon and averaged NLS equations. Although a general averaging procedure of nonlocal Poisson brackets has not yet been proposed, these important examples support the evidence that such a theory should exist. Some recent results in this connection are due to V. L. Alekseev and M. V. Pavlov [37, 38, 46].

Acknowledgements

I would like to express my sincere gratitude to S. P. Novikov, B. A. Dubrovin, H. Gümral, N. H. Ibragimov, V. G. Mikhalev, O. I. Mokhov, Y. Nutku, M. V. Pavlov, M. B. Sheftel, S. P. Tsarev, and A. P. Veselov for their encouragement and helpful discussions.

The research described in this publication was partially supported by Grant No. RKR000 from the International Science Foundation.

References

1. B. A. Dubrovin and S. P. Novikov, *Hamiltonian formalism of one-dimensional systems of hydrodynamic type and the Bogolyubov–Whitham averaging method*, Dokl. Akad. Nauk SSSR **270** (1983), no. 4, 781–785; English transl., Soviet Math. Dokl. **27** (1983), 665–669.
2. _____, *On Poisson brackets of hydrodynamic type*, Dokl. Akad. Nauk SSSR **279** (1984), no. 2, 294–297; English transl., Soviet Math. Dokl. **30** (1984), 651–654.
3. _____, *Hydrodynamics of weakly deformed soliton lattices. Differential geometry and Hamiltonian theory*, Uspekhi Mat. Nauk **44** (1989), no. 6, 29–98; English transl., Russian Math. Surveys **44** (1989), no. 6, 35–124.
4. _____, *Hydrodynamics of soliton lattices*, Soviet Sci. Rev. Sect. C: Math. Phys. Rev. **9** (1993), no. 4, 1–136.
5. S. P. Tsarev, *On Poisson brackets and one-dimensional Hamiltonian systems of hydrodynamic type*, Dokl. Acad. Nauk SSSR **282** (1985), no. 3, 534–537; English transl., Soviet Math. Dokl. **31** (1985), 488–491.
6. _____, *The geometry of Hamiltonian systems of hydrodynamic type. The generalized hodograph method*, Izv. Acad. Nauk SSSR Ser. Mat. **54** (1990), no. 5, 1048–1068; English transl., Math. USSR-Izv. **37** (1991), no. 2, 397–419.
7. O. I. Mokhov, *Poisson brackets of the Dubrovin-Novikov type (DN-brackets)*, Funktsional. Anal. i Prilozhen. **22** (1988), no. 4, 92–93; English transl., Functional Anal. Appl. **22** (1988), 336–338.
8. G. V. Potemin, *On Poisson brackets of differential-geometric type*, Dokl. Akad. Nauk SSSR **286** (1986), no. 1, 39–42; English transl., Soviet Math. Dokl. **33** (1986), 30–33.
9. P. W. Doyle, *Differential geometric Poisson bivectors in one space variable*, J. Math. Phys. **34** (1993), no. 4, 1314–1338.
10. N. I. Grinberg, *Poisson brackets of hydrodynamic type with degenerate metric*, Uspekhi Mat. Nauk **40** (1985), no. 4, 217–218; English transl., Russian Math. Surveys **40** (1985), no. 4, 231–232.
11. M. V. Polyak, *One-dimensional Hamiltonian systems of hydrodynamic type with an explicit dependence on the spatial variable*, Uspekhi Mat. Nauk **42** (1987), no. 3, 195–196; English transl., Russian Math. Surveys **42** (1987), no. 3, 229–230.
12. I. Ya. Dorfman, *Dirac structures for integrable evolution equations*, Doctorial Dissertation, Moscow, 1988.
13. A. A. Balinskiĭ and S. P. Novikov, *Poisson brackets of hydrodynamic type, Frobenius algebras and Lie algebras*, Dokl. Akad. Nauk SSSR **283** (1985), no. 5, 1036–1039; English transl., Soviet Math. Dokl. **32** (1985), 228–231.
14. O. I. Mokhov and E. V. Ferapontov, *Nonlocal Hamiltonian operators of hydrodynamic type associated with constant curvature metrics*, Uspekhi Mat. Nauk **45** (1990), no. 3, 191–192; English transl. in Russian Math. Surveys **45** (1990).
15. O. I. Mokhov, *Hamiltonian systems of hydrodynamic type and constant curvature metrics*, Phys. Lett. A **166** (1992), no. 3–4, 215–216.
16. E. V. Ferapontov, *Hamiltonian systems of hydrodynamic type and their realization on hypersurfaces of a pseudoeuclidean space*, Geom. Sbornik **22** (1990), 59–96; English transl., J. Soviet Math. **55** (1991), 1970–1995.
17. _____, *Differential geometry of nonlocal Hamiltonian operators of hydrodynamic type*, Funktsional. Anal. i Prilozhen. **25** (1991), no. 3, 37–49; English transl. in Functional Anal. Appl. **25** (1991).

18. E. V. Ferapontov and M. V. Pavlov, *Quasiclassical limit of Coupled KdV equations. Riemann invariants and multi-Hamiltonian structure*, Phys. D **52** (1991), 211–219.
19. E. V. Ferapontov, *Dirac reduction of the Hamiltonian operator $\delta^{ij}d/dx$ to a submanifold of the Euclidean space with flat normal connection*, Funktsional. Anal. i Prilozhen. **26** (1992), no. 4, 83–86; English transl. in Functional Anal. Appl. **26** (1992).
20. V. V. Sokolov, *On Hamiltonian property of Krichever-Novikov equation*, Dokl. Akad. Nauk SSSR **277** (1984), no. 1, 48–50; English transl. in Soviet Math. Dokl. **30** (1984).
21. Y. Nutku, *On a new class of completely integrable nonlinear wave equations. Multi-Hamiltonian structure*, J. Math. Phys. **28** (1987), 2579–2585.
22. P. Olver and Y. Nutku, *Hamiltonian structures for systems of hyperbolic conservation laws*, J. Math. Phys. **29** (1988), 1610–1619.
23. H. Gümral and Y. Nutku, *Multi-Hamiltonian structure of equations of hydrodynamic type*, J. Math. Phys. **31** (1990), 2606–2611.
24. P. W. Doyle, *Symmetry classes of 2-component hyperbolic systems*, Preprint.
25. M. Arik, et al., *Multi-Hamiltonian structure of the Born–Infeld equation*, J. Math. Phys. **30** (1989), 1338–1344.
26. B. A. Dubrovin, *Differential geometry of the space of orbits of a Coxeter group*, Preprint SISSA (1993), Trieste.
27. M. V. Pavlov, *Discrete symmetry and local Hamiltonian structures of hydrodynamic-type systems*, Uspekhi Mat. Nauk **48** (1993), no. 6, 167–168; English transl. in Russian Math. Surveys **48** (1993).
28. Yu. N. Sidorenko, *Elliptic bundle and generating operators*, Zap. Nauchn. Sem. Leningrad. Otdel. Mat. Inst. Steklov. (LOMI) **161** (1987), 76–87; English transl. in J. Soviet Math. **46** (1989), no. 5.
29. P. Dirac, *Principles of quantum mechanics*, Clarendon Press, Oxford, 1947.
30. I. Ishimory, *A relationship between the Ablowitz–Kaup–Newell–Segur and Wadati–Konno–Ishikawa schemes of the inverse scattering method*, J. Phys. Soc. Japan **51** (1982), no. 9, 3036–3041.
31. K. Konno, *Relationship among modified WKI equation, equation of motion of a thin vertex filament and that of a continuous Heisenberg spin system*, J. Phys. Soc. Japan **59** (1990), no. 40, 3417–3420.
32. E. V. Ferapontov, *Nonlocal matrix Hamiltonian operators. Differential geometry and applications*, Teoret. Mat. Fiz. **91** (1992), no. 3, 452–462; English transl. in Theoret. and Math. Phys. **91** (1992).
33. _____, *Dupin hypersurfaces and integrable Hamiltonian systems of hydrodynamic type, which do not possess Riemann invariants*, Diff. Geometry and Appl. (1994) (to appear).
34. _____, *Reciprocal transformations and their invariants*, Differentsial'nye Uravneniya **25** (1989), no. 7, 1256–1265; English transl., Differential Equations **25** (1989), no. 7, 898–905.
35. _____, *Reciprocal autotransformations and hydrodynamic symmetries*, Differentsial'nye Uravneniya **27** (1991), no. 7, 1250–1263; English transl. in Differential Equations **27** (1991), no. 7, 885–895.
36. C. Rogers, *Reciprocal transformations and their applications*, in "Nonlinear Evolutions", Proc. 5th Workshop on Nonlinear Evolution Equations and Dynamical Systems (1987), France, 109–123.
37. V. L. Alekseev, *private communication*.
38. M. V. Pavlov, *private communication*.
39. S. P. Tsarev, *Differential-geometric methods of integration of hydrodynamic type systems*, Doctorial Thesis (1993), Moscow.
40. M. Antonowicz and A. P. Fordy, *Coupled KdV equations with multi-Hamiltonian structures*, Phys. D **28** (1987), 345–357.
41. H. Gümral and Y. Nutku, *Bi-Hamiltonian structures of d-Boussinesq and Benney–Lax equations*, Phys. Lett. A: Math. Gen. **27** (1994), 193–200.
42. M. Antonowicz and A. P. Fordy, *Coupled Harry-Dym equations with multi-Hamiltonian structures*, J. Phys. A: Math. Gen. **21** (1988), 269–275.
43. H. Knörrer, *Geodesics on quadrics and a mechanical problem of C. Newmann*, J. Reine Angew. Math. **334** (1982), 69–78.
44. S. P. Tsarev, *Semi-Hamiltonian formalism of diagonal systems of hydrodynamic type, and integrability of the equations of chromatography and electrophoresis*, Preprint No. 106, LIIAN, Leningrad (1989). (Russian)

45. M. V. Pavlov, *Hamiltonian formalism of the equations of electrophoresis. Integrable equations of hydrodynamics*, Preprint No. 17, Landau Inst. Theor. Phys., Acad. Sci. USSR (1987), Moscow. (Russian)
46. V. L. Alekseev and M. V. Pavlov, *Hamiltonian structures of the Whitham equations*, Proceedings of the conference on NLS, Chernogolovka (1994) (to appear).

Translated by THE AUTHOR

INSTITUTE FOR MATHEMATICAL MODELING, MOSCOW, RUSSIA

Nonselfintersecting Magnetic Orbits on the Plane. Proof of the Overthrowing of Cycles Principle

P. G. GRINEVICH AND S. P. NOVIKOV

§1. Introduction. Overthrowing of cycles. Unsolved problems

Since 1981, one of the present authors (S. Novikov) has published a series of papers [1–4] (partly in collaboration with I. Schmelzer and I. Taimanov) developing an analog of Morse theory for closed 1-forms-multivalued functions and functionals on finite-dimensional and infinite-dimensional manifolds (*Morse–Novikov Theory*). This theory was extensively developed for finite-dimensional manifolds (many people worked in this direction later). The notion of "*multivalued action*" was understood and "*topological quantization of the coupling constant*" for such functionals was formulated by Novikov in 1981 as a consequence of the requirement that the *Feynman amplitude should be one-valued on the space of fields–maps*, by Deser–Jackiv–Templeton in 1982 for the special case of Chern–Simons functional, and by Witten in 1983. This idea found very important applications in quantum field theory. A very beautiful analog of this theory appeared also in the late eighties in symplectic geometry and topology, when the so-called Floer homology theory was discovered.

The very first topological idea of this theory, formulated in the early eighties, was the so-called *principle of overthrowing of cycles*. It led to results that were not proved rigorously until now. Our goal is to prove some of them.

We recall that Novikov studied, in particular, an important class of classical Hamiltonian systems of different physical origin formally equivalent to the motion of a charged particle on Riemannian manifolds M^n in an external magnetic field Ω, the latter being a closed 2-form on the manifold (see [3]). In terms of symplectic geometry, these Hamiltonian systems on phase spaces like $W^{2n} = T^*(M^n)$ are generated by the standard Hamiltonian functions (as are the so-called "natural systems" in classical mechanics) corresponding to the nonstandard symplectic structure determined by the external magnetic field. In the most interesting cases, our symplectic form is topologically nontrivial (i.e., it may have a nontrivial cohomology class in $H^2(W^{2n}, R)$).

Periodic orbits are the extremals of the (possibly multivalued) action functional S on the space $L(M^n)$ of closed loops (i.e., smooth or piecewise smooth mappings

---

1991 *Mathematics Subject Classification*. Primary 58E05.

of the circle in the manifold M^n):

$$S\{\gamma(t)\} = \oint_\gamma \frac{1}{2}\left(\frac{d\gamma}{dt}\right)^2 + e \oint_\gamma d^{-1}(\Omega).$$

This quantity is not well defined in general as a functional, but its *variation* δS *is well defined as a closed* 1-*form on the space of closed loops* $L(M^n)$ (this is the situation of the *Dirac monopole*).

Even if the closed 1-form δS is exact, its integral S may be unbounded from below. In such cases standard Morse theory does not work. For a fixed energy E, we replace the action functional by the "*Maupertuis–Fermat*" *functional* with the same extremals:

$$F_E(\gamma) = (2E)^{1/2}l(\gamma) + e\oint_\gamma d^{-1}\Omega.$$

This functional is also multivalued in general. Here $l(\gamma)$ is the ordinary Riemannian length. Let the charge e be equal to 1 and the form Ω be exact $\Omega = dA$ and small enough. The functional above is positive. We have a very nice special case of the Finsler metric (its geometry was investigated by E. Cartan many years ago). We may apply the ordinary Morse–Lusternik–Shnirel′ man theory in this case.

DEFINITION 1. The functional F_E is said to be *not everywhere positive* if the form $p^*\Omega$ is exact on the universal covering $p\colon M \to M^n$ and there exists a closed curve γ on the universal covering (or a curve homotopic to zero in M^n) such that $F_E(\gamma) < 0$. By definition, $F_E(\gamma_0) = 0$ for any constant curve γ_0.

DEFINITION 2. The functional F_E is said to be *essentially multivalued* if the form $p^*\Omega$ is not exact on the universal covering M, but is well defined as a functional on some regular nontrivial free abelian covering space $\widehat{L} \to L(M)$ with discrete fiber Z^k over the loop space of the manifold M.

In this case there is a natural embedding of the (trivial) covering space over the one-point curves $M \times Z^k \subset \widehat{L}$ such that $F_E(M \times 0) = 0$ for some selected point $0 \in Z^k$ and $F_E(M \times j) \neq 0$ for $j \neq 0$.

In the last case the functional F_E is obviously not everywhere positive on the covering space $\widehat{L}$. There exists an index $j \in Z^k$ such that

$$F_E(M \times j) < 0.$$

There is a natural free action of the group $\pi_1(M^n)$ on the loop space $L(M)$ such that the quotient space is isomorphic to the space $L_0(M^n)$. Here $L_0(X)$ denotes the space of loops homotopic to zero on X. This action extends naturally to the space $\widehat{L}$ and we get the quotient space $\widehat{L}_0$ of the space $\widehat{L}$ by the group $\pi_1(M^n)$.

On the last space $\widehat{L}_0$, our functional F_E is well defined and not everywhere positive. There is an embedding

$$M^n \times Z^k \subset \widehat{L}_0$$

such that

$$F_E(M^n \times 0) = 0, \qquad F_E(M^n \times j) < 0$$

for the same indices as above. This makes sense only if our functional is essentially multivalued. For single-valued functionals, we have $k = 0$ and for $k = 0$, Z^k contains only one point 0.

The following two lemmas are trivial, but important.

LEMMA 1. *All embeddings $M^n \times j \to \widehat{L}_0$ are homotopic to an embedding of index 0.*

LEMMA 2. *Our functional has nondegenerate manifolds of local minima for all one-point families $M_j = M^n \times j \subset \widehat{L}_0$.*

We now come to the following important definition.

DEFINITION 3. By an *overthrowing of a cycle* (set) $Z \subset M^n$ (in the negative domain) for the given multivalued or not everywhere positive functional F_E, we mean any continuous map

$$f: Z \times [0, 1] \to \widehat{L}_0$$

such that $f(Z \times 0) = Z \subset M_0$ and $F_E(Z \times 1) < 0$.

The existence of such an overthrowing was pointed out in the early eighties by Novikov as a main topological reason for the existence of periodic orbits homotopic to zero in the magnetic fields. There are two important examples.

EXAMPLE 1. In the case of essentially multivalued functionals, we may take $Z = M^n$. Here overthrowing is a homotopy between $M^n \times 0$ and $M^n \times j$ as above.

EXAMPLE 2. For the case of single-valued but not everywhere positive functionals, we may take Z to be a single point in M^n. Taimanov proved in [5] that there exists an overthrowing with $Z = M^n$ for any not everywhere positive functional.

As a consequence of overthrowing, an *analog of the Morse inequalities* was formulated. Let all critical points be nondegenerate. For the number of such points with Morse index equal to i and a *positive value of the functional*, we have the inequality

$$m_i(F_E) \geqslant b_{i-1}(M^n), \qquad i \geqslant 1.$$

Here the integers b_i are Betti numbers or any of their improved versions of the Smale type. Critical points may be degenerate or they may be multiples of one smaller extremal. Therefore we expect to prove the existence of one periodic extremal by using this argument. However, there exists an important difficulty (pointed out by Bolotin many years ago):

We can prove the existence of positive critical values $c_s > 0$ for the functional F_E by minimax arguments, but actual critical points may not exist. Our functional violates the important compactness principle.

The critical value $F_E = c_s > 0$ may be realized by an infinitely long curve γ that satisfies the Euler-Lagrange equation and may be approximated by a locally convergent sequence of closed curves $\gamma_i(t) \to \gamma(t)$ such that

$$F_E(\gamma_i) \to c_s + 0, \qquad l(\gamma_i) \to \infty.$$

At present we do not know any examples of such infinitely long extremals obtained via the overthrowing of cycles.

In the present paper we shall prove the overthrowing principle completely for the important case $M^2 = T^2$ with Euclidean metric and arbitrary nonzero magnetic field.

We may think about the Euclidean plane $\mathbb{R}^2$ with an everywhere nonzero double-periodic magnetic field directed along z-axis. *All four periodic extremals for any generic energy E (with Morse indices* 1, 2, 2, 3 *of the Maupertuis–Fermat functional) will be found as convex nonselfintersecting curves. Therefore they are geometrically distinct.* Of course, we obtain other extremals from them by discrete translations on periods. In principle, homological arguments do not give anything else.

REMARK 1. After a long history, in the paper [4] a correct criterion was found (in Theorem 1) for the existence of nonselfintersecting extremals of multivalued and not everywhere positive functionals on the 2-sphere. The idea of the proof was incomplete for the essentially multivalued case. Later it was completed and finally proved by Taimanov (see the proof and the history of this problem in the survey article [6]).

REMARK 2. It is clear for us now that no analogs of Morse type theory can be constructed for the space of immersions. Therefore Theorem 2 of the paper [4] is not natural. Its most general form is probably wrong. It should be replaced by a stronger result, namely, by the main theorem of the present paper for nonzero magnetic fields.

REMARK 3. In the very interesting papers [7, 8], V. Ginzburg proved the existence of periodic orbits with sufficiently small energy and sufficiently large energy, using perturbations of the limiting pictures. In particular, he pointed out to us that Theorem 3, announced without proof in the paper [4], is wrong. In fact, it contradicts the example of constant negative curvature and magnetic field equal to the Gaussian 2-form (with the appropriate sign) for which the extremals are exactly the horocycles: there is no periodic horocycles on compact surfaces. This mistake is interesting: Theorem 3 was deduced from Lemma 3 in [4], which claims that our functional is bounded from below in any free homotopy class of loops, if it is true for the trivial one. This lemma is wrong. It is true for the homotopy classes of mappings $(S^1, s) \to (M^2, x)$ representing any element of the fundamental group $\pi_1(M^2, x)$, but may be wrong for some free homotopy classes containing an infinite number of elements of the fundamental group. This is exactly what is going on in this counterexample.

For finite conjugacy classes our theorem might still be true. However, we have no proof: the absence of compactness leads to the same difficulties as above.

We may find some finite critical value $c > -\infty$, but the corresponding extremal may be infinitely long as before. What kind of extremal may we get? For surfaces with negative curvature and horocycles, we have $c = -\infty$, so this case is outside the scope of our arguments.

We get a nontrivial example on the 2-torus T^2 with exact magnetic field $\Omega = dA$ such that our functional F_E is positive on the space of closed curves homotopic to zero.

This property is always true for energy larger than some critical value E_0. *In many cases an interval of energies exists such that the Maupertuis–Fermat functional is not a well-defined Finsler metric but is positive on the space of loops homotopic to zero.*

There is another interesting example for $n = 3$. Let the manifold M^n be a fiber bundle with fiber isomorphic to the circle $S^1 \subset M^3$. For magnetic fields Ω with homology classes from the base, we may require that the periodic extremals be homotopic to the fiber. For the 3-torus this restriction means that our magnetic field has at most one rationally independent flux over the integral 2-cycles.

§2. Nonzero double periodic magnetic fields on the plane. Proof of the overthrowing of cycles for convex polygons

Now we consider a nonzero smooth double periodic magnetic field on the Euclidean plane $\mathbb{R}^2$ directed perpendiculary to this plane:

$$B(x+1, y) = B(x, y+1) = B(x, y) > 0.$$

For an energy level such that $(2E)^{1/2} = 1$, we denote the Maupertuis–Fermat functional by $F = F_E$. We shall consider only this case (without any loss of generality).

Consider the space of closed convex curves oriented in such a way that

$$F\{\gamma\} = l(\gamma) - \iint_K B(x, y)\, dx\, dy, \qquad x^1 = x,\ x^2 = y.$$

In this formula K denotes a positively oriented domain inside the curve γ, the magnetic field B is positive. We shall say that the second term is the *magnetic area*. It comes with a minus sign.

In this section we consider the functional F on the *space P_N of the rectilinear convex "parametrized" polygons $\gamma \in P_N$ containing exactly N equal line segments of any length L*. By definition, "parametrized polygon" means "polygon with some natural numeration of vertices"

$$AB\ldots CDA = A_1 A_2 \ldots A_{N-1} A_N A_{N+1}, \qquad A_{N+1} = A_1.$$

Cyclic permutations of this numeration leads to the free action of the group $\mathbb{Z}/N$ on the space P_N. We denote the quotient space by $\bar{P}_N$.

Let $B_{\min}$, $B_{\max}$ denote the minimum and maximum of $B(x, y)$ on the torus T^2. We introduce the following parameters:

$$N_0 = \left[\frac{8 B_{\max}}{B_{\min}}\right] + 1,$$

$$\alpha_0 = \min\left\{\frac{1}{1000 N},\ \frac{2}{N} \arctan\left(\frac{9 B_{\min}}{20 B_{\max}(2N^3 + N/2)}\right)\right\},$$

$$L_0 = \frac{4N}{\sin(\alpha_0/2) B_{\min}}.$$

These parameters depend on N, $B_{\min}$, $B_{\max}$.

We shall consider the spaces P_N for $N > N_0$ only. For $N \to \infty$ we have $\alpha_0 \to 0$ and $L_0 \to \infty$. Let AB, BC be neighboring edges of the convex polygon. At the point B, we have the *external angle* α and the *internal angle* β such that $\alpha + \beta = \pi$.

DEFINITION 4. A convex closed polygon from the space P_N is called *admissible* if all its external angles are larger than $\alpha_0/2$ and $L < 2L_0$.

On the subspace of admissible polygons P_N^a, we define the *corrected functional*

$$F_a(\gamma) = F(\gamma) + \sum_{k=1}^{N} \phi(\alpha_0/\alpha_k) + L_0 \psi(L/L_0).$$

Here α_k denotes the kth external angle of our admissible convex polygon γ, ϕ and ψ are real nonnegative functions on the closed interval $[0, 2]$ such that

$$\phi(x) = \psi(x) = 0, \qquad x \leq 1,$$
$$d\phi/dx > 0, \quad d\psi/dx > 0, \qquad x > 1,$$
$$\phi(x) \to +\infty, \quad \psi(x) \to 1, \qquad x \to 2,$$

and both x-derivatives of this functions converge to $+\infty$ as $x \to 2$.

THEOREM 1. *Let $\gamma \in P_N$ be a convex polygon with $N > N_0$. It is an extremal for the functional F on this space such that $F(\gamma) > 0$ if and only if γ is an admissible curve, is also an extremal for the functional F_a, and $F(\gamma) = F_a(\gamma) > 0$.*

The following obvious geometric facts are true:

LEMMA 3. *Let γ_i, $i = 1, 2$, be two convex polygons such that γ_1 lies completely inside γ_2. Then $l(\gamma_1) < l(\gamma_2)$.*

LEMMA 4. *Let $\gamma \in P_N$ be a convex polygon with total length NL such that there exist two internal angles in it less than $\pi/3$. It follows that the distance between the corresponding two vertices (say A and B) is at least $LN/4$.*

For the proof of the last lemma, we observe that the vertices A and B cannot be neighboring in γ. Our entire curve γ belongs to the interior of the rumbus whose two opposite vertices are exactly A, B with the corresponding internal angles equal to $2\pi/3$. The perimeter of the rumbus is less than $4|AB|$. We conclude therefore that $LN < 4|AB|$ from Lemma 3. Lemma 4 is proved.

LEMMA 5. *Let D be any convex subset in Euclidean space $\mathbb{R}^2$ bounded by the polygon $\gamma \in P_N$, $l(\gamma) = NL$, and C be any point inside it. After the rotation of this set by a small angle $\delta\alpha$ about the point C, we get a domain D_1 whose magnetic area satisfies the inequality*

$$\left| \iint_{D_1} B \, dx \, dy - \iint_{D} B \, dx \, dy \right| < N^2 L^2 (B_{\max} - B_{\min}) \frac{\delta\alpha}{2}.$$

For the proof of this lemma, we note that after the rotation by the small angle $\delta\alpha$, the total set D_1 minus the original one will have an area not exceeding $N^2 L^2 \delta\alpha/2$. Combining this with obvious estimates for the integral, we get our final estimate. Lemma 5 is proved.

LEMMA 6. *Let $\gamma \in P_N$, $N > N_0$ and let there exist two different vertices (say, A and B) such that the corresponding internal angles are less than $\pi/3$. There exists a small deformation γ_t of the curve $\gamma = \gamma_0$ in the space P_N (of fixed length) such that the magnetic area increases in the linear approximation and all external angles do not decrease. Therefore the curve γ cannot be an extremal for either of the functionals F, F_a.*

PROOF OF THE LEMMA. Let the segment AB be horizontal in our picture. It divides γ into two pieces $AC \ldots DB$ (upper piece) and $AE \ldots FB$ (lower piece). Our deformation will be such that the lower piece $E \ldots F$ does not move $E_t = E, \ldots, F_t = F$ and the upper piece $C \ldots D$ moves up perpendicularly to the segment AB parallel to itself by the distance t and $(C_t \ldots D_t) \parallel (C \ldots D)$.

The position of the vertices A_t, B_t is defined from this completely, because the length L does not change.

This deformation has the desired properties (see the elementary trigonometric calculation in the Appendix). Lemma 6 is proved.

So we cannot have two internal angles less than $\pi/3$ for the extremals.

LEMMA 7. *No polygon $\gamma \in P_N$ can have all external angles, except possibly one, say α_N, less than α_0.*

The proof of this statement follows immediately from the definition of α_0 and elementary geometric facts: the total sum of all external angles is equal to 2π, each of them is less than π. Therefore we have

$$\sum_{k=1}^{N-1} \alpha_k = 2\pi - \alpha_N < (N-1)\alpha_0 \leqslant \frac{N-1}{1000N}.$$

At the same time we have $\alpha_N < \pi$. This leads to a contradiction, which proves our lemma.

Now consider a curve (a polygon from the space P_N) containing at least two vertices with external angles greater than α_0. Let these vertices be A, B and all vertices between them (from one side) have "small" external angles (i.e., less than α_0). We construct a deformation γ_t of this curve $\gamma = \gamma_0$ in the space P_N with fixed length. Let $[AB]$ be the segment between these two points and C, D be the vertices with orthogonal projections on the segment $[AB]$ closest to the center. Here C belongs to the arc with small external angles and D belongs to the other arc of γ. Let $A_t = A$ and B_t be obtained from B by a small shift $\delta x = t$ along the segment $[AB]$ in the direction of A. We rotate the arcs $A \ldots C$ and $A \ldots D$ around the point $A = A_t$. We shift the arcs $B \ldots C$ and $B \ldots D$ parallel to themselves by the same distance as the point B. After that we rotate them about the point B_t. Finally we find the points C_t, D_t as crossing points. The following lemma is true for this deformation.

LEMMA 8. *The deformation γ_t described above does not change the length. It is such that all external angles (except the angles at the vertices A, B with external angles more than α_0) do not decrease; the t-derivative of the magnetic area for $t = 0$ is nonzero. Therefore the curve γ cannot be an extremal for the functionals F, F_a; any* *curve which is an extremal for each of them is such that all external angles are more than α_0.*

The proof of this lemma uses the lemmas above. It is based on elementary trigonometric calculations using the values of parameters α_0, N_0, L_0 fixed at the beginning of this section (see the Appendix for details).

LEMMA 9. *Let the curve $\gamma \in P_N$ be such that all external angles are larger than α_0 and $L > L_0$. In this case we have $F(\gamma) < 0$. If the curve is admissible ($L < 2L_0$), we have $F_a(\gamma) < 0$.*

For the proof of this lemma it suffices to estimate the magnetic area of any triangle ABC based on two edges AB, BA of our polygon. The external angle is larger than α_0. Therefore its area S is larger than S_0:

$$S > S_0 = (1/2)L^2 \sin(\alpha_0/2),$$

and its magnetic area is larger than $B_{\min} S_0 > 2NL$. We have $l(\gamma) = NL$. As a consequence, we get the inequality

$$F(\gamma) < NL - 2NL < 0.$$

All external angles are larger than α_0. So the contribution of the function ϕ to the value of the functional $F_a(\gamma)$ is equal to zero. By definition, we always have $\psi \leqslant 1$. Therefore, we conclude that for $L > L_0$,

$$F_a(\gamma) = F(\gamma) + L_0 \psi(L/L_0) < -NL + L_0 < 0.$$

Lemma 9 is proved.

The proof of Theorem 1 now follows from the lemmas. Theorem 1 is proved.

We shall now construct a natural analog of Morse theory for the functional F_a on the space P_N^a of admissible polygons, or, more precisely, on the space $\overline{P}_N^a$ of admissible polygons completed naturally by one-point curves and quotiented by the discrete group $\mathbb{Z}^2 \times (\mathbb{Z}/N)$ generated by the basic translations of the plane $\mathbb{R}^2$ and cyclic permutations of the vertices.

This space is homotopy equivalent to the torus T^2 (i.e., to the subspace of one-point curves). This space without one-point curves is homotopy equivalent to the 3-torus $T^2 \times S^1$.

We shall use Morse type estimates "modulo the subspace P^0" for which the functional F_a is less than or equal to zero, i.e.,

$$P^0 = \{F_a \leqslant 0\}.$$

The following easy lemma is true:

LEMMA 10. *The space P^0 is not connected. It contains at least two components. One of them is exactly the set of all one-point curves $T_0^2 \subset P^0$. The other one, P_1^0, contains all N-polygons γ with equal angles such that the length of the edges L is big enough (but less than $2L_0$).*

We already know that the set of one-point curves is a local minimum for the functional F. By definition, the value of the functional F_a on it is equal to zero.

Since $F_a \geqslant F$ for any admissible curve, this set is a local minimum for F_a also. Therefore the set of one-point curves is isolated in P^0. The curves with equal angles have all external angles equal to $2\pi/N > \alpha_0$. For $L = L_0$ and large $N > N_0$, we have $F = F_a$ for these curves and $F < 0$.

Lemma 10 is proved.

For any point in the plane, we construct an overthrowing of this point, continuously depending on this point.

DEFINITION 5. By definition, the image of our *initial overthrowing* is the set of all N-polygons with the center at this point, with equal angles, and edges of length $L < L_0$. This set determines a map

$$f : (T^2 \times S^1) \times [0, 1] \to \overline{P}_N^a$$

such that

$$f(T^2 \times S^1 \times 0) = T_0^2 \subset P_0, \qquad f(T^2 \times S^1 \times 1) \subset P_1^0.$$

The parameter along the circle S^1 here is exactly the angle parametrizing all polygons with the same center and same length, the parameter in the interval $[0, 1]$ coincides with the radius divided by the maximal radius, so that the whole image of $f(T^2 \times S^1 \times 1)$ belongs to the negative values of our functionals for all central points in the plane.

The following obvious lemma is true:

LEMMA 11. *The overthrowing*

$$f : (T^2 \times S^1 \times [0, 1], T^2 \times S^1 \times 0 \cup T^2 \times S^1 \times 1) \to (\overline{P}_N^a, P^0)$$

generates monomorphisms in the homology groups:

$$H_{i-1}(T^2 \times S^1) \to H_i(\overline{P}_N^a, P^0), \qquad i \geqslant 1.$$

Our functional F_a generates a cell decomposition of the space $\overline{P}_N^a$ modulo P_0, corresponding to the critical points such that $F_a(\gamma) > 0$. This is a corollary of our lemmas, because this space is invariant under the gradient flow (all gradient lines go inside it). So we can apply standard arguments of Morse theory to this space modulo the negative subspace P_0. Combining this fact with the previous lemma, we get the following theorem:

THEOREM 2. *For any value $N > N_0$ there exist at least two different extremals of the Maupertuis–Fermat functional F in the space P_N^a of admissible convex polygons. If the critical points are nondegenerate, there exists at least eight of them in the same space with Morse indices equal to* 1, 2, 2, 2, 3, 3, 3, 4.

PROOF OF THE THEOREM. By the minimax principle, we always have at least one extremal in this space. Suppose we have only one critical point. After the long gradient deformation starting from the initial overthrowing process

$$f = f_0 : T^2 \times S^1 \times [0, 1] \to \overline{P}_N^a,$$

we get the new overthrowing process f_1 in which almost the entire image is below the critical level and the remaining part is concentrated in a small neighborhood of the critical point. After removing some small neighborhood of the critical point from the space $\overline{P}_N^a$, the new overthrowing will split into some pieces (at least two) such that for each piece the images of its boundaries $T^2 \times S^1 \times 0$ and $T^2 \times S^1 \times 1$ belong to different components.

This follows from the fact that any new overthrowing of one point must pass through the same small neighborhood as the new overthrowing of the entire torus T^2. By any new overthrowing of the point we have in mind any curve $f_1 \colon \tau(t) \to \overline{P}_N^a$, where $\tau(t)$ is a continuous curve in $T^2 \times S^1 \times [0,1]$ such that $\tau(0) \in T^2 \times S^1 \times 0$ and $f_1(\tau(1)) \in P^0$.

Our space P_N^a is locally contractible. Using this, we deform its identity map onto itself in such a way that after the deformation all small neighborhoods of the critical point collapse to this point. Finally, we construct a deformation of the set of one-point curves to one (critical) point in the space P_N. However this set is nonhomotopic to zero in the space P_N. Thus we have a contradiction. So we have at least two extremals.

The other part of this theorem is a standard corollary from the handle decomposition generated by the critical points of the functional F_a. It follows from the lemmas above that we may apply the standard arguments of Morse theory. Theorem 2 is proved.

§3. Compactness property for $N \to \infty$. Main results

DEFINITION 6. For any convex polygon $\gamma \in P_N$ by its *maximal diameter* $D_{\max}$ we mean the maximal distance between two points of this polygon. By the *diameter in the direction* ϕ, we mean the maximal distance D_ϕ between two straight lines parallel to the direction ϕ that have nontrivial intersection with γ. The maximal and minimal diameters $D_{\max}$, $D_{\min}$ are defined as the maximum and minimum of the function D_ϕ, corresponding to directions denoted by $\phi_{\max}$, $\phi_{\min}$.

THEOREM 3. *Let $\gamma_N \in P_N$ be an extremal of the functional F such that $F(\gamma_N) > 0$. The following estimates for its maximal diameter and for its maximal length $\mathcal{L}_0$ hold*:

$$D_{\max} \leqslant \frac{8(3 + B_{\max} B_{\min}^{-1})}{B_{\min}(1 - 8N^{-1})}, \qquad \mathcal{L}_0 \leqslant 4 D_{\max}.$$

The proof of this theorem follows from the next lemma.

LEMMA 12. *For any extremal γ of the functional F on the space P_N such that $F(\gamma) > 0$, the following estimate is valid*:

$$\frac{D_{\max}}{D_{\min}} \leqslant \frac{B_{\max} B_{\min}^{-1} + 3}{1 - 8N^{-1}}.$$

PROOF OF THE LEMMA. We describe a deformation that preserves the length of γ and changes the magnetic area in the linear approximation if the inequality is not true. The y-axis is exactly the direction $\phi_{\min}$ in our picture.

Let AB, HG be the left most edges such that their angles with x-axis are no more than $\pi/4$ and CD, FE are the right most edges with the same property. The arcs $AB\ldots CD$ and $HG\ldots FE$ belong to the upper and lower parts of γ. The points A and H or D and E may coincide, but this is not important. Our deformation γ_t, $\gamma_0 = \gamma$, is such that $A_t\ldots H_t = A\ldots H$ the arc $D_t\ldots E_t$ is obtained from $D\ldots E$ by a parallel shift by the distance $\delta x = t$ to the left. The arc $B_t\ldots C_t$ is obtained by a parallel shift of the arc $B\ldots C$ upwards (by the distance δy_1) and to the left (by the distance δx_1). The arc $G_t\ldots F_t$ is obtained from $G\ldots F$ by a parallel shift downwards (by the distance δy_2) and left (by the distance δx_2). The value of all these parameters as functions of the variable t are specified by the requirement that all lengths be the same and the new polygon γ_t be closed.

The detailed proof of the lemma follows from trigonometric calculations presented in the Appendix.

PROOF OF THEOREM 3. Let A, B, C, D be the left most, upper most, right most and lowest vertices in γ. The polygon $ABCD$ with four edges belongs completely to the interior of γ. Therefore its area is at least $D_{\min}D_{\max}/2$ and the magnetic area Q is at least $D_{\min}D_{\max}B_{\min}/2$. Combining this with the trivial estimate $l(\gamma) \leqslant 4D_{\max}$ and $F(\gamma) > 0$, we get the inequality

$$0 < F = l - Q < 4D_{\max} - D_{\min}D_{\max}B_{\min}/2,$$

or finally

$$D_{\min} < 8B_{\min}^{-1}.$$

Using the lemma above, we obtain the desired inequality for $D_{\max}$. Theorem 3 is proved.

Now consider a sequence γ_N of extremals of the functional F on the spaces P_N with $F > 0$ and $N > N_0$. In fact we shall only consider the subsequence $N_k = N_1 2^k$.

For large N, small external angles and bounded total length NL, the polygons κ and $p_N(\kappa)$ are very close to each other.

From the theorem above, we conclude that *there exists a subsequence $k_j \to \infty$ such that $\kappa_j = \gamma_{N_{k_j}} \to \gamma$, where γ is a continuous curve*, because the whole family of our extremal N_{k_j}-polygons κ_j with positive values of the functional F is precompact.

THEOREM 4. *The limiting curve γ is a periodic smooth extremal of the functional F for which is positive, $F(\gamma) > 0$.*

LEMMA 13. *Let $\kappa_j \in P_N$, $N = N_{k_j}$ as above be a sequence of "relative extremals" with positive values of the functional F and γ be the limiting continuous curve. For $N \to \infty$, the entire family of external angles of the curves κ_j converges to zero as $O(1/N)$.*

Let A be a vertex with largest external angle α, and let B be the "opposite" vertex such that the arc $A\ldots B$ contains exactly $N/2$ edges. Consider the following deformation γ_t of the polygon $\kappa_j = \gamma_0$: $B_t = B$, all vertices except A move along their own edges towards B in such a way that the distance between any two vertices becomes exactly $L - \delta L$, $\delta L = t$. The shift of the vertex A will be completely determined by the requirement that the new polygon has equal edges of length $L - t$.

For the variation of the functional F at the point $t = 0$, we get the inequality (see the elementary trigonometric calculation in the Appendix)

$$|\delta F| > \delta L(N - \pi \mathcal{L}_0 B_{\max} - \mathcal{L}_0 B_{\max}(\sin \alpha)^{-1}).$$

At the same time we remember that $\delta F = 0$ for $t = 0$. Finally, we get the inequality

$$\sin \alpha < \left(\frac{N}{\mathcal{L}_0 B_{\max}} - \pi\right)^{-1}.$$

Therefore, we have proved that for N large enough there exists constant c such that $\alpha < cN^{-1}$. Lemma 13 is proved.

LEMMA 14. *The limiting curve γ belongs to the class C^1.*

We have already found an upper estimate for the length of the "relative extremals" in the lemma above. It is also easy to find a lower estimate for this length. Consider any point x inside γ. We perform the homothety with the center at this point and coefficient $1 + p$. For the variation we have

$$0 = \delta F = \delta l(\gamma) - \delta \iint_K B \, dx \, dy > pl(\gamma) - pl(\gamma) d_{\max} B_{\max}.$$

Here $d_{\max}$ denotes the maximal distance from the point x to γ. We deduce from this the following inequality:

$$d \geqslant B_{\max}^{-1}, \qquad NL = l(\gamma) \geqslant l_0 = 2B_{\max}^{-1}.$$

Consider any arc $P \ldots Q \ldots R \ldots S$ on the extremal γ, containing n edges, $n \leqslant N/2$. By the previous lemma, the angle ϕ between the lines PQ and RS at their intersection point (outside γ) is of order $O(nN^{-1})$:

$$\phi \leqslant nc/N.$$

For the length of the arc $P \ldots S$ we have $l(P \ldots S) = nL \geqslant nl_0 N^{-1}$. Combining this with the previous inequality, we get

$$\phi \leqslant cl(P \ldots S) l_0^{-1}.$$

For the limiting curve $N \to \infty$, we have an upper estimate for the angle between two "tangent" lines at the points P, S

$$\phi \leqslant c(B_{\min}, B_{\max}) l(P, S)$$

(the distance l is along the curve).

DEFINITION 7. By a *tangent line* to any convex curve, we mean any straight line which has our entire curve on one side and intersects it. For vertices of convex polygons, a tangent line has only one common point with our polygon.

Our curve is convex because it is the limit of convex curves. Lemma 14 follows from this estimate.

The polygons from our sequence κ_j belong to the spaces $P_{2^{k_j}N_1}$. We fix a numeration such that all the vertices $P_{j,s}$ with numbers $2^{k_j}s$ converge to some points P_s on the limiting curve for fixed values of s, $j \to \infty$, $s = 1, \ldots, N_1$.

For sufficiently large N_1 and any two vertices $P_{0,s} = R_0$, $P_{0,r} = Q_0$ on this curve, we have two sequences

$$P_{j,s} = R_j \to R, \qquad P_{j,t} = Q_j \to Q.$$

LEMMA 15. *We have the following estimate for the angle ϕ between two straight lines tangent to the polygons κ_j at the vertices R_j, Q_j:*

$$\phi = \int_{R_j}^{Q_j} B(\kappa_j(t))\, dl(t) + O\left(\frac{1}{2^{k_j}N_1}\right), \qquad j \to \infty.$$

Here the integral is taken along the curve κ_j using the natural parameter l.

PROOF. Consider the arc $TR_jU\ldots A\ldots VQ_jS$, where A is the "central" vertex between R_j and Q_j (or one of two central vertices, if the number of edges in the arc is odd). Let B be the "central" vertex of the opposite arc $R_jT\ldots B\ldots SQ_j$ in the same sense, let T_1, T_2 denotes two tangent lines at the vertices R_j, Q_j and ϕ denote external angle at their crossing point.

We construct a deformation γ_t, $\gamma_0 = \kappa_j$, of this extremal such that the entire arc $R_jT\ldots B\ldots SQ_j$ does not move.

We move the points R_j, Q_j inside the curve γ_0 by the small distance $\delta s = t$ in the direction perpendicular to the edges R_jU, Q_jV. We denote the new vertices by R^1, Q^1. We move all edges of the arc $R_jU\ldots VQ_j \to R^1U^1\ldots V^1Q^1$ inside by the distance δs in the directions perpendicular to each edge and construct from their pieces a new arc $R^1U^1\ldots V^1Q^1$ with slightly smaller edges (not necessary equal to each other).

After that we make all edges equal by a deformation such that the point B does not move, all vertices from the arc $A^1\ldots Q^1\ldots B^1$ move along their own edges on this arc in the direction towards B, all vertices on the arc $A^1\ldots R^1\ldots B$ move along their own edges on this arc towards B, the vertex A moves along one of the two edges (which of the two is uniquely determined by the condition that we finally get equal edges).

At the first step of the deformation, we have the following estimate for the length:

$$\delta l = \phi\, \delta s + O\left(\frac{1}{2^{k_j}N_1}\right)$$

(from Lemma 13), and the following estimate for the magnetic area:

$$\delta \iint_K B\, dx\, dy = \int_{R_j}^{Q_j} B\, dl + O\left(\frac{1}{2^{k_j}N_1}\right).$$

From the same Lemma 13, we conclude that the total product over all vertices is equal to one plus something small

$$\cos\alpha_1 \cdots \cos\alpha_{2^{k_j}N_1} = 1 + O\left(\frac{1}{2^{k_j}N_1}\right)$$

for the polygon $\kappa_j = \gamma_0$. Changing the length of one edge on the arc $A^1 R^1 B$ by the value δL, we have a shift of the point A^1 by the distance

$$\delta L \prod_l \cos\alpha_l = \delta L\left(1 + O\left(\frac{1}{2^{k_j}N_1}\right)\right).$$

Therefore, the variation of length at the second step is small enough, and a shift of any vertex is no more than $c\delta s$, where c is some constant independent of j. So the variation of the magnetic area is small enough at the second step.

We get the total variation of the functional by summing all the contributions:

$$0 = \delta F = \phi - \int_{R_j}^{Q_j} B\, dl + O\left(\frac{1}{2^{k_j}N_1}\right).$$

Lemma 15 is proved.

PROOF OF THEOREM 4. From Lemma 15 above applied to the limiting (as $j \to \infty$) curve, we get the statement of the theorem. Theorem 4 is proved.

THEOREM 5. *For any smooth positive doubly periodic magnetic field on the Euclidean plane (directed along the third axis orthogonal to the plane) there exist at least two different periodic convex extremals for which the value of the Maupertuis–Fermat functional is positive.*

PROOF. Let us assume that we have only one extremal for the functional F after passing to the limit as $j \to \infty$. Consider the overthrowing process

$$f_j: T^2 \times [0,1] \to P^a_{2^{k_j}N_1}$$

after it was subjected to a long gradient deformation determined by the corrected functional F^a. For large j, we have several (at least two) different extremals in this space, which have the same limit as $j \to \infty$.

Therefore our "relative extremals" $\kappa_{j,p}$, $p = 1, 2, \ldots$, for all large values of j belong to the same very small contractible neighborhood W of the limiting extremal κ in the space of convex piecewise smooth curves.

After the long gradient deformation (mentioned above) the new overthrowing process belongs to the negative subspace $P^0 \subset P^a_{2^{k_j}N_1}$ everywhere outside the neighborhood W. As above in the proof of Theorem 2, any overthrowing process of the point determined by the map f_j must pass through this set W. For this reason, the embedding of the manifold of all one-point curves is contractible in the space of nonparametrized closed convex curves (i.e., is the quotient by the action of the group SO_2, changing the initial point in the natural parametrization). But this is an obvious contradiction. Theorem 5 is proved.

THEOREM 6. *Let all periodic convex extremals with a positive value of Maupertuis–Fermat functional be nondegenerate in the sense of Morse in the space of nonparametrized curves. In this case there exist at least four periodic convex extremals for any fixed value of energy whose Morse indices are equal to* $(1, 2, 2, 3)$.

For the proof of this theorem, we shall use Theorem 3 and the comparison of Morse indices of periodic extremals with the Morse indices of the "relative extremals" κ_j for all sufficiently large values of j. This comparison (which looks easy) was never carried out rigorously for our spaces. So we shall finish the complete proof later. Here we present the idea of the proof. The very first question is: *what is the Morse index for the Maupertuis–Fermat functional on the space of all smooth curves*?

For the definition of this quantity, we can introduce some specific recipe for parametrizing the curves, because our functional does not depend on it. The natural parameter (length) is sufficient for our goals.

After that we consider the Morse index on the space P of the convex curves with the natural parametrization. This index is finite. There is a trivial "nullity" of this critical point equal to 1. This nullity corresponds to the choice of the initial point on the curve. Our functional is invariant under the free action of the group SO_2 on the space P, as mentioned above. The quotient space $\bar{P}$ of the space P by this action is homotopy equivalent to the torus T^2. Our functional in the generic case has only nondegenerate critical points in $\bar{P}$.

In the process of approximation, we use the spaces P_N with the natural action of the group $\mathbb{Z}/N \subset SO_2$. After the approximation, our functional is only $\mathbb{Z}/N$-invariant. The corresponding quotient spaces $\bar{P}_N$ have the homotopy type of $T^2 \times S^1$. We found more critical points in Theorem 3 than are needed for the limit $N \to \infty$. In fact, pairs of critical points with neighboring indices $i, i+1$ must have the same limit. Returning to the spaces P_N with discrete parametrization, we get free $\mathbb{Z}/N$-critical orbits instead of points. These orbits converge to the SO_2-orbits as $N \to \infty$. Each nondegenerate critical SO_2-orbit with Morse index i generates at least two nondegenerate critical $\mathbb{Z}/N$-orbits with Morse indices $i, i+1$ for obvious homological reasons *if we shall be able to prove that this approximation is really equivalent to a small perturbation of our functional in the C^2 norm.*

Following the classical papers of Marston Morse, we introduce a *finite-dimensional approximation* of the space P in a small neighborhood of the given closed extremal γ with Morse index equal to i. It is convenient for us to use the same spaces P_N of polygons with N equal edges and total length not far from the length NL of our extremal γ.

The approximation of the functional used by Morse is the following.

For L small enough, we join all pairs of neighboring vertices by the unique small extremal and construct thereby the *extremal polygon* with the same vertices as any given polygon from the space P_N. This space is canonically isomorphic to the space P_N, but the value of our functional on the extremal polygon is different from its value on the rectilinear polygon with the same vertices. We denote this functional on the space P_N by F^e. *Its extremals are exactly the same as the smooth extremals γ on the space of all smooth curves.*

Consider a small neighborhood of the extremal γ. All external angles of polygons in this neighborhood have order $O(N^{-1})$. We shall compare the functionals F^e

and F in this part of the space P_N. An easy trigonometric estimate shows that the difference between these functionals is of order $O(N^{-2})$.

More precisely, consider a small line segment AC with length equal to L and a small piece of extremal with the same vertices A, C (it looks like an arc of the circle of radius $R_0 = B^{-1}$ in our approximation, for very small values of L). Here B is value of the magnetic field at the midpoint of the interval AC. Calculating the terms of order $O(N^{-3})$ for the length of the extremal arc and the magnetic area of this small domain, we shall also need the first derivatives of the field B at the same point. Using some elementary trigonometry, we obtain the following lemma.

LEMMA 16. *Let our magnetic field belong to the class C^2 on the torus. The value of the Maupertuis–Fermat functional for all such "local" geometric figures bounded by a small line segment and a small piece of the extremal is less than or equal to the quantity $O(L^3)$ with the coefficient depending on the maximal values of the magnetic field B and its first derivatives on the torus.*

Combining this result with some elementary statements of Euclidean geometry, we see that any small variation of the local geometric figures leads to a variation of the functional of order $O(L^3)$ and $O(L^2 \delta L)$ in the variable N^{-1}. If we consider any variation of the polygon from the small neighborhood of the extremal γ of length l, we know *a priori* that $L \sim O(N^{-1})$ and $\delta L \sim O(N^{-1})$.

We now use the additivity property of our functional: the difference between the functionals $F - F^e$ is equal to the sum of N "local" terms corresponding to the individual edges described in the lemma above. Therefore we obtain the following result:

LEMMA 17. *In a small neighborhood of our extremal γ in the space P_N, all derivatives of any order of the difference $F - F^e$ are magnitudes of order $O(N^{-2})$.*

Now we complete the proof of the theorem. In the process of approximating the space of curves by extremal polygons, we consider the sequence $N = 2^{k_j} N_1$ as above, for which we have the convergence of the "relative" extremals of the functional F on the subspaces P_N to the smooth extremal γ. From old results of Morse, we know that for all large values of N, we have the same curve γ as extremal of the functional F^e, i.e., of the same functional on the space of extremal polygons. We know also that the tangent spaces to P_N at the point γ for all values of $N = 2^k N_1$ may be regarded as finite-dimensional subspaces T_j of the same Hilbert space T. This sequence T_j converges in the sense that all T_j with larger numbers "almost" contain the previous ones: there exist a natural projector

$$\pi_{j,j+s} \colon T_j \to T_{j+s}$$

such that $\|\pi_{j,j+s}(u) - u\| \to 0$ as $j \to \infty$ homogeneously in s and for all unit vectors u. We shall identify the subspaces T_j and $\pi_{j,j+s}(T_j)$ in our notations.

The second variation of the functional F^e also converges. This means that this second variation is strictly positive on the subspaces orthogonal to the image of T_j in T_{j+s} with the lowest eigenvalue converging to $+\infty$ as $j \to \infty$. In the image of T_j, all the lower eigenvalues and eigenvectors converge to their values in the space of normal vector fields along the curve γ. Therefore we may use the finite spaces P_N for the calculation of the Morse index. Our theorem now follows from the lemmas above, because the Morse index is stable under the perturbations of the function F^e in the spaces P_N provided the perturbations are small together with their first, second (and third) derivatives at all points of their neighborhoods under investigation. The role of one-dimensional nullity is the following: it leads generically to the splitting of one nondegenerate critical circle and the creation of some nonzero even number of nondegenerate critical points in the spaces P_N in the process of approximation: half of them with index i and the other half with index $i + 1$; all of them converge to our extremal circle, which is a single point γ in the space of the nonparametrized curves.

Theorem 6 is proved.

§4. Appendix. Trigonometric calculations. Proof of the lemmas

PROOF OF LEMMA 6. It is convenient to decompose the deformation shown in Figure 1 into two steps.

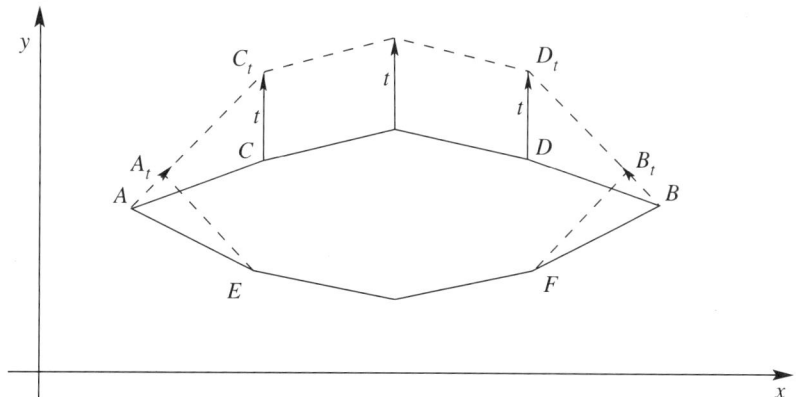

FIGURE 1

These two steps are shown in Figures 2a and 2b.

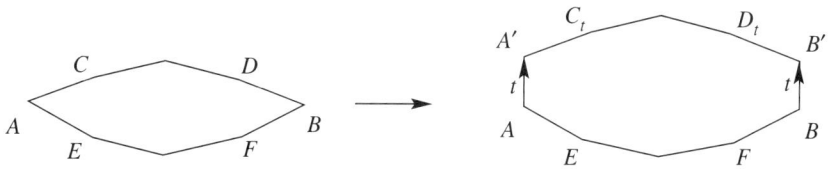

FIGURE 2A

Step 1: We shift the arc $ACDB$ up by the distance t. The images of the points A, B, C, D under the shifts are denoted by A', B_t, C_t, D'. We also add small vertical segments connecting the points A, A' and B, B'.

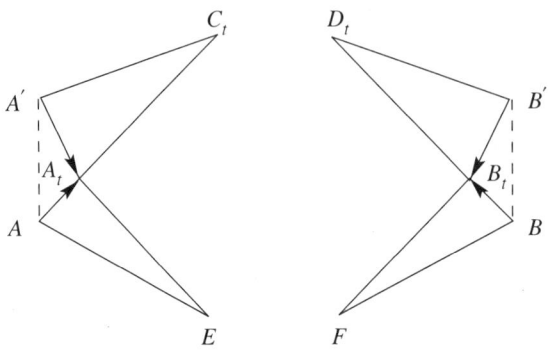

FIGURE 2B

Step 2: We rotate the segments AE, $A'C_t$, BF, $B'D_t$ around the points E, C_t, F, D_t by angles such that the images of the points A, A' coincide and the images of B, B' coincide as well. We shall denote them by A_t and B_t respectively.

Let us denote the variation of the magnetic area at the first step and at the second one by $\delta_1 Q$ and $\delta_2 Q$ respectively. We have

$$\delta_1 Q \geqslant B_{\min} |AB| t,$$

$$\delta_2 Q \geqslant -B_{\max} \cdot \tfrac{1}{2}(|AE||AA_t| + |C_t A'||A'A_t| + |BF||BB_t| + |D_t B'||B'B_t|).$$

The infinitesimal triangles $AA'A_t$ and $BB'B_t$ have the magnetic area $O(t^2)$ and may be neglected in the first-order calculations.

The angles $AA_t A'$ and $BB_t B'$ are greater than $2\pi/3$, thus

$$|AA_t| < |AA'| = t, \quad |A'A_t| < t, \quad |BB_t| < t, \quad |B'B_t| < t,$$

$$\delta_2 Q \geqslant -B_{\max} 2Lt, \quad |AB| \geqslant LN/4 \geqslant LN_0/4 > 2(B_{\max}/B_{\min})L.$$

Combining all these estimates, we get

$$\delta Q = \delta_1 Q + \delta_2 Q > 2B_{\max} Lt - 2B_{\max} Lt = 0.$$

Lemma 6 is proved.

PROOF OF LEMMA 8. It is convenient to decompose the deformation into two steps.

Step 1: We rotate the entire arcs AC, BC, AD, BD around the points A and B by the following angles

$$\delta\varphi_1 = \frac{1}{\cos\varphi_1} \frac{1}{\tan\varphi_1 + \tan\varphi_2} \frac{\delta x}{|AC|}, \quad \delta\varphi_2 = \frac{1}{\cos\varphi_2} \frac{1}{\tan\varphi_1 + \tan\varphi_2} \frac{\delta x}{|BC|},$$

$$\delta\varphi_3 = \frac{1}{\cos\varphi_3} \frac{1}{\tan\varphi_3 + \tan\varphi_4} \frac{\delta x}{|BD|}, \quad \delta\varphi_4 = \frac{1}{\cos\varphi_4} \frac{1}{\tan\varphi_3 + \tan\varphi_4} \frac{\delta x}{|AD|}.$$

Here φ_1, φ_2, φ_3, φ_4 are the angles between the x-axis and the intervals AC, BC, BD, AD respectively, $|AC|$ denotes the length of the chord AC.

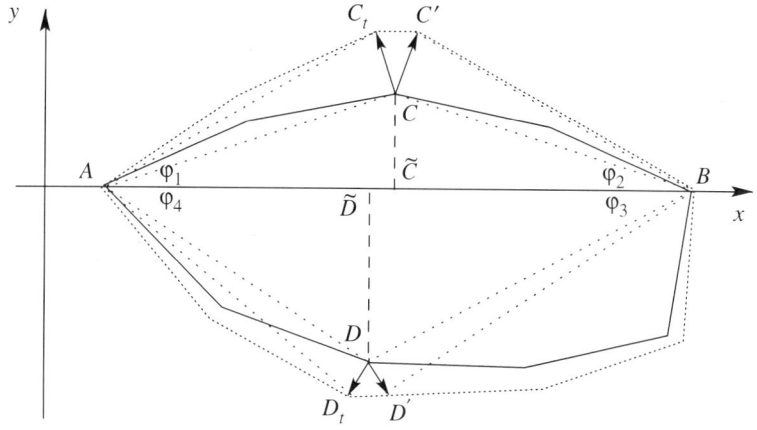

FIGURE 3

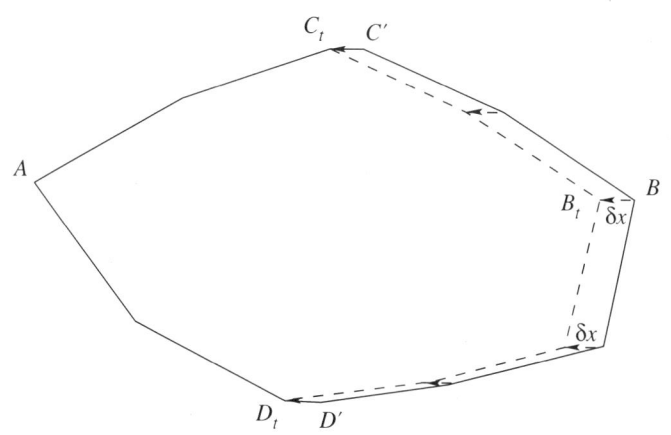

FIGURE 4

Let us denote the images of the points C, D after the rotations about the point A by C_t, D_t, the images of the points C, D after the rotations about the point B by C', D', the orthogonal projections of the points C and D on the interval AB by $\tilde{C}$, $\tilde{D}$. Finally we add small horizontal intervals connecting C_t and C', D_t and D'.

Step 2: We shift the entire arc $C'BD'$ to the left by the distance δx.

Let us denote the magnetic area above the line AB and below the line AB by Q_+ and Q_- respectively, let $Q = Q_+ + Q_-$, Q_{AD} be the magnetic area of the polygon bounded by the arc AD and by the segment AD, let Q_{BD} be the magnetic area of the polygon bounded by the arc BD and by the segment BD, δ_1 be the variation at the first step, δ_2 be the variation at the second step. Then

(1) $\quad \delta_1 Q_+ \geq \frac{1}{2} B_{\min} |AC| |CC_t| + \frac{1}{2} B_{\min} |BC| |CC'|,$

$\delta_1 Q_- \geq \delta_1 Q_{AD} + \delta_1 Q_{BD} + B_{\min} \cdot \{\text{area of triangle } ADD_t\}$

(2) $\quad\quad\quad\quad\quad\quad\quad + B_{\min} \cdot \{\text{area of triangle } BDD'\},$

(3) $\quad \delta_2 Q \geq -\frac{1}{2} NL B_{\max} \delta x.$

We shall use the following estimates:

1) $\varphi_1 + \varphi_2 < N\alpha_0/2$, $\varphi_1 < N\alpha_0/2$, $\varphi_2 < N\alpha_0/2$, $\cos \varphi_1 > 0.9$, $\cos \varphi_2 > 0.9$,

$$\tan \varphi_1 < \frac{9 B_{\min}}{20 B_{\max} (2N^3 + N/2)}, \quad \tan \varphi_2 < \frac{9 B_{\min}}{20 B_{\max} (2N^3 + N/2)}.$$

2) The angles between the x-axis and all the segments of the arc AB are less than $N\alpha_0/2$, their cosines are greater then 0.9.

3) $|AB| > 1.8L$, $|A\widetilde{C}| > 0.4L$, $|A\widetilde{D}| > 0.4L$, $|B\widetilde{C}| > 0.4L$, $|B\widetilde{D}| > 0.4L$, $|AC| > 0.4L$, $|AD| > 0.4L$, $|BC| > 0.4L$, $|BD| > 0.4L$.

4) $|D\widetilde{D}| > L/2N$.

The estimates 1)–3) follow directly from the definition of α_0. Let us prove estimate 4).

Lemma 6 implies that if γ is an extremal then at least one of the arcs AD or BD has no internal angles less than $\pi/3$. For the sake of concreteness, let us assume that the arc AD has this property. Then there are two possibilities.

1) Moving from the point A to the point D along the arc AD, we always move to the right. Let us denote by Q the neighboring vertex to A in the arc AD. The angle between the x-axes and AQ is greater than $\pi/3 - N\alpha_0/2 > \pi/3 - 0.001$.

2) Moving from the point A to the point D along the arc AD, we move to the left and then to the right. Let T be the "turning vertex", P and Q be the preceding and succeeding vertices. Then the projection of the interval PQ on the y-axis is greater than $\sqrt{3}L/2$.

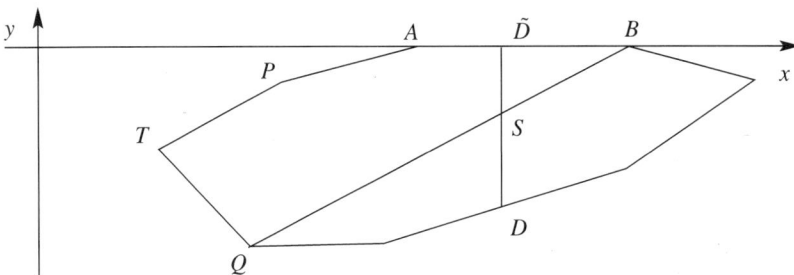

FIGURE 5

In both cases the distance between Q and the line AB is greater than $(\sqrt{3}/2 - 0.001)L$. Let us connect Q with B and denote the crossing point of the line segments QB and $D\widetilde{D}$ by S. Since $|QB| < NL/2$, $|SB| > 0.4$, we have

$$|D\widetilde{D}| > |SD| > \frac{|SB|}{|QB|} \left(\frac{\sqrt{3}}{2} - 0.001 \right) L > \frac{0.4L}{NL/2} \left(\frac{\sqrt{3}}{2} - 0.001 \right) L > \frac{L}{2N}.$$

For the infinitesimal intervals CC_t, CC', we have

$$|CC_t| = |AC|\delta\varphi_1 = \frac{1}{\cos\varphi_1}\frac{1}{\tan\varphi_1 + \tan\varphi_2}\delta x > \frac{1}{\tan\varphi_1 + \tan\varphi_2}\delta x,$$

thus

$$|CC_t| > \frac{10B_{\max}(2N^3 + N/2)}{9B_{\min}}\delta x, \qquad |CC'| > \frac{10B_{\max}(2N^3 + N/2)}{9B_{\min}}\delta x.$$

Next, $|AC| + |BC| > |A\widetilde{C}| + |B\widetilde{C}| > 1.8L$, thus

(4) $\quad \delta_1 Q_+ > 0.9 B_{\min} L \dfrac{10B_{\max}(2N^3 + N/2)}{9B_{\min}}\delta x = (2N^3 + N/2) L B_{\max}\delta x,$

$$\delta\varphi_3 = \frac{1}{\cos\varphi_3}\frac{1}{\tan\varphi_3 + \tan\varphi_4}\frac{\delta x}{|BD|} < \frac{1}{\cos\varphi_3}\frac{1}{\tan\varphi_3}\frac{\delta x}{|BD|} = \frac{1}{\sin\varphi_3}\frac{\delta x}{|BD|}.$$

Finally, $\sin\varphi_3 |BD| = |D\widetilde{D}|$,

$$\delta\varphi_3 < (2N/L)\delta x, \qquad \delta\varphi_4 < (2N/L)\delta x.$$

Applying Lemma 5 and taking into account the fact that the magnetic areas of the triangles ADD_t and BDD' are positive, we obtain

(5) $\qquad\qquad\qquad \delta_1 Q_- \geqslant -2N^3 L B_{\max}\delta x.$

Combining (3), (4), and (5), we get

$$\delta Q = \delta_1 Q_+ + \delta_1 Q_- + \delta_2 Q > 0.$$

This completes the proof of Lemma 8.

PROOF OF LEMMA 12. To estimate the variation of the magnetic area Q under this deformation, it is convenient to decompose this deformation in two steps.

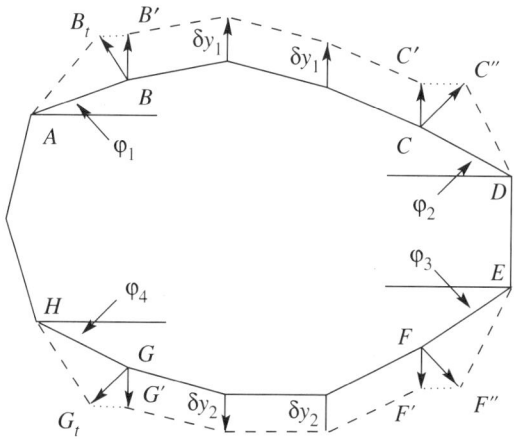

FIGURE 6

Step 1: We shift the arc BC up by the distance δy_1, the arc FG down by the distance δy_2, and we rotate the segments AB, CD, EF, GH about the points A, D, E, H by the angles

$$\delta\varphi_1 = \frac{\delta y_1}{L\cos\varphi_1}, \quad \delta\varphi_2 = \frac{\delta y_1}{L\cos\varphi_2}, \quad \delta\varphi_3 = \frac{\delta y_2}{L\cos\varphi_3}, \quad \delta\varphi_4 = \frac{\delta y_2}{L\cos\varphi_4},$$

where φ_1, φ_2, φ_3, φ_4 are the angles between the real line and the segments AB, CD, EF, GH, $|\tan\varphi_k| \leq 1$, $k = 1, \ldots, 4$. Let us denote the images of the points B, C, F, G under the shifts by B', C', F', G', the images under the rotations by B_t, C'', F'', G_t.

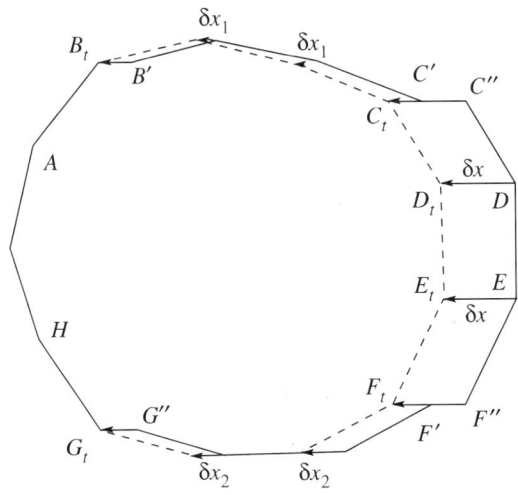

FIGURE 7

Step 2: We shift the arc $B'C'$ to the left by the distance $\delta x_1 = L\sin\varphi_1\delta\varphi_1 = \tan\varphi_1 \delta y_1$, the arc $F'G'$ to the left by the distance $\delta x_2 = L\sin\varphi_4\delta\varphi_4 = \tan\varphi_4\delta y_2$, and the arc $C''DEF''$ to the left by the distance

(6) $$\delta x = \delta y_1(\tan\varphi_1 + \tan\varphi_2) = \delta y_2(\tan\varphi_3 + \tan\varphi_4).$$

From (6) it follows that $\delta y_1 = t/(\tan\varphi_1 + \tan\varphi_2)$, $\delta y_2 = t/(\tan\varphi_3 + \tan\varphi_4)$.

Let us denote by BC_x and FG_x the lengths of the projections of the arcs BC and FG on the x-axis.

For the change of magnetic area in Step 1, we have

$$\delta_1 Q \geq B_{\min}(\delta y_1 BC_x + \delta y_2 FG_x) = B_{\min}\left[\frac{t\, BC_x}{\tan\varphi_1 + \tan\varphi_2} + \frac{t\, FG_x}{\tan\varphi_3 + \tan\varphi_4}\right],$$

thus $\delta_1 Q \geq B_{\min} t \min(BC_x, FG_x)$.

The angles between the x-axis and all segments of the arcs AH and DE are greater then $\pi/4$; thus the projection of these arcs on the x-axis are smaller than $D_{\min}$ and

$$\min(BC_x, FG_x) \geq D_x - 2D_{\min} - 2L,$$

where D_x denotes the diameter of γ in the direction x. It is easy to show that
$$D_x \geqslant D_{\max} - D_{\min}, \qquad L \leqslant (4/N) D_{\max}.$$
Thus
$$\delta_1 Q \geqslant B_{\min} t [(1 - 8/N) D_{\max} - 3 D_{\min}].$$
For the variation of the magnetic area in the second step, we have
$$\delta_2 Q \geqslant - B_{\max} D_{\min} t.$$
Therefore
$$\delta Q = \delta_1 Q + \delta_2 Q \geqslant \{ B_{\min}[(1 - 8/N) D_{\max} - 3 D_{\min}] - B_{\max} D_{\min} \} t.$$
If
$$\frac{D_{\max}}{D_{\min}} > \frac{B_{\max} B_{\min}^{-1} + 3}{1 - 8 N^{-1}},$$
then
$$\delta Q \geqslant \{ B_{\min}[B_{\max} B_{\min}^{-1} D_{\max} + 3 D_{\min} - 3 D_{\min}] - B_{\max} D_{\min} \} t > 0.$$
Lemma 12 is proved.

PROOF OF LEMMA 13.

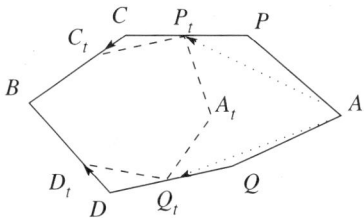

FIGURE 8

The variation of the magnetic area under this deformation consists of two parts:

1) The magnetic area of the small triangles near all the vertices except A and B (we denote it by $\delta_1 Q$).

2) The magnetic area of the small quadrangle $AP_t A_t Q_t$ near the vertex A. Here we denote the neighbors of A by P and Q, the shifts of the points A, P, Q by A_t, P_t, Q_t, the magnetic area of the quadrangle by $\delta_2 Q$.

We shall use the following estimate. Let F be a vertex of our polygon, suppose the arc BF contains k segments. Then the shift of the vertex F under the deformation FF_t is less than kt. It is easy to prove this by induction.

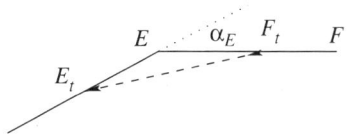

FIGURE 9

Let E be the neighbor of F on the arc BF, α_E be the external angle at the vertex E, let E_t be the shifted point E. Then

$$|FF_t| = |EE_t|\cos\alpha_E + t < |EE_t| + t\,.$$

Then the shifts of all the vertices except A are by less than $Nt/2$. For the area of the triangle near the vertex E we have

$$\{\text{area of the triangle } E_t EF_t\} = \tfrac{1}{2}|EE_t||EF_t|\sin\alpha_E < \tfrac{1}{2}NLt\alpha_E < \tfrac{1}{2}\mathcal{L}_0 t\alpha_E\,.$$

The sum of all external angles is equal to 2π; hence the sum of areas of all triangles is less than $\pi\mathcal{L}_0 t$ and the magnetic area of these triangles is less than $B_{\max}\pi\mathcal{L}_0 t$.

The distances $|AA_t|$ and $|P_t Q_t|$ can be estimated by

$$|AA_t| \leqslant 2\frac{Nt}{2}\frac{1}{\sin\alpha}\,,\qquad |P_t Q_t| < 2\frac{\mathcal{L}_0}{N}\,,$$

where α is the external angle at the vertex A. Then for the area of the quadrangle $AP_t A_t Q_t$ we have

$$\{\text{area of the quadrangle } 6AP_t A_t Q_t\} \leqslant \tfrac{1}{2}|P_t Q_t||AA_t| < \mathcal{L}_0 t (\sin\alpha)^{-1}\,.$$

Finally we get

$$\delta F = \delta l(\gamma) - \delta Q < -Nt + B_{\max}\pi\mathcal{L}_0 t + \mathcal{L}_0 B_{\max} t(\sin\alpha)^{-1}\,.$$

References

1. S. P. Novikov, *Multivalued functions and functionals. An analog of Morse theory*, Dokl. Akad. Nauk SSSR **260** (1981), no. 1, 31–35; English transl. in Soviet Math. Dokl. **24** (1981).
2. S. P. Novikov and I. Schmelzer, *Periodic solution of the Kirchhoff equations*, Funktsional. Anal. i Prilozhen. **15** (1981), no. 3, 54–66; no. 4, 37–52; English transl. in Functional Anal. Appl. **15** (1981).
3. S. P. Novikov, *The Hamiltonian formalism and a multivalued analog of Morse theory*, Uspekhi Mat. Nauk **37** (1982), no. 5, 33–49; English transl. in Russian Math. Surveys **37** (1982).
4. S. Novikov and I. A. Taimanov, *Periodic extremals of multivalued and not everywhere positive functionals*, Dokl. Akad. Nauk SSSR **274** (1984), no. 1, 26–28; English transl. in Soviet Math. Dokl. **29** (1984).
5. I. A. Taimanov, *Principle of the overthrowing of cycles in Morse–Novikov theory*, Dokl. Akad. Nauk SSSR **268** (1983), no. 1, 46–50; English transl. in Soviet Math. Dokl. **27** (1983).
6. _____, *Closed extremals on two-dimensional manifolds*, Uspekhi Mat. Nauk **47** (1992), no. 2, 143–185; English transl. in Russian Math. Surveys **37** (1982).
7. V. Ginzburg, *New generalization of Poincaré's geometric theorem*, Funktsional. Anal. i Prilozhen. **21** (1987), no. 2, 16–22.
8. _____, *On the existence and non-existence of closed trajectories for some Hamiltonian flows*, Stanford University Preprint, April 1994.

Translated by THE AUTHORS

Landau Institute for Theoretical Physics, Kosygina 2, Moscow, 117940, Russia
E-mail address: pgg@landau.ac.ru

Landau Institute for Theoretical Physics, Kosygina 2, Moscow, 117940, Russia
E-mail address: novikov@landau.ac.ru

Spin Generalization of the Calogero–Moser System and the Matrix KP Equation

I. KRICHEVER, O. BABELON, E. BILLEY, AND M. TALON

ABSTRACT. Complete solutions of the spin generalization of the elliptic Calogero Moser systems are constructed. They are expressed in terms of Riemann theta-functions. Analogous constructions for the trigonometric and rational cases are also presented.

§1. Introduction

The elliptic Calogero–Moser system [1, 2] is a system of N identical particles on a line interacting with each other via the potential $V(x) = \wp(x)$, where $\wp(x) = \wp(x|\omega, \omega')$ is the Weierstrass elliptic function with periods 2ω, $2\omega'$. This system (and its quantum version, as well) is a completely integrable system [3]. The complete solution of the elliptic Calogero–Moser model was constructed by algebro-geometrical methods in [4]. The degenerate cases where $V(x) = 1/\sinh^2 x$ or $V(x) = 1/x^2$ are also of interest, and admit nice interpretations as reductions of geodesic motions on symmetric spaces [3, 16]. A similar interpretation for the elliptic case was recently given in [5].

In this work, we consider the spin generalization of the Calogero–Moser model, which was defined in [6]. Again this model exists in the elliptic, trigonometric and rational versions, each one being of its own interest. In particular the hidden symmetry of the model changes from a current algebra type in the rational case, to a Yangian type in the trigonometric case [6–8]. Our main goal is to construct the action-angle type variables for these spin generalizations of the Calogero–Moser system, and to solve the equations of motion in terms of Riemann theta-functions. The algebro-geometric constructions of the solutions substantially differ in the three cases and we shall present them in parallel.

Let us consider the classical Hamiltonian system of N particles on a line, with coordinates x_i and momenta p_i, and internal degrees of freedom described for each particle by an l-dimensional vector $a_i = (a_{i,\alpha})$ and an l-dimensional covector $b_i^+ = (b_i^\alpha)$, $\alpha = 1, \ldots, l$. The Hamiltonian has the form

$$(1.1) \qquad H = \frac{1}{2}\sum_{i=1}^{N} p_i^2 + \frac{1}{2}\sum_{i\neq j}(b_i^+ a_j)(b_j^+ a_i) V(x_i - x_j),$$

1991 *Mathematics Subject Classification.* Primary 58F07.

©1995, American Mathematical Society

where $(b_i^+ a_j)$ stands for the corresponding scalar product

$$(1.2) \qquad (b_i^+ a_j) = (b_i^\alpha a_{i,\alpha}),$$

and the potential $V(x)$ is one of the functions $\wp(x)$, $1/\sinh^2 x$, or $1/x^2$. The nontrivial Poisson brackets between the dynamical variables x_i, p_i, b_i^α, $a_{i,\alpha}$ are

$$(1.3) \qquad \{p_i, x_j\} = \delta_{ij}, \qquad \{b_i^\beta, a_{j,\alpha}\} = -\delta_{i,j}\delta_\alpha^\beta.$$

The equations of motion have the form

$$(1.4) \qquad \ddot{x}_i = \sum_{j \neq i}(b_i^+ a_j)(b_j^+ a_i) V'(x_i - x_j), \qquad V'(x) = \frac{dV(x)}{dx},$$

$$(1.5) \qquad \dot{a}_i = -\sum_{j \neq i} a_j (b_j^+ a_i) V(x_i - x_j),$$

$$(1.6) \qquad \dot{b}_i^+ = \sum_{j \neq i} b_j^+ (b_i^+ a_j) V(x_i - x_j).$$

From (1.5), (1.6) it follows that $(b_i^+ a_i)$ are integrals of motion. We restrict the system to the invariant submanifold

$$(1.7) \qquad (b_i^+ a_i) = c = 2.$$

REMARK. The reduction of system (1.1) to the invariant submanifold defined by the constraint (1.7) is a completely integrable Hamiltonian system for any value of the constant c. Changing the value of c amounts to a rescaling of the time variable. In the following, we shall assume $c = 2$ for definiteness.

Let us introduce the quantities

$$(1.8) \qquad f_{ij} = (b_i^+ a_j).$$

The Poisson brackets (1.2) imply

$$(1.9) \qquad \{f_{ij}, f_{kl}\} = \delta_{jk} f_{il} - \delta_{il} f_{kj}.$$

The Hamiltonian (1.1) in terms of these new variables has the form

$$(1.10) \qquad H = \frac{1}{2}\sum_{i=1}^N p_i^2 + \frac{1}{2}\sum_{i \neq j} f_{ij} f_{ji} V(x_i - x_j).$$

The system (1.10) with f_{ij} as dynamical variables satisfying the relations (1.9) was introduced in [6] and was called the Euler–Calogero–Moser system. When $l < N$, relations (1.8) give a parametrization of special symplectic leaves of the system (1.9), (1.10).

Let us count the number of nontrivial degrees of freedom. We start with $2N+2Nl$ dynamical variables corresponding to the x_i, p_i, $a_{i,\alpha}$, b_i^α. The Hamiltonian (1.10) has a symmetry under rescaling

$$(1.11) \qquad a_i \to \lambda_i a_i, \qquad b_i \to \lambda_i^{-1} b_i$$

(notice that f_{ij} is noninvariant but $f_{ij}f_{ji}$ is invariant, and the Poisson brackets are also invariant). The corresponding momentum is given by the collection of $b_i^+ a_i$ and we fix it to the values $b_i^+ a_i = 2$, which makes N conditions. The stabilizer of this momentum consists in the whole group, so that the reduced system is defined by N more constraints, e.g., $\sum_\alpha b_i^\alpha = 1$, leaving us with a phase space of dimension $2Nl$. Moreover the Hamiltonian and the symplectic structure are invariant under a further symmetry

$$(1.12) \qquad a_i \to W^{-1} a_i, \qquad b_i^+ \to b_i^+ W,$$

where W is any matrix in $GL(r, \mathbb{R})$ independent of the label i, preserving the above condition on the b_i's. This means that W must leave the vector $v = (1, \ldots, 1)$ invariant. Hence this group is of dimension $l^2 - l$. Fixing the momentum, $\mathcal{P}$, gives $l^2 - l$ conditions. The stabilizer of a generic momentum is trivial: this is because such a generic element can be diagonalized as $\mathcal{P} = m^{-1} \Lambda m$. Its stabilizer under the adjoint action consists of the matrices of the form $g = m^{-1} D m$ with D diagonal. The condition $gv = v$ translates into $Dmv = mv$ which implies generically $D = 1$, i.e., $g = 1$. We have proved the

PROPOSITION 1.1. *The dimension of the reduced phase space $\mathcal{M}$ is*

$$(1.13) \qquad \dim \mathcal{M} = 2[Nl - l(l-1)/2].$$

Our method for the construction of the solutions of system (1.1) is a generalization of the approach that was used for the classical Calogero–Moser system. In [9] a remarkable connection between the Calogero–Moser system and the motion of poles of the rational and the elliptic solutions of the KdV equation was found. It turned out that the corresponding relation becomes an isomorphism in the case of the rational or the elliptic solutions of the Kadomtsev–Petviashvili equation. In [10] and [11] this isomorphism in the rational case was used in opposite directions.

In [11], using the known solutions of the rational Calogero–Moser system, rational solutions of the KP equation were constructed.

In [10], the construction of rational solutions for various partial differential equations admitting a zero-curvature representation was proposed. The application of this result to the KP equation yielded an alternative way of solving the Calogero–Moser system. This approach was generalized in [4], where the action-angle variables for the elliptic Calogero–Moser system were constructed and an exact formula for elliptic solutions of the KP equation was obtained. (Further developments in the theory of so-called elliptic solitons are presented in the special issue of Acta Applicandae Mathematicae **35** (1994) dedicated to the memory of J. L. Verdier).

§2. Relationship with the matrix KP equation

The zero-curvature representation for the KP equation

$$\tfrac{3}{4} u_{yy} = (u_t - \tfrac{3}{2} u u_x + \tfrac{1}{4} u_{xxx})_x \tag{2.1}$$

has the form ([12, 13])

$$[\partial_y - L, \partial_t - M] = 0, \tag{2.2}$$

where

$$L = \partial_x^2 - u(x, y, t), \qquad M = \partial_x^3 - \tfrac{3}{2} u \partial_x + w(x, y, t). \tag{2.3}$$

In this scalar case ($l = 1$), assuming that u is an elliptic function of the variable x, the comparison of singular terms in the expansion of the right- and left-hand sides of (2.1) near the poles of u gives that:

1) Any elliptic (in the variable x) solution of the KP equation has the form

$$u(x, y, t) = 2 \sum_{i=1}^{N} \wp(x - x_i(y, t)) + \text{const}. \tag{2.4}$$

2) The dependence of the poles $x_i(y, t)$ on the variable y coincides with the elliptic Calogero–Moser system and their dependence on the variable t is described by the "third integral" of this system.

Let us consider the same equations (2.2) in the case when the operators (2.3) have $(l \times l)$ matrix coefficients. They are equivalent to the system

$$w_x = \tfrac{3}{4} u_y, \qquad w_y = u_t - \tfrac{3}{4}(u u_x + u_x u) + \tfrac{1}{4} u_{xxx} - [u, w], \tag{2.5}$$

that we call the *matrix KP equation*.

In the matrix case, we do not know the complete classification of all elliptic solutions of (2.5). It turns out that the system (1.1) is isomorphic to the special elliptic solutions of matrix KP equation of the form

$$u(x, y, t) = \sum_{i=1}^{N} p_i(y, t) \wp(x - x_i(y, t)), \tag{2.6}$$

$$w(x, y, t) = \sum_{i=1}^{N} (A_i(y, t) \zeta(x - x_i(y, t)) + B_i(y, t) \wp(x - x_i(y, t))), \tag{2.7}$$

where p_i is a rank-one matrix-valued function depending on y and t,

$$p_i = a_i b_i^+, \quad \text{i.e.,} \quad p_{i,\alpha}^\beta = a_{i,\alpha} b_i^\beta. \tag{2.8}$$

The precise relationship is provided by

THEOREM 2.1. *Let us consider the functions*

$$(2.9) \qquad V(x) = \wp(x), \qquad \Phi(x, z) = \frac{\sigma(z-x)}{\sigma(z)\sigma(x)} e^{\zeta(z)x}.$$

The equations

$$(2.10) \qquad \left(\partial_t - \partial_x^2 + \sum_{i=1}^N a_i(t) b_i^+(t) V(x - x_i(t))\right) \Psi = 0,$$

$$(2.11) \qquad \Psi^+ \left(\partial_t - \partial_x^2 + \sum_{i=1}^N a_i(t) b_i^+(t) V(x - x_i(t))\right) = 0$$

(where we define $\Psi^+ \partial \equiv -\partial \Psi^+$*) have solutions* Ψ, Ψ^+ *of the form*

$$(2.12) \qquad \Psi = \sum_{i=1}^N s_i(t, k, z) \Phi(x - x_i(t), z) e^{kx + k^2 t},$$

$$(2.13) \qquad \Psi^+ = \sum_{i=1}^N s_i^+(t, k, z) \Phi(-x + x_i(t), z) e^{-kx - k^2 t},$$

where s_i and s_i^+ are l-dimensional vector $s_i = (s_{i,\alpha})$ and covector $s_i^+ = (s_i^\alpha)$, respectively, if and only if $x_i(t)$ satisfy the equations (1.4) and the vectors a_i, b_i^+ satisfy the constraints (1.7) and the system of equations

$$(2.14) \qquad \dot{a}_i = -\sum_{j \neq i} a_j (b_j^+ a_i) V(x_i - x_j) - \lambda_i a_i,$$

$$(2.15) \qquad \dot{b}_i^+ = \sum_{j \neq i} b_j^+ (b_i^+ a_j) V(x_i - x_j) + \lambda_i b_i^+,$$

where $\lambda_i = \lambda_i(t)$ are scalar functions.

REMARK 1. System (1.4), (2.14), (2.15) is "gauge equivalent" to system (1.4)–(1.6). This means that if (x_i, a_i, b_i^+) satisfy the equations (1.4), (2.14), (2.15), then x_i and the vector-valued functions

$$(2.16) \qquad \hat{a}_i = a_i q_i, \quad \hat{b}_i^+ = b_i q_i^{-1}, \quad q_i = \exp\left(\int^t \lambda_i(t) \, dt\right)$$

are solutions of system (1.4)–(1.6).

REMARK 2. In the scalar case, the ansatz (2.10) was introduced in [4]. Its particular form was inspired by the well-known formula for the solution of the Lamé equation

$$(2.17) \qquad \left(\frac{d^2}{dx^2} - 2\wp(x)\right) \Phi(x, z) = \wp(z) \Phi(x, z).$$

PROOF. Inserting (2.12) into equation (2.10), we find the condition

$$A \equiv \sum_{i=1}^{N} \left\{ \dot{s}_i \Phi(x - x_i, z) - (\dot{x}_i + 2k) s_i \Phi'(x - x_i, z) \right.$$
$$\left. - s_i \Phi''(x - x_i, z) + \sum_{j=1}^{N} a_j (b_j^+ s_i) \wp(x - x_j) \Phi(x - x_i, z) \right\} = 0,$$

where $\Phi' = \partial_x \Phi$ and so on.

The vanishing of the triple pole $(x - x_i)^{-3}$ gives the condition

$$(2.18) \qquad a_i (b_i^+ s_i) = 2 s_i.$$

Using this condition and the Lamé equation (2.17), we can identify the double pole $(x - x_i)^{-2}$. Its vanishing gives the condition

$$(2.19) \qquad s_i (\dot{x}_i + 2k) + \sum_{j \neq i} a_i b_i^+ s_j \Phi(x_i - x_j, z) = 0.$$

We finally identify the residue at the simple pole and obtain the condition

$$(2.20) \quad \dot{s}_i + \left(\sum_{j \neq i} a_j b_j^+ \wp(x_i - x_j) - \wp(z) \right) s_i + a_i \sum_{j \neq i} (b_i^+ s_j) \Phi'(x_i - x_j, z) = 0.$$

Now inserting equations (2.18)–(2.20) into the expression for A, one sees that A vanishes identically due to the functional equation

$$(2.21) \quad \Phi'(x, z) \Phi(y, z) - \Phi(x, z) \Phi'(y, z) = (\wp(y) - \wp(x)) \Phi(x + y, z).$$

We have shown that the function ψ given by (2.12) satisfies equation (2.10) if and only if the conditions (2.18)–(2.20) are fulfilled.

Equation (2.18) implies that the vector s_i is proportional to the vector a_i. Hence

$$(2.22) \qquad s_{i,\alpha}(t, k, z) = c_i(t, k, z) a_{i,\alpha}(t).$$

Moreover, (2.18) implies that the constraints (1.7) are fulfilled.

Equation (2.19) can then be rewritten as a matrix equation for the vector $C = (c_i)$:

$$(2.23) \qquad (L(t, z) + 2k I) C = 0,$$

where the Lax matrix $L(t, z)$ with spectral parameter z is given by

$$(2.24) \qquad L_{ij}(t, z) = \dot{x}_i \delta_{ij} + (1 - \delta_{ij}) f_{ij} \Phi(x_i - x_j, z).$$

We can rewrite equation (2.20) as

$$(2.25) \qquad \dot{a}_i = -\lambda_i a_i - \sum_{j \neq i} a_j (b_j^+ a_i) \wp(x_i - x_j),$$

where we have defined

$$\lambda_i = \frac{\dot{c}_i}{c_i} - \wp(z) + \sum_{j\neq i}(b_i^+ a_j)\Phi'(x_i - x_j, z)\frac{c_j}{c_i}.$$

But this last equation can be rewritten as

(2.26) $$(\partial_t + M)C = 0,$$

where the second element M of the Lax pair is given by

(2.27) $$M_{ij}(t, z) = (-\lambda_i - \wp(z))\delta_{ij} + (1 - \delta_{ij})f_{ij}\Phi'(x_i - x_j, z).$$

The same arguments show that the existence of a solution Ψ^+ of the form (2.13) implies (cancellation of the triple pole)

(2.28) $$s_i^\alpha = c_i^+ b_i^\alpha,$$

and the covector $C^+ = (c_i^+)$ satisfies the equation (cancellation of the double pole)

(2.29) $$C^+(L(z) + 2k) = 0.$$

Finally looking at the simple pole, one gets

(2.30) $$\dot{b}_i^+ = \lambda_i^+ b_i^+ + \sum_{j\neq i}(b_i^+ a_j)b_j^+ \wp(x_i - x_j)$$

with a new scalar λ_i^+ given by

$$\lambda_i^+ = -\frac{\dot{c}_i^+}{c_i^+} - \wp(z) + \sum_{j\neq i}\frac{c_j^+}{c_i^+}(b_j^+ a_i)\Phi'(x_j - x_i).$$

The equations (2.25), (2.30) are compatible with $f_{ii} = b_i^+ a_i = 2$ only when $\lambda_i^+ = \lambda_i$. Finally we can rewrite the definition of λ_i^+ as

(2.31) $$\partial_t C^+ - C^+ M = 0.$$

To complete the proof of the theorem, we must establish that all $x_i(t)$ satisfy equation (1.4). To do this, we use the compatibility conditions between equation (2.23), (2.26) and between equations (2.29), (2.31) which read respectively

$$(\dot{L} + [M, L])C = 0, \qquad C^+(\dot{L} + [M, L]) = 0.$$

Computing $\dot{L} + [M, L]$, we see that the off-diagonal elements vanish identically due to equations (2.14), (2.15), while the diagonal elements are precisely the equations of motion of the x_i. The computation uses equation (2.21) again and we have therefore established the Lax form of the equations of motion

(2.32) $$\dot{L} = [L, M].$$

Theorem 2.1 is proved. □

REMARK. In [1] it was proved that the Lax equation (2.32) with the matrices L and M given by formulas (2.24), (2.27) (with $f_{ij} = 2$, $\lambda_i = 0$) is equivalent to the equations of motion of the Calogero–Moser system if and only if the functional equation (2.21) is fulfilled. In [1], the particular solutions of the functional equation corresponding to the values $z = \omega_l$ was found. The proof of the fact that this equation is valid for arbitrary values of the spectral parameter z was given in [4].

Let us comment on the trigonometric and rational limits of the above formulas. The trigonometric limit is obtained when one of the periods $\omega \to \infty$. We choose the other one as $i\pi$. In this limit the function Φ becomes

$$\Phi(x, z) = (\coth x - \coth z) e^{x \coth z}.$$

The exponential factor in Φ comes from the factor $\exp(\zeta(z)x)$ in the elliptic case and is necessary to induce the double periodicity of Φ in z. In the trigonometric case, however, it can be absorbed into a redefinition of k and s_i of the form

$$k \to k - \coth z, \qquad s_i \to s_i \exp(x_i(t) \coth z + 2kt \coth z - t \coth^2 z)$$

and similarly for the dual quantities. In the following we shall therefore remove this exponential factor from the definition of Ψ. The definitions of the functions $V(x)$ and $\Phi(x, z)$ become

(2.33) $\qquad V(x) = 1/\sinh^2(x), \qquad \Phi(x, z) = \coth x - \coth z.$

With these new functions, the above theorem remains valid, but due to the redefinition of the function $\Phi(x, z)$, the expression of the Lax matrices is slightly modified and now reads

(2.34)
$$L_{ij}(t, z) = (\dot{x}_i - 2\coth z)\delta_{ij} + (1 - \delta_{ij}) f_{ij} \Phi(x_i - x_j, z),$$

(2.35) $\qquad M_{ij}(t) = -\lambda_i \delta_{ij} - (1 - \delta_{ij}) f_{ij} V(x_i - x_j).$

The rational limit is obtained in a straightforward way from the trigonometric limit by sending the second period $\omega' \to \infty$. The functions $V(x)$ and $\Phi(x, z)$ become

(2.36) $\qquad V(x) = 1/x^2, \qquad \Phi(x, z) = 1/x - 1/z,$

and of course $\coth z \to 1/z$ in (2.34).

Notice that, as compared to the elliptic case, there is a decoupling between the spectral parameter z and the x_i's in the Lax matrix (2.34).

Part I. The Direct Problem

§3. The spectral curve

Due to equation (2.23), the parameters k and z are constrained to obey the identity

(3.1) $\qquad R(k, z) \equiv \det(2kI + L(t, z)) = 0.$

This defines a curve Γ, which is time-independent due to the Lax equation (2.32). This curve plays a fundamental role in the subsequent analysis. Its properties are different in the elliptic, trigonometric and rational cases. Notice, moreover, that Γ is invariant under the symmetries (1.11), (1.12).

PROPOSITION 3.1. *In the elliptic case we have*

$$(3.2) \qquad R(k, z) = \sum_{i=0}^{N} r_i(z) k^i,$$

where the $r_i(z)$ are elliptic functions of z, independent of t and having the form

$$(3.3) \qquad r_i(z) = I_i^0 + \sum_{s=0}^{N-i-2} I_{i,s} \partial_z^s \wp(z).$$

In a neighborhood of $z = 0$ the function $R(k, z)$ can be represented in the form

$$(3.4) \qquad R(k, z) = 2^N \prod_{i=1}^{N} (k + v_i z^{-1} + h_i(z)),$$

where the $h_i(z)$ are regular functions of z and

$$(3.5) \qquad v_i = 1, \qquad i > l.$$

PROOF. The matrix elements (2.24) are double periodic functions of the variable z with essential singularity at $z = 0$, but the functions $r_i(z)$ are meromorphic because $L(t, z)$ can be represented in the form

$$(3.6) \quad L(t, z) = G(t, z) \widetilde{L}(t, z) G^{-1}(t, z), \qquad G_{ij} = \delta_{ij} \exp(\zeta(z) x_i(t)),$$

where $\widetilde{L}_{ij}(t, z)$ are meromorphic functions of the variable z in a neighborhood of the point $z = 0$. In fact we have

$$(3.7) \qquad \widetilde{L}(t, z) = -z^{-1}(F(t) - 2I) + O(z^0),$$

where $F(t)$ is the matrix of elements $f_{ij}(t)$. Therefore $r_i(z)$ are elliptic functions having poles of degree $N - i$, at most, at the point $z = 0$. Hence they can be represented in the form (3.3) as a linear combination of the function $\wp(z)$ and its derivatives. The coefficients I_i^0, $I_{i,s}$ of this expansion are the integrals of motion of the system (1.1). Each set of given values of these integrals defines an algebraic curve Γ.

Since around $z = 0$ the function $r_i(z)$ has a pole of order $N - i$, a factorization of the form (3.4) holds. Due to equation (3.7), the coefficients $-2v_i$ in (3.4) are the eigenvalues of the matrix $F - 2I$. From (1.8) we see that F is of rank l, hence the eigenvalue $v_i = 1$ has multiplicity $N - l$. Moreover, the corresponding $(N - l)$-dimensional subspace of eigenvectors $C = (c_1, \ldots, c_N)$ is defined by the equations

$$(3.8) \qquad \sum_{j=1}^{N} c_j a_{j,\alpha} = 0, \qquad \alpha = 1, \ldots, l. \quad \square$$

REMARK. The conditions (3.5) imply a full set of linear relations on the integrals I_i^0, I_{is} of the system (1.1). Let us take any polynomial (in k) $R(k, z)$ of the form (3.2) with $r_i(z)$ of the form (3.3). This polynomial depends on $N(N+1)/2$ parameters I_i^0, I_{is}.

Let us introduce the variable $\tilde{k} = k + z^{-1}$. Then the polynomial in this variable $\widetilde{R}(\tilde{k}, z) = R(\tilde{k} - z^{-1}, z)$ for a generic set of variables I_i^0, I_{is} can be represented in the form

$$(3.9) \qquad \widetilde{R}(\tilde{k}, z) = \sum_{i=0}^{N} \widetilde{R}_i(\tilde{k}) z^{-i} + \mathcal{R}(z, \tilde{k}),$$

where $\widetilde{R}_i$ are polynomials in $\tilde{k}$ of degree $\deg \widetilde{R}_i = N - i$ and $\mathcal{R}(z, \tilde{k}) = O(z)$ is a regular series in z with coefficients that are polynomials in $\tilde{k}$ of degree $N - 1$. The conditions (3.5) imply that

$$(3.10) \qquad \widetilde{R}_i(\tilde{k}) = 0, \qquad i > l.$$

The coefficients of $\widetilde{R}_i$ are linear combinations of the parameters I_i^0, I_{is}. Therefore, (3.10) is equivalent to a set of $(N-l)(N-l+1)/2$ linear equations on these parameters. The total number of independent parameters is therefore equal to $Nl - l(l-1)/2$, which is exactly half the dimension of the reduced phase space.

In the trigonometric and rational cases the parametrization of the corresponding spectral curve is even more explicit.

PROPOSITION 3.2. *In the trigonometric case, we have*

$$(3.11) \qquad R(k, z) = R_0(k) + \coth z R_1(k) + \cdots + \coth^l z R_l(k),$$

where the $R_m(k)$ are polynomials in k of degree $\deg_k R_m = N - m$ and

$$(3.12) \qquad R(k, z = -\infty) = R(k+2, z = +\infty).$$

In a neighborhood of $z = 0$, the function $R(k, z)$ can be factorized in the form (3.4), *where now $v_i = 0$, $i > l$.*

PROOF. The matrix $L(t, z)$ depends on z only via the term $\coth zF$. Since F is of rank l, $R(k, z)$ is of the form (3.11). To prove relation (3.12), it suffices to remark that

$$(3.13) \qquad L(t, -\infty) + 2kI = e^{2X}(L(t, +\infty) + 2(k+2)I)e^{-2X}$$

with $X = \text{Diag}(x_i(t))$. The conditions $v_i = 0$, $i > l$ follow from the fact that around $z = 0$ we now have

$$(3.14) \qquad L(t, z) = -z^{-1}F + O(z^0). \qquad \square$$

PROPOSITION 3.3. *In the rational case, we have*

(3.15) $$R(k, z) = R_0(k) + z^{-1}R_1(k) + \cdots + z^{-l}R_l(k),$$

where the $R_m(k)$ are polynomials in k of degree $\deg_k R_m = N - m$ and

(3.16) $$R_1(k) = -dR_0(k)/dk.$$

In a neighborhood of $z = 0$, the function $R(k, z)$ can be factorized in the form (3.4), *where now $v_i = 0$, $i > l$.*

PROOF. Since

(3.17) $$[X, L(t, \infty)] = F - 2I,$$

we have

(3.18)
$$L(t, z) + 2kI = L(t, \infty) - z^{-1}[X, L(t, \infty)] + 2(k - z^{-1})I$$
(3.19) $$= (I - z^{-1}X)(L(t, \infty) + 2(k - z^{-1})I)(I - z^{-1}X)^{-1} + O(z^{-2}),$$

hence $R(k, z) = R_0(k - z^{-1}) + O(z^{-2})$, so that $R_1 = -R_0'$. □

As a consequence, we can count the number of parameters occurring in the spectral curve. Each R_m depends on $N - m + 1$ parameters, but relations (3.12) or (3.16) remove N parameters and the leading term of R_0 is already given, so that we get $Nl - l(l-1)/2$ parameters, which is exactly half the dimension of the reduced phase space. These parameters can be identified with the action variables of our model and are in involution, since there exists an r-matrix for L [8].

We now compute the genus of the spectral curve Γ.

PROPOSITION 3.4. *For generic values of the action variables, the genus of the spectral curve is given by*

(3.20) $$\text{Elliptic case}: g = Nl - l(l+1)/2 + 1,$$
(3.21)
$$\text{Trigonometric and rational cases}: g = N(l-1) - l(l+1)/2 + 1.$$

PROOF. Equation (3.1) allows us to present the compact Riemann surface Γ as an N-sheeted branched covering of the base curve of the variable z, i.e., the completed plane in the trigonometric and rational cases and the torus in the elliptic case. The sheets are the N roots in k. By the Riemann–Hurwitz formula, we have $2g - 2 = N(2g_0 - 2) + v$, where g_0 is the genus of the base curve, i.e., $g_0 = 0$ in the trigonometric and rational cases, $g_0 = 1$ in the elliptic case. Here v is the number of branch points, i.e., the number of values of z for which $R(k, z)$ has a double root in k. This is the number of zeros of $\partial_k R(k, z)$ on the surface $R(k, z) = 0$. But $\partial_k R(k, z)$ is a meromorphic function on the surface, hence it has as many zeros as poles. The poles are located above $z = 0$ or $k = \infty$, which is the same, and are easy to count.

Let P_i be the points of Γ lying on the different sheets over the point $z = 0$. In the neighborhood of P_i the function k has the expansion

$$k_i = -v_i z^{-1} - h_i(z). \tag{3.22}$$

Hence, the function $\partial R/\partial k$ in the neighborhood of P_i has the form

$$\frac{\partial R}{\partial k} = 2^N \prod_{j \neq i}((v_j - v_i)z^{-1} + (h_j(z) - h_i(z))). \tag{3.23}$$

From this we see that on each of the l sheets $(k_i(z), z)$ $(i = 1, \ldots, l)$ we have one pole of order $N - 1$. On each of the $N - l$ sheets $(k_i(z), z)$, $i = l+1, \ldots, N$, we have one pole of order l. Finally $v = l(N - 1) + (N - 1)l$ in either case. Inserting this value in the Riemann–Hurwitz theorem yields the result. $\square$

§4. Analytic properties of the eigenvectors of the Lax matrix

For a generic point P of the curve Γ, i.e., for a pair $(k, z) = P$ that satisfies equation (3.2), there exists at time $t = 0$ a unique eigenvector $C(0, P)$ of the matrix $L(0, z)$ normalized by the condition $c_1(0, P) = 1$. In fact the unnormalized components $c_i(0, P)$ can be taken as $\Delta_i(0, P)$, where $\Delta_i(0, P)$ are suitable minors of the matrix $L(0, z) + 2kI$, and are thus holomorphic functions on Γ outside the points above $z = 0$. After normalizing the first component, all the other coordinates $c_j(0, P)$ are meromorphic functions on Γ outside the points P_i above $z = 0$. The poles of $c_j(0, P)$ are the zeros on Γ of the first minor of the matrix $L(0, z) + 2kI$, i.e., they are defined by the system of the equation (3.2) and the equation

$$\det(2k\delta_{ij} + L_{ij}(0, z)) = 0, \quad i, j > 1. \tag{4.1}$$

Thus the position of these poles depends only on the initial data.

In the trigonometric and rational cases, nothing particular happens above $z = 0$. In the elliptic case, however, one has to be careful because of the essential singularity.

PROPOSITION 4.1. *In the elliptic case, in the neighborhood of the point P_i, the coordinate $c_j(0, P)$ has the form*

$$c_j(0, P) = (c_j^{(i)}(0) + O(z)) \exp[\zeta(z)(x_j(0) - x_1(0))], \tag{4.2}$$

where $c_j^{(i)}(t)$ is the eigenvector of the matrix $F(t)$ corresponding to the nonzero eigenvalue $2(1 - v_i)$, i.e.,

$$\sum_{j=1}^{N} f_{kj}(t) c_j^{(i)}(t) = 2(1 - v_i) c_k^{(i)}(t). \tag{4.3}$$

PROOF. From equation (3.6), we have $C(0, P) = G(0, z)\widetilde{C}(0, P)$, where $\widetilde{C}(0, P)$ is an eigenvector of $\widetilde{L}(0, z)$. Using equation (3.7), we get $\widetilde{C}(0, P) = \widetilde{C}^{(i)} + O(z)$, where $\widetilde{C}^{(i)}$ is an eigenvector of $F - 2I$. Therefore we have $c_j(0, P) = (c_j^{(i)}(0) + O(z)) \exp(\zeta(z) x_j(0))$. Normalizing $c_1(0, P) = 1$ yields the result. $\square$

We can now compute the number of poles of C on Γ. This number is the same in all cases, although its relation to the genus of Γ differs in the elliptic and other cases.

PROPOSITION 4.2. *The number of poles of $C(0, P)$ is*

(4.4) $\quad m = Nl - l(l+1)/2 = g - 1 \quad$ (*elliptic case*),

(4.5) $\quad m = Nl - l(l+1)/2 = g + N - 1 \quad$ (*trigonometric and rational cases*).

PROOF. Let us introduce the function W of the complex variable z defined by

$$W(z) = (\mathrm{Det}\,|c_i(M_j)|)^2,$$

where the M_j's are the N points above z. It is well defined on the base curve, since Det^2 does not depend on the order of the M_j's.

In the trigonometric and rational cases, W is a meromorphic function, hence has the same number of zeros and poles. In the elliptic case, it has an essential singularity at $z = 0$ of the form $\exp 2\zeta(z) \sum(x_i(0) - x_1(0))$. This does not affect the property that the number of poles is equal to the number of zeros. Clearly W has a double pole for values of z such that there exists a point M above z at which $C(M)$ has a simple pole.

We now show that $W(z)$ has a simple zero for values of z corresponding to a branch point of the covering, hence $m = v/2$.

First notice that $W(z)$ only vanishes at branch points, where there are at least two identical columns. Indeed, let $M_i = (k_i, z)$ be the N points above z. Then the $C(M_i)$ are the eigenvectors of $L(z)$ corresponding to the eigenvalues $-2k_i$, hence are linearly independent when all the k_i's are different. Therefore $W(z)$ cannot vanish at such a point. Let us assume now that z corresponds to a branch point which is generically of order 2. At such a point, $W(z)$ has a simple zero. Indeed, let ξ be an analytical parameter on the curve around the branch point. The covering projection $M \to z$ can be expressed as $z = z_0 + z_1 \xi^2 + O(\xi^3)$. The determinant vanishes to order ξ^1, hence W vanishes to order ξ^2, but this is precisely proportional to $z - z_0$. $\square$

At this point the analysis of the elliptic and the trigonometric and rational cases begin to differ substantially. We treat them separately.

4.1. The elliptic case. In this case we compute the time evolution of the above eigenvectors.

PROPOSITION 4.3. *The coordinates $c_j(t, P)$ of the vector-valued function $C(t, P)$ are meromorphic functions on Γ except at the points P_i. Their poles $\gamma_1, \ldots, \gamma_{g-1}$ do not depend on t. In the neighborhood of P_i they have the form*

(4.6) $\quad c_j(t, P) = c_j^{(i)}(t, z) \exp(\zeta(z)(x_j(t) - x_1(0)) + \mu_i(z)t),$

where the $c_j^{(i)}(t, z)$ are regular functions of z,

(4.7) $\quad c_j^{(i)}(t, z) = c_j^{(i)}(t) + O(z)$

and

(4.8) $\quad \mu_i(z) = (1 - 2v_i)z^{-2} - 2h_i(0)z^{-1} + O(z^0).$

PROOF. The fundamental matrix $S(t, z)$ of solutions to the equation

$$(4.9) \qquad (\partial_t + M(t, z))S(t, z) = 0, \qquad S(0, z) = 1,$$

is a holomorphic function of the variable z for $z \neq 0$. At $z = 0$, however, it has an essential singularity.

We have $L(t, z) = S(t, z)L(0, z)S^{-1}(t, z)$. Therefore the vector $C(t, z) = S(t, z)C(0, z)$ is the common solution to (2.26) and to the equation

$$(4.10) \qquad (L(t, z) + 2kI)C(t, P) = 0, \qquad P = (k, z) \in \Gamma.$$

Since $S(t, z)$ is regular for $z \neq 0$, we see that $C(t, P)$ has the same poles as $C(0, P)$.

Let us consider the vector $\widetilde{C}(t, P)$ defined by

$$(4.11) \qquad C(t, P) = G(t, z)\widetilde{C}(t, P),$$

where $G(t, z)$ is the same as in (3.6). This vector is an eigenvector of the matrix $\widetilde{L}(t, z)$ and satisfies the equation

$$(4.12) \qquad (\partial_t + \widetilde{M}(t, z))\widetilde{C}(t, P) = 0, \qquad \widetilde{M} = G^{-1}\partial_t G + G^{-1}MG.$$

From (2.24), (2.27) it follows that

$$(4.13) \qquad \widetilde{M}(t, z) = -z^{-2}I + z^{-1}\widetilde{L}(t, z) + O(z^0).$$

This relation implies that around P_i we have

$$\partial_t \widetilde{C}(t, z) = (\tilde{\mu}_i(t, z) + O(z^0))\widetilde{C}(t, P),$$

where

$$(4.14) \qquad \tilde{\mu}_i(t, z) = z^{-2} + 2k_i(z)z^{-1} = (1 - 2v_i)z^{-2} - 2h_i(0)z^{-1} + O(z^0).$$

From this, we deduce that around P_i, we have

$$\widetilde{C}(t, P) = e^{\mu_i(z)t}\widehat{C}(t, P),$$

where the vector $\widehat{C}(t, P)$ is regular around P_i. Multiplying by $G(t, z)$ and normalizing $c_1(0, P) = 1$, we get the result. □

4.2. The trigonometric and rational cases. In these cases M is constant on the curve and we can choose

$$(4.15) \qquad C(t, P) = S(t)C(0, P),$$

where $S(t)$ is defined as above and is independent of the point of the curve. Hence $C(t, P)$ is a meromorphic vector with the same poles as $C(0, P)$. Moreover, since $C(0, P)$ is regular above $z = 0$, the same is true for $C(t, P)$.

However, there appear new points at infinity that play the major role.

In the trigonometric case, we have two series of such points above $z = \pm\infty$. Let us denote these points by $Q_j(k = \chi_j, z = -\infty)$, $j = 1, \ldots, N$ and $T_j(k = \chi_j + 2, z = +\infty)$, $j = 1, \ldots, N$ (the k-coordinate of T_j is $\chi_j + 2$ because of equation (3.12)).

In the rational case, we have only one series of such points above $z = \infty$. We denote them by $Q_j(k = \chi_j, z = \infty)$, $j = 1, \ldots, N$.

In the trigonometric case, the base curve is in fact a cylinder, i.e., a sphere with two marked points, while in the rational case it is a sphere with only one marked point.

We study the solutions of the equation

$$L(t, P)C(t, P) \equiv (L(t, z) + 2kI)C(t, P) = 0$$

around these points.

PROPOSITION 4.4. *In the trigonometric case, the eigenvectors at the points Q_j and T_j are related by*

(4.16) $$C(t, T_j) = \mu_j e^{-4(\chi_j + 1)t} e^{-2X} C(t, Q_j).$$

In the rational case, at the point Q_j we have

(4.17) $$\partial_k C(t, Q_j) = -(X + 2\chi_j t - \mu_j)C(t, Q_j).$$

The parameters μ_j are constants and $X = \mathrm{Diag}(x_i(t))$. Moreover, with the normalization $c_1(0, P) = 1$, all the μ_j's are equal to $e^{2x_1(0)}$ or $x_1(0)$ respectively.

PROOF. Let us first establish (4.16). From equation (3.13) we see that

$$C(t, T_j) = \mu_j(t) e^{-2X} C(t, Q_j).$$

To compute $\mu_j(t)$, we exploit the Lax equation $\dot{C} = -MC$ at the points Q_j, T_j, using the fact that M is independent of the point on Γ. Using the relation

$$e^{-2X} M(t) e^{2X} = M(t) + 2L(t, +\infty) - 2\dot{X} + 4I,$$

we find $\dot{\mu}_j = -(4\chi_j + 4)\mu_j$.

The proof of (4.17) is slightly more complicated. First of all, around a point Q_j, the curve has the equation

$$R_0(k) - z^{-1} R_0'(k) + O(z^{-2}) = 0,$$

which implies

(4.18) $$\left.\frac{1}{z^2}\frac{dz}{dk}\right|_{Q_j} = -1 \implies \frac{1}{z} = (k - \chi_j) + O(k - \chi_j)^2,$$

hence k is an analytic parameter around Q_j.

Next we consider the equation $[L(t, \infty) + 2kI - z^{-1}F]C(t, P) = 0$. It gives

$$(4.19) \qquad L(t, Q_j)\partial_k C(t, Q_j) = (F - 2I)C(t, Q_j).$$

To solve this equation, we remark that by virtue of equation (3.17), we have

$$L(t, Q_j)(-XC(t, Q_j)) = [X, L(t, Q_j)]C(t, Q_j) = (F - 2I)C(t, Q_j),$$

therefore the general solution of equation (4.19) is of the form

$$(4.20) \qquad \partial_k C(t, Q_j) = -XC(t, Q_j) + \mu_j(t)C(t, Q_j).$$

To find the functions $\mu_j(t)$, we use the evolution equation $\dot{C} = -MC$, which implies

$$\dot{\mu}_j C(t, Q_j) = (\dot{X} - [X, M])C(t, Q_j),$$

but it is straightforward to check that $\dot{X} - [X, M] = L(t, \infty)$. Therefore $\dot{\mu}_j = -2\chi_j$. Applying equation (4.17) for $i = 1$ at $t = 0$, we get $c_1 = 1$, $\partial_k c_1 = 0$, hence all the μ_j's are equal to $x_1(0)$. □

Similarly, for the covector C^+ we have

PROPOSITION 4.5. *At the point Q_j, we have in the trigonometric case*

$$(4.21) \qquad C^+(t, T_j) = \mu_j^+ e^{4(\chi_j+1)t} C^+(t, Q_j) e^{2X}$$

and in the rational case

$$(4.22) \qquad \partial_k C^+(t, Q_j) = C^+(t, Q_j)(X + 2\chi_j t + \mu_j^+).$$

Moreover, with the normalization $c_1^+(0, P) = 1$, all the μ_j^+'s are equal to $e^{-2x_1(0)}$, or $-x_1(0)$ in the rational case.

§5. The analytic properties of Ψ and Ψ^+

In this section we encode the previous results on the eigenvectors of the Lax matrix into analyticity properties of Ψ and Ψ^+. We treat the three cases separately.

5.1. The elliptic case.

THEOREM 5.1. *The components $\Psi_\alpha(x, t, P)$ of the solution $\Psi(x, t, P)$ to the nonstationary matrix Schrödinger equation (2.10) are defined on the N-sheeted covering Γ of the initial elliptic curve. They are meromorphic on Γ outside l points P_i, $i = 1, \ldots, l$. For general initial conditions, the curve Γ is smooth, its genus equals $g = Nl - l(l+1)/2 + 1$, and the functions Ψ_α have $g - 1$ poles $\gamma_1, \ldots, \gamma_{g-1}$, which do not depend on the variables x, t. In a neighborhood of P_i, $i = 1, \ldots, l$, the function Ψ_α has the form*

$$(5.1) \qquad \Psi_\alpha(x, t, P) = \left(\chi_0^{\alpha i} + \sum_{s=1}^{\infty} \chi_s^{\alpha i}(x, t) z^s\right) e^{\lambda_i(z)x + \lambda_i^2(z)t} \Psi_1(0, 0, P),$$

where

(5.2) $$\lambda_i(z) = z^{-1} + k_i(z) = (1 - v_i)z^{-1} - h_i(0) + O(z)$$

and the $\chi_0^{\alpha i}$ are constants independent of t.

PROOF. We recall the relationship between the function Ψ and the eigenvectors of the Lax matrix

$$\Psi(x, t, P) = \sum_{j=1}^{N} s_j(t, P)\Phi(x - x_j(t), z)e^{kx+k^2 t},$$

$$s_j(t, P) = c_j(t, P)a_j(t).$$

It is obvious that the $g - 1$ poles γ_k of the c_i's are time-independent poles of Ψ. To study the behavior of Ψ above $z = 0$, we use the expansion of Φ at $z = 0$

(5.3) $$\Phi(x, z) = (-z^{-1} + \zeta(x) + O(z))e^{\zeta(z)x}$$

and the expansion of the eigenvectors as in equation (4.6). We get
(5.4)
$$\Psi_\alpha = \sum_{j=1}^{N}(-z^{-1} + \zeta(x - x_j(t)) + O(z))a_{j,\alpha}(t)c_j^{(i)}(t, z)e^{\lambda_i(z)x + \lambda_i^2(z)t}e^{-z^{-1}x_1(0)}.$$

On the $N - l$ branches $i > l$, we see that $\lambda_i(z)$ is regular at $z = 0$ due to equation (3.5), so that Ψ has no essential singularity at the P_i for $i > l$ apart from the irrelevant constant factor $\exp(-z^{-1}x_1(0))$. Moreover, there is no pole at these points. This is because $\sum_j a_j^\alpha(t)c_j^{(i)}(t, z) = O(z)$ due to equation (3.8).

It only remains to prove that the leading term of the expansion of the first factor in the right-hand side of (5.1) does not depend on t. (It does not depend on x because the singular part of Φ at $z = 0$ does not depend on x.) The substitution of the right-hand side of (5.1) with $\chi_0^{\alpha j} = \chi_0^{\alpha j}(t)$ into the equation

(5.5) $$(\partial_t - \partial_x^2 + u(x, t))\Psi(x, t, P) = 0,$$

gives

(5.6) $$u = 2(\partial_x \chi_1)\Lambda\chi_0^{-1} - (\partial_t \chi_0)\chi_0^{-1},$$

where χ_s is a matrix with entries $\chi_s^{\alpha j}$ and Λ is a diagonal matrix $\Lambda^{\alpha j} = (1 - v_j)\delta^{\alpha j}$. From (5.4) it follows that χ_1 has the form

(5.7) $$\chi_1 = \sum_{i=1}^{N} R_i(t)\zeta(x - x_i(t)).$$

Therefore, for a potential of the form $u = \sum p_i(t)\wp(x - x_i(t))$, relation (5.7) implies that

(5.8) $$p_i = -2R_i(t)\Lambda\chi_0^{-1}, \qquad (\partial_t \chi_0)\chi_0^{-1} = 0.$$

Dividing (5.4) by the normalization factor $\Psi_1(0, 0, P)$, which plays no role in the fact that $\Psi(x, t, P)$ satisfies the differential equation (2.10), we get the final result. One can express χ_0 in terms of the $c_i^{(j)}(t)$ defined in (4.3)

$$\chi_0^{\alpha j} = \sum_{i=1}^{l} c_i^{(j)}(0) a_{i,\alpha}(0). \quad \square \tag{5.9}$$

The same arguments show that:

THEOREM 5.2. *The components* $\Psi^{+,\alpha}(x, t, P)$ *of the solution* $\Psi^+(x, t, P)$ *to the nonstationary matrix Schrödinger equation* (2.11) *are defined on the same curve* Γ. *They are meromorphic on* Γ *outside the l punctures* P_i, $i = 1, \ldots, l$. *In general* $\Psi^{+,\alpha}$ *have* $g - 1$ *poles* $\gamma_1^+, \ldots, \gamma_{g-1}^+$ *which do not depend on the variables* x, t. *In a neighborhood of* P_i, $i = 1, \ldots, l$, *the function* $\Psi^{+,\alpha}$ *has the form*

$$\Psi^{+,\alpha}(x, t, P) = \left(\chi_0^{+,\alpha i} + \sum_{s=1}^{\infty} \chi_s^{+,\alpha i}(x, t) z^s\right) e^{-\lambda_i(z)x - \lambda_i^2(z)t} \Psi^{+,1}(0, 0, P), \tag{5.10}$$

where the $\chi_0^{+,\alpha j}$ *are constants.*

REMARK. Theorem 5.1 states, in particular, that the solution Ψ of equation (2.10) is (up to normalization) a Baker–Akhiezer vector-valued function ([14]). In the next section we show that this function is uniquely defined by the curve Γ, its poles γ_s, the matrix χ_0 and the value $x_1(0)$. All these values are defined by the initial Cauchy data and do not depend on t. At the same time it is absolutely necessary to emphasize that part of them depend on the choice of the normalization point t_0 that we choose as $t_0 = 0$. Let us be more accurate. Any point of the phase space $\{x_i, p_i, a_i, b_i^+ \mid (b_i^+, a_i) = 2\}$ defines the matrix L with the help of formulas (2.24). The characteristic equation (3.2) defines an algebraic curve Γ. The equation (4.1) defines a set of $g - 1$ points γ_s on Γ. Therefore, we may define a map

$$\{x_i, p_i, a_i, b_i^+ \mid (b_i^+, a_i) = 2\} \mapsto \{\Gamma, D \in J(\Gamma)\}, \tag{5.11}$$

$$D = \sum_{s=1}^{g-1} A(\gamma_s) + x_1 U^{(1)}, \tag{5.12}$$

where $A: \Gamma \to J(\Gamma)$ is an Abel map and $U^{(1)}$ is a vector depending on Γ only (see (6.9)). The coefficients of the equation (3.2) are integrals of the Hamiltonian system (1.1). As we shall see in the next section, the second part of the data (5.11) define angle-type variables, i.e., the vector $D(t)$ evolves linearly $D(t) = D(t_0) + (t - t_0) U^{(2)}$ if a point in phase space evolves according to equations (1.4)–(1.6). These equations have the obvious symmetries:

$$a_i, b_i^+ \to \lambda_i a_i, \lambda_i^{-1} b_i^+, \quad a_i, b_i^+ \to W^{-1} a_i, b_i^+ W, \tag{5.13}$$

where q_i are constants and W is an arbitrary constant matrix. In the next section we prove that to the data Γ, D one can associate a unique point in the phase space reduced under the symmetry (5.13).

5.2. The trigonometric and rational cases.

THEOREM 5.3. *The components $\Psi_\alpha(x, t, P)$ of the solution $\Psi(x, t, P)$ to the nonstationary matrix Schrödinger equation (2.10) are defined on an N-sheeted covering Γ of the completed complex plane. They are meromorphic on Γ outside l points P_i, $i = 1, \ldots, l$. For general initial conditions the curve Γ is smooth, its genus equals $g = N(l-1) - l(l+1)/2 + 1$ and the Ψ_α have $g + N - 1$ poles $\gamma_1, \ldots, \gamma_{g+N-1}$ which do not depend on the variables x, t. In a neighborhood of P_i, $i = 1, \ldots, l$, the function Ψ_α has the form*

$$(5.14) \qquad \Psi_\alpha(x, t, P) = \left(\chi_0^{\alpha i} z^{-1} + \sum_{s=1}^\infty \chi_s^{\alpha i}(x, t) z^{s-1} \right) e^{k_i(z)x + k_i^2(z)t},$$

where

$$(5.15) \qquad k_i(z) = -v_i z^{-1} - h_i(0) + O(z)$$

and the $\chi_0^{\alpha i}$ are constants.

In the trigonometric case, at the points T_j and Q_j above $z = \pm\infty$, we have

$$(5.16) \qquad \Psi(x, t, T_j) = \mu_j \Psi(x, t, Q_j).$$

In the rational case, at the points Q_j above $z = \infty$, we have

$$(5.17) \qquad \partial_k \Psi(x, t, Q_j) = \mu_j \Psi(x, t, Q_j),$$

where the μ_j's are defined in equations (4.16), (4.17).

PROOF. In the trigonometric case, the result follows immediately from (4.16). In the rational case, we use (2.12) with $s_i(t, P) = a_i(t) c_i(t, P)$, obtaining

$$\partial_k \Psi = \sum_i a_i \left\{ -\frac{1}{z} c_i x - \frac{1}{z} \partial_k c_i + \left(1 + \frac{1}{z^2} \partial_k z - \frac{2kt}{z} \right) c_i \right. $$
$$\left. + \frac{\partial_k c_i + (x_i + 2kt) c_i}{x - x_i} \right\} e^{kx + k^2 t}.$$

When $z = \infty$, we have $1 + z^{-2} \partial_k z = 0$ and

$$\partial_k \Psi|_{Q_j} = \left(\sum_i a_i \frac{\partial_k c_i + (x_i + 2kt) c_i}{x - x_i} \right) e^{kx + k^2 t} \bigg|_{Q_j}.$$

The result follows from equation (4.17). □

Similarly, for Ψ^+ we have the following result.

THEOREM 5.4. *The components $\Psi^{+,\alpha}(x, t, P)$ of the solution $\Psi^+(x, t, P)$ to the nonstationary matrix Schrödinger equation* (2.11) *are defined on an N-sheeted covering Γ of the completed complex plane. They are meromorphic on Γ outside l points P_i, $i = 1, \ldots, l$ and have $g + N - 1$ poles $\gamma_1^+, \ldots, \gamma_{g+N-1}^+$, which do not depend on the variables x, t. In a neighborhood of P_i, $i = 1, \ldots, l$, the function $\Psi^{+,\alpha}$ has the form*

$$(5.18) \quad \Psi^{+,\alpha}(x, t, P) = \left(\chi_0^{+,\alpha i} z^{-1} + \sum_{s=1}^{\infty} \chi_s^{+,\alpha i}(x, t) z^{s-1} \right) e^{-k_i(z)x - k_i^2(z)t},$$

where the $\chi_0^{+,\alpha i}$ are constants. In addition, at the points T_j, Q_j above $z = \pm\infty$, in the trigonometric case

$$(5.19) \quad \Psi^+(T_j) = \mu_j^+ \Psi^+(Q_j)$$

and in the rational case we have at the points Q_j above $z = \infty$

$$(5.20) \quad \partial_k \Psi^+(x, t, Q_j) = \mu_j^+ \Psi^+(x, t, Q_j),$$

where the μ_j^+'s are defined in equations (4.21), (4.22).

REMARK. Obviously, one can multiply the functions Ψ_α by a meromorphic function $f(P)$ on Γ without affecting the Schrödinger equation (2.10). We have already used this property in equation (5.1) to factor out $\Psi_1(0, 0, P)$. The $m + l$ poles of the resulting Baker–Akhiezer function are now in arbitrary position. In the trigonometric and rational cases, we can use the same feature in order to match the normalizations used in the elliptic case. Let us define $f(P)$ with $m + l$ zeros at the points $\gamma_1, \ldots, \gamma_m$ and $P_1, \ldots, P_l$, with N poles at the points $Q_1, \ldots, Q_N$ and $l - 1$ poles $\gamma_1', \ldots, \gamma_{l-1}'$ at some arbitrary prescribed positions. This defines a divisor of degree g, and such a function f is uniquely determined. It has g extra poles $\gamma_l', \ldots, \gamma_{g+l-1}'$. The function

$$(5.21) \quad \psi' = f\psi$$

now has $m + l$ poles at the points Q_j and γ_k', and satisfies the differential equation (2.10). In the following sections we shall use ψ' (denoted by ψ) and γ_k' (denoted by γ_k).

Part II. The Inverse Problem

§6. The elliptic case

6.1. The Baker–Akhiezer functions. At the beginning of this section we present some information on finite-gap theory [14] needed in the sequel.

THEOREM 6.1. *Let Γ be a smooth algebraic curve of genus g with fixed local coordinates $w_i(P)$ in neighborhoods of l punctures P_i, $w_i(P_i) = 0$, $i = 1, \ldots, l$. Then for each set of $g + l - 1$ points $\gamma_1, \ldots, \gamma_{g+l-1}$ in general position there exists a unique function $\psi_\alpha(x, t, P)$ such that:*

$1°$. *The function ψ_α of the variable $P \in \Gamma$ is meromorphic outside the punctures and has at most simple poles at points γ_s (if all of them are distinct).*

$2°$. *In the neighborhood of the puncture P_j the function ψ_α has the form*

$$(6.1) \quad \psi_\alpha(x, t, P) = e^{w_j^{-1}x + w_j^{-2}t}\left(\delta_{\alpha j} + \sum_{s=1}^{\infty} \xi_s^{\alpha j}(x, t) w_j^s\right), \quad w_j = w_j(P).$$

PROOF. The existence follows from the explicit formula given below in terms of Riemann theta functions. Uniqueness follows from the Riemann–Roch theorem applied to the ratio of two such functions. □

We now give a fundamental formula expressing the Baker–Akhiezer functions in terms of Riemann theta functions. According to the Riemann–Roch theorem, for any divisor $D = \gamma_1 + \cdots + \gamma_{g+l-1}$ in general position there exists a unique meromorphic function $h_\alpha(P)$ such that the divisor of its poles coincides with D and such that

$$(6.2) \quad h_\alpha(P_j) = \delta_{\alpha j}.$$

Using the results recalled in Appendix B, this function can be written as follows:

$$(6.3) \quad h_\alpha(P) = \frac{f_\alpha(P)}{f_\alpha(P_\alpha)}, \quad f_\alpha(P) = \theta(A(P) + Z_\alpha) \frac{\prod_{j \neq \alpha} \theta(A(P) + R_j)}{\prod_{i=1}^{l} \theta(A(P) + S_i)},$$

where

$$(6.4) \quad R_j = -\mathcal{K} - A(P_j) - \sum_{s=1}^{g-1} A(\gamma_s), \quad j = 1, \ldots, l,$$

$$(6.5) \quad S_i = -\mathcal{K} - A(\gamma_{g-1+i}) - \sum_{s=1}^{g-1} A(\gamma_s),$$

$$(6.6) \quad Z_\alpha = Z_0 - A(P_\alpha), \quad Z_0 = -\mathcal{K} - \sum_{i=1}^{g+l-1} A(\gamma_i) + \sum_{j=1}^{l} A(P_j).$$

Let $d\Omega^{(i)}$ be the unique meromorphic differential holomorphic on Γ outside the punctures P_j, $j = 1, \ldots, l$, that has the form

$$(6.7) \quad d\Omega^{(i)} = d(w_j^{-i} + O(w_j))$$

near the punctures, and is normalized by the conditions (see Appendix B for the definition of the cycles a_k^0, b_k^0)

$$(6.8) \quad \oint_{a_k^0} d\Omega^{(i)} = 0.$$

It defines a vector $U^{(i)}$ with coordinates

$$(6.9) \quad U_k^{(i)} = \frac{1}{2\pi i} \oint_{b_k^0} d\Omega^{(i)}.$$

THEOREM 6.2. *The components of the Baker–Akhiezer function $\psi(x, t, P)$ are equal to*

(6.10)
$$\psi_\alpha(x, t, P) = h_\alpha(P) \frac{\theta(A(P) + U^{(1)}x + U^{(2)}t + Z_\alpha)\theta(Z_0)}{\theta(A(P) + Z_\alpha)\theta(U^{(1)}x + U^{(2)}t + Z_0)} e^{(x\Omega^1(P) + t\Omega^{(2)}(P))},$$

(6.11)
$$\Omega^{(i)}(P) = \int_{q_0}^{P} d\Omega^{(i)}.$$

PROOF. It suffices to check that the function defined by formula (6.10) is well defined (i.e., it does not depend on the path of integration between q_0 and P) and has the desired analytical properties. (The ratio of the product of theta-functions at $P = P_\alpha$ equals 1 by (6.6).) □

From the exact theta-function formula (6.10), it follows that ψ can be represented in the form

(6.12)
$$\psi(x, t, P) = r(x, t)\widehat{\psi}(x, t, P),$$

where $\widehat{\psi}(x, t, P)$ is an entire function of the variables x, t and $r(x, t)$ is a meromorphic function.

We now give the definition of the dual Baker–Akhiezer function. For any set of $g + l - 1$ points in general position there exists a unique meromorphic differential $d\Omega$ that has poles of the form

(6.13)
$$d\Omega = \frac{dw_j}{w_j^2} + \frac{\lambda_j\, dw_j}{w_j} + O(1)\, dw_j$$

at the punctures P_j, and equals zero at the points γ_s

(6.14)
$$d\Omega(\gamma_s) = 0.$$

In addition to γ_s, this differential has $g + l - 1$ other zeros, which we denote γ_s^+.

The *dual Baker–Akhiezer function* is the unique function $\psi^+(x, t, P)$ with coordinates $\psi^{+,\alpha}(x, t, P)$ such that:

1°. The function $\psi^{+,\alpha}$ as a function of the variable $P \in \Gamma$ is meromorphic outside the punctures and has at most simple poles at the points γ_s^+ (if all of them are distinct);

2°. In the neighborhood of the puncture P_j the function $\psi^{+,\alpha}$ has the form

(6.15)
$$\psi^{+,\alpha}(x, t, P) = e^{-w_j^{-1}x - w_j^{-2}t}\left(\delta_{\alpha j} + \sum_{s=1}^{\infty} \xi_s^{+,\alpha j}(x, t) w_j^s\right).$$

Let $h_\alpha^+(P)$ be the function that has poles at the points of the dual divisor $\gamma_1^+, \ldots, \gamma_{g+l-1}^+$ and is normalized by (6.2) (i.e., $h_\alpha^+(P_j) = \delta_{\alpha j}$). It can be written in the form (6.3) with γ_s replaced by γ_s^+. It follows from the definition of the dual divisors that

(6.16)
$$\sum_{s=1}^{g+l-1} A(\gamma_s) + \sum_{s=1}^{g+l-1} A(\gamma_s^+) = K_0 + 2\sum_{j=1}^{l} A(P_j),$$

where K_0 is the canonical class (i.e., the equivalence class of the divisor of zeros of a holomorphic differential).

THEOREM 6.3. *The components of the dual Baker–Akhiezer function $\psi^+(x, t, P)$ are equal to*

(6.17)
$$\psi_\alpha^+(x, t, P) = h_\alpha^+(P) \frac{\theta(A(P) - U^{(1)}x - U^{(2)}t + Z_\alpha^+)\theta(Z_0^+)}{\theta(A(P) + Z_\alpha^+)\theta(U^{(1)}x + U^{(2)}t - Z_0^+)} e^{-(x\Omega^1(P) + t\Omega^{(2)}(P))},$$

where

(6.18) $$Z_0^+ = Z_0 - 2\mathcal{K} - K_0, \qquad Z_\alpha^+ = Z_0^+ - A(P_\alpha).$$

The above results are valid for any curve Γ. We now consider a more specific setting, which will correspond to the elliptic model.

THEOREM 6.4. *Let Γ be a smooth algebraic curve defined by an equation of the form*

(6.19) $$R(k, z) = k^N + \sum_{i=0}^{N} r_i(z) k^i = 0,$$

where $r_i(z)$ are elliptic functions, holomorphic outside the point $z = 0$, such that the covering $P \to z$ has no branching points over $z = 0$ (i.e., the function $k(P)$ has N simple poles $P_1, \ldots, P_N$ on Γ which are preimages of $z = 0$). Let us assume also that the residues v_j of $k(P)$ at the poles defined by the expansion of $R(z, k)$ near $z = 0$

(6.20) $$R(k, z) = \prod_{i=1}^{N}(k + v_i z^{-1} + O(z^0))$$

satisfy (3.5), $v_j = 1$, $j > l$. *Then there exists a function $\varphi_i(P)$ on Γ such that the Baker–Akhiezer function ψ corresponding to the curve Γ and the local parameters $w_j = (k_j(z) + \zeta(z))^{-1}$ at the puncture P_j obeys*

(6.21) $$\psi(x + 2\omega_i, t, P) = \varphi_i(P)\psi(x, t, P).$$

PROOF. Consider the functions

(6.22) $$\varphi_i(P) = \exp(2(k(P) + \zeta(z))\omega_i - 2\eta_i z), \qquad i = 1, 2,$$

where $2\omega_i$ are periods of the base elliptic curve and $\eta_i = \zeta(\omega_i)$. From the monodromy properties of the ζ-function and relation $2(\eta_1\omega_2 - \eta_2\omega_1) = \pi i$, it follows that $\varphi(P)$ is a well-defined function on the curve Γ. It is holomorphic outside the points $P_1, \ldots, P_l$. In the neighborhood of P_j it has the form

(6.23) $$\varphi_i(P) = (1 + O(z)) \exp(2(k_j(z) + \zeta(z))\omega_i).$$

Let $\psi(x, t, P)$ be the Baker–Akhiezer vector function corresponding to Γ, P_j, $w_j(P)$ and to any divisor D of degree $g + l - 1$. Then equation (6.21) follows from the fact that the right- and left-hand sides have the same analytical properties. □

COROLLARY 6.1. *The vector Baker–Akhiezer function $\psi(x, t, P)$ with components ψ_α, $\alpha = 1, \ldots, l$, can be written in the form*

$$(6.24) \quad \psi(x, t, P) = \sum_{i=1}^{m} s_i(t, P) \Phi(x - x_i(t), z) e^{kx+k^2 t}, \qquad P = (k, z).$$

PROOF. Let $x_i(t), i = 1, \ldots, m$, be the set of poles of the function $\psi(x, t, P)$ (as a function of the variable x) in the fundamental domain of the lattice with periods 2ω, $2\omega'$. It follows from (6.12) that these poles do not depend on P. Let us assume that they are simple poles. Then their exist vectors $s_i(t, P)$ such that the function

$$\mathcal{F}(x, t, P) = \psi(x, t, P) - \sum_{i=1}^{m} s_i(t, P) \Phi(x - x_i(t), z) e^{kx+k^2 t}$$

is *holomorphic* in x in this fundamental domain. This function has the same monodromy properties (6.21) as the function ψ. Any nontrivial function satisfying (6.21) has at least one pole in the fundamental domain. Hence, $\mathcal{F} = 0$. □

Let us remark that for the specific curve Γ above, the Baker–Akhiezer function is exactly of the form postulated in equation (2.12). The same arguments show that the dual Baker–Akhiezer function has the form (2.13).

6.2. The potential. The following theorem is a particular case of a general statement from [14].

THEOREM 6.5. *Let $\psi(x, t, P)$ be a vector-valued function with the coordinates $\psi_\alpha(x, t, P)$ defined above. Then it satisfies the equation*

$$(6.25) \quad (\partial_t - \partial_x^2 + u(x, t)) \psi(x, t, P) = 0,$$

where the entries of the matrix-valued function u are equal to

$$(6.26) \quad u^{\alpha i}(x, t) = 2 \partial_x \xi_1^{\alpha i}(x, t).$$

The potentials $u(x, t)$ corresponding to some Baker–Akhiezer vector-valued function are called *algebro-geometrical* or *finite-gap* potentials.

PROOF. This follows directly from the fact that $(\partial_t - \partial_x^2) \psi_\alpha$ has the same analytic properties as ψ but for the normalizations at the P_i's, so can be expanded in the ψ_β with coefficients $-u^{\alpha\beta}$. □

THEOREM 6.6. *The dual Baker–Akhiezer function satisfies the equation*

$$(6.27) \quad \psi^+(x, t, P)(\partial_t - \partial_x^2 + u(x, t)) = 0,$$

where $u(x, t)$ is the same as in (6.25).

PROOF. To show that the potentials are the same as in (6.25), we consider the form $\psi_\alpha \psi^{+\beta} d\Omega$, where $d\Omega$ is defined by equation (6.13) and conditions (6.14). This is a meromorphic 1-form on Γ with poles only at the P_j's. Around P_j we have

$$\psi_\alpha \psi^{+\beta} d\Omega = \left[\frac{\delta_{\alpha j} \delta_{\beta j}}{w_j^2} + (\delta_{\alpha j} \delta_{\beta j} \lambda_j + \delta_{\alpha j} \xi_1^{+\beta j} + \delta_{\beta j} \xi_1^{\alpha j}) \frac{1}{w_j} + O(1) \right] dw_j.$$

Since the sum of residues must vanish, we get

$$\xi_1^{+\beta\alpha} + \xi_1^{\alpha\beta} = -\lambda_\alpha \delta_{\alpha\beta}.$$

This implies the result, since $u^{\alpha\beta} = 2\partial_x \xi_1^{\alpha\beta}$, $u^{+\alpha\beta} = -2\partial_x \xi_1^{+\beta\alpha}$ and λ_α is independent of x. □

In general position, the algebro-geometrical potentials are quasi-periodic functions of all arguments. In [4] the necessary conditions on the algebraic geometrical data $\{\Gamma, P_1, \ldots, P_l, w_1(P), \ldots, w_l(P)\}$ were found in order that the corresponding potentials be elliptic functions of the variable x. Here we have

PROPOSITION 6.1. *Let Γ be a specific curve as in Theorem 6.4. Then the algebro-geometrical potential $u(x, t)$ corresponding to the curve Γ, to the points $P_1, \ldots, P_l$, and to the local coordinates $w_j(z) = (k_j(z) + \zeta(z))^{-1}$ is an elliptic function. In general position this potential has the form*

$$(6.28) \qquad u(x, t) = \sum_{i=1}^{N} a_i(t) b_i^+(t) \wp(x - x_i(t)).$$

PROOF. The potential is elliptic because of equation (6.21). Since the Baker–Akhiezer function has the form (6.24), u has only double poles. Hence the potential is of the form $u = \sum \rho_i(t) \wp(x - x_i(t))$. We show that the matrices $\rho_i(t)$ are of rank one. It follows from (6.10) that the poles $x = x_i(t)$ of the Baker–Akhiezer function correspond to the solutions of the equation

$$(6.29) \qquad \theta(U^{(1)} x + U^{(2)} t + Z) = 0.$$

From (6.6) it follows that for such a pair $(x_i(t), t)$, the first factor in the numerator of formula (6.10) vanishes at P_α. At a point P_β, $\beta \neq \alpha$, it is the function $h_\alpha(P)$ which vanishes. Therefore, the residue $\psi_{\alpha,i}^0(t, P)$ of $\psi_\alpha(x, t, P)$ at $x = x_i(t)$, as a function of the variable P, has the following analytical properties:

1°. It is a meromorphic function outside the punctures P_j and has the same set of poles as ψ.

2°. In a neighborhood of the puncture P_j, it has the form

$$(6.30) \qquad \psi_{\alpha,i}^0(t, P) = \exp(w_j^{-1} x_i(t) + w_j^{-2} t) O(w_j).$$

Hence, it has the same analytical properties as the Baker–Akhiezer function but with one difference. Namely, the regular factor of its expansion at *all* the punctures vanishes. For generic x, t there are no such functions. For special pairs $(x = x_i(t), t)$ such a function $\psi_{i0}(t, P)$ exists and is *unique* up to a constant (in P) factor (it is unique in the generic case when $x_i(t)$ is a simple root of (6.29)). Therefore, the components of the Baker–Akhiezer function can be represented in the form

$$(6.31) \qquad \psi_\alpha(x, t, P) = \frac{\phi_\alpha(t) \psi_{i0}(t, P)}{x - x_i(t)} + O((x - x_i(t))^0).$$

The last equality implies that the residue $\tilde{\rho}_i(t)$ of the matrix $\xi_1(x, t)$ with entries $\xi_1^{\alpha j}(x, t)$,

$$\xi_1(x, t) = \frac{\tilde{\rho}_i(t)}{x - x_i(t)} + O((x - x_i(t))^0) \tag{6.32}$$

is of rank 1. It follows from (6.26) that $\rho_i(t) = -2\tilde{\rho}_i(t)$. Hence this matrix is of rank one too, and therefore has the form (6.28). □

Now let us examine the effect of replacing the divisor by an equivalent one. Let $D = \gamma_1 + \cdots + \gamma_{g+l-1}$ and $D^{(1)} = \gamma_1^{(1)} + \cdots + \gamma_{g+l-1}^{(1)}$ be two equivalent divisors (i.e., there exists a meromorphic function $h(P)$ on Γ such that D is a divisor of its poles and $D^{(1)}$ is a divisor of its zeros). Then for the corresponding Baker–Akhiezer vector-valued functions $\psi(x, t, P)$ and $\psi^{(1)}(x, t, P)$ we have the equality

$$H\psi(x, t, P) = \psi^{(1)}(x, t, P) h(P). \tag{6.33}$$

Here H is a diagonal matrix

$$H^{\alpha j} = h(P_j) \delta^{\alpha j}. \tag{6.34}$$

The proof of (6.33) follows from the uniqueness of the Baker–Akhiezer functions, because the left-hand and right-hand sides of (6.33) have the same analytical properties.

COROLLARY 6.2. *The algebraic-geometrical potentials $u(x, t)$ and $u^{(1)}(x, t)$ corresponding to Γ, P_j, $w_j(P)$, and to equivalent effective divisors D and $D^{(1)}$, respectively, are gauge equivalent*

$$u^{(1)}(x, t) = H u(x, t) H^{-1}, \qquad H^{\alpha j} = h_j \delta^{\alpha j}. \tag{6.35}$$

COROLLARY 6.3. *A curve Γ in general position satisfies the conditions of Theorem 6.2 if and only if it is the spectral curve (3.2) of a Lax matrix L defined by formula (2.24), where x_i, p_i are arbitrary constants and f_{ij} are defined by vectors a_i, b_i^+ satisfying (1.7), with the help of (1.8). The corresponding vectors are defined uniquely up to the transformation (5.13).*

Notice that the Baker–Akhiezer function $\Psi_\alpha(x, t, P)/\Psi_1(0, 0, P)$ in (5.1) is related to the normalized Baker–Akhiezer function $\psi_\alpha(x, t, P)$ appearing in equation (6.1) by

$$\frac{\Psi_\alpha(x, t, P)}{\Psi_1(0, 0, P)} = \sum_\beta \tilde{\chi}_0^{\alpha\beta} \psi_\beta(x, t, P).$$

This induces a similarity transformation on the potential, and we have the following

COROLLARY 6.4. *If $a_i(t)$, $b_i(t)$, $x_i(t)$ give a solution of the equation of motion of the Hamiltonian system (1.1), then*

$$\sum_{i=1}^N a_i(t) b_i^+(t) \wp(x - x_i(t)) = \chi_0 u(x, t) \chi_0^{-1}, \tag{6.36}$$

where $u(x, t)$ is the algebro-geometrical potential corresponding to the data described in Theorem 6.2 and to the normalized Baker–Akhiezer functions.

COROLLARY 6.5. *The correspondence*

(6.37) $$a_i(t), b_i^+(t), x_i(t) \mapsto \{\Gamma, [D]\},$$

where $[D]$ is the equivalence class of the divisor D (i.e., the corresponding point of the Jacobian), is an isomorphism up to the equivalence (5.13).

The curve Γ is time-independent. At the same time $[D]$ depends on the choice of the normalization point $t_0 = 0$. From the exact formulas for the solutions (see below) it follows that the dependence of $D(t_0)$ is linear onto the Jacobian.

6.3. Reconstruction formulas.

THEOREM 6.7. *Let Γ be a curve defined by an equation of the form* (3.2) *and $D = \gamma_1, \ldots, \gamma_{g+l-1}$ be a set of points in general position. Then the formulas*

(6.38) $$\theta(U^{(1)}x_i(t) + U^{(2)}t + Z_0) = 0,$$

(6.39) $$a_{i,\alpha}(t) = Q_i^{-1}(t) h_\alpha(q_0) \frac{\theta(U^{(1)}x_i(t) + U^{(2)}t + Z_\alpha)}{\theta(Z_\alpha)},$$

(6.40) $$b_i^\alpha(t) = Q_i^{-1}(t) h_\alpha^+(q_0) \frac{\theta(U^{(1)}x_i(t) + U^{(2)}t - Z_\alpha^+)}{\theta(Z_\alpha^+)},$$

where

(6.41) $$Q_i^2(t) = \frac{1}{2} \sum_{\alpha=1}^{l} h_\alpha^+(q_0) h_\alpha(q_0)$$
$$\times \frac{\theta(U^{(1)}x_i(t) + U^{(2)}t - Z_\alpha)\theta(U^{(1)}x_i(t) + U^{(2)}t - Z_\alpha^+)}{\theta(Z_\alpha)\theta(Z_\alpha^+)}$$

define the solutions of the system (1.4), (2.14), (2.15). *Any solution of the system* (1.1) *can be obtained from the solutions* (6.38)–(6.40) *using the symmetries* (5.13).

REMARK. Relation (6.38) may be reformulated as follows: the vector $U^{(1)}$ defines an embedding of the elliptic curve into the Jacobian $J(\Gamma)$. Therefore the restriction of the theta function to the corresponding elliptic curve is a product of elliptic σ-Weierstrass functions, i.e., of the vectors

(6.42) $$\theta(U^{(1)}x + U^{(2)}t + Z_0) = \text{const} \prod_{i=1}^{N} \sigma(x - x_i(t)).$$

It should be mentioned that exactly the same formula was obtained in [4] for the solutions of the elliptic Calogero–Moser system. The difference for various spins l is encoded in the set of corresponding curves.

From (2.12), (2.22) it follows that the vector $a_i(t)$ is proportional to the singular part of the expansion of $\psi(x, t, P)$ near $x = x_i(t)$ and up to a scalar factor this singular part does not depend on P. Therefore, using the formula (6.10) for $P = q_0$, we obtain (6.39). Theorem 6.7 is proved. $\square$

§7. The rational and trigonometric cases

7.1. Baker–Akhiezer functions. Let Γ be the smooth algebraic curve of genus g defined by equation (3.11) or (3.15). The function $k(P)$ has l simple poles above $z=0$ denoted by $P_1, \ldots, P_l$. In the vicinity of P_i, one has $k = k_i(z) \equiv -v_i/z - h_i(z)$ with h_i regular. We take $1/k_i(z)$ as the local parameter around P_i. We shall consider the space of Baker–Akhiezer functions with $g+l-1$ poles γ_k and N poles at the points Q_j above $z = -\infty$, which are of the form $\exp(kx + k^2t)$ times a meromorphic function. The space of such functions is of dimension $N+l$, and imposing the N conditions (5.16) or (5.17) one ends up with a space of dimension l. One can define a basis ψ_α of this space by choosing the normalization in a neighborhood of the points P_j above $z=0$ with local parameters (z, k_j) as

$$(7.1) \qquad \psi_\alpha(P, x, t) = \left(\delta_{\alpha j} + \sum_{s=1}^\infty \frac{\xi_s^{\alpha j}(x, t)}{k_j^s} \right) e^{k_j x + k_j^2 t}.$$

Now we have at our disposal $g+l-1$ parameters γ_k and the N coefficients μ_j, which adds up to $m + l$ parameters, i.e., half of the dimension of the phase space, just as in the elliptic case.

One can construct the Baker functions ψ_α directly from the above analyticity properties. One first chooses a convenient basis to expand them.

There exists a unique function $g_j(P)$ with $g+l-1$ poles γ_k, one pole at Q_j with residue 1 (i.e., of the form $1/(k - \chi_j)$) and l zeros at the P_α's. Then we can write

$$(7.2) \qquad \psi_\alpha(P, x, t) = \left(h_\alpha(P) + \sum_{j=1}^N g_j(P) r_{j\alpha}(x, t) \right) e^{kx + k^2 t},$$

where $h_\alpha(P)$ is the function defined in equation (6.3), so that conditions (7.1) are fulfilled.

It remains only to express the N conditions (5.16) or (5.17). This yields the

THEOREM 7.1. *The components of the Baker–Akhiezer function $\psi(x, t, P)$ are given by*

$$(7.3) \qquad \psi_\alpha(x, t, P) = \frac{\mathrm{Det}\begin{pmatrix} h_\alpha(P) & g_j(P) \\ h_\alpha(T_i) & \Theta_{ij}(x, t) \end{pmatrix}}{\mathrm{Det}(\Theta_{ij}(x, t))} e^{kx + k^2 t},$$

where in the trigonometric case Θ is the matrix with elements

$$(7.4) \qquad \Theta_{ii} = -\sigma_i e^{-2x - 4(\chi_i + 1)t} + g_i(T_i), \qquad \Theta_{ij} = g_j(T_i), \quad i \neq j.$$

In the rational case one must replace $h_\alpha(T_i)$ by $h_\alpha(Q_i)$ in (7.3) and define

$$(7.5) \qquad \Theta_{ii} = x + 2\chi_i t - \sigma_i + g_i^{(1)}, \qquad \Theta_{ij} = g_j(Q_i), \quad i \neq j,$$

where $g_j(P) = 1/(k - \chi_j) + g_j^{(1)} + O(k - \chi_j)$.

PROOF. We first express the conditions on ψ arising from conditions (5.16) or (5.17) on ψ/f, where f is the meromorphic function introduced in equation (5.21). As a matter of fact, in the trigonometric case near the point Q_j we have

$$\psi_\alpha(x, t, P) = \frac{R_{j\alpha}^{(-1)}(x, t)}{k - \chi_j} + O((k - \chi_j)^0),$$

while around T_j we have

$$\psi_\alpha(x, t, P) = R_{j\alpha}^{(0)}(x, t) + O(k - \chi_j).$$

The relations (5.16) take the form

(7.6) $\qquad R_{j\alpha}^{(0)}(x, t) = \sigma_j R_{j\alpha}^{(-1)}(x, t) \quad \text{with } \sigma_j = \mu_j f(T_j)/f_j^{(0)},$

where around Q_j the function f appearing in equation (5.21) has the following expansion

$$f(P) = \frac{f_j^{(0)}}{k - \chi_j} + f_j^{(1)} + O(k - \chi_j)$$

and σ_j is independent of x and t. In the rational case, near Q_j we have

$$\psi_\alpha(P, x, t) = \frac{R_{j\alpha}^{(-1)}(x, t)}{k - \chi_j} + R_{j\alpha}^{(0)}(x, t) + O(k - \chi_j)$$

and conditions (5.17) on ψ are equivalent to

(7.7) $\qquad R_j^{(0)}(x, t) = \sigma_j R_j^{(-1)}(x, t) \quad \text{with } \sigma_j = \mu_j + f_j^{(1)}/f_j^{(0)}.$

Using the expression (7.2) for ψ, these conditions take the form (in the rational case T_j is replaced by Q_j below)

(7.8) $\qquad \sum_k \Theta_{jk}(x, t) r_{k\alpha} = -h_\alpha(T_j).$

Solving this linear system by Cramer's rule yields the result. □

PROPOSITION 7.1. *The Baker–Akhiezer function given in* (7.3) *can be written in the form*

(7.9) $\qquad \psi = \sum_{i}^{N} s_i(t, k, z) \Phi(x - x_i(t), z) e^{kx + k^2 t}.$

PROOF. Let us give the proof in the trigonometric case. In the rational case, the proof is similar and even simpler. From (7.3) we see that one can write

(7.10)
$$\psi_\alpha(x, t, P) = \left(h_\alpha(P) - \sum_{i=1}^{N} \frac{2e^{-2x_i} s_{i,\alpha}(t, P)}{e^{-2x} - e^{-2x_i}} \right) e^{kx + k^2 t}$$

$$= \left(h_\alpha(P) + \sum_{i=1}^{N} s_{i,\alpha}(t, P) + \sum_{i=1}^{N} s_{i,\alpha}(t, P) \coth(x - x_i) \right) e^{kx + k^2 t}.$$

We must show that

$$h_\alpha(P) = -(1 + \coth z) \sum_{i=1}^N s_{i,\alpha}(t, P).$$

But the function $h_\alpha(P)/(1 + \coth z)$ vanishes at the points P_i above $z = 0$ and has poles at the points Q_j above $z = -\infty$. Hence, we can write

$$\frac{h_\alpha(P)}{1 + \coth z} = \sum_{j=1}^N \frac{h_\alpha(Q_j)}{\alpha_j} g_j(P),$$

where the constants α_j are defined by $1 + \coth z = \alpha_j(k - \chi_j) + O(k - \chi_j)^2$ around Q_j. Using this formula at $P = T_i$, we get in particular

(7.11) $$h_\alpha(T_i) = 2 \sum_{j=1}^N \frac{h_\alpha(Q_j)}{\alpha_j} g_j(T_i).$$

On the other hand, using (7.10), we find

$$2 \sum_{i=1}^N s_{i,\alpha}(t, P) = \psi_\alpha(x, t, P) e^{-kx - k^2 t}\Big|_{x=+\infty} - \psi_\alpha(x, t, P) e^{-kx - k^2 t}\Big|_{x=-\infty}$$

$$= \frac{\mathrm{Det}\begin{pmatrix} 0 & g_j(P) \\ h_\alpha(T_i) & g_j(T_i) \end{pmatrix}}{\mathrm{Det}(g_j(T_i))}.$$

Expanding the determinant in the numerator along the first row, and using equation (7.11) to evaluate $h_\alpha(T_j)$, we get

$$\sum_{i=1}^N s_{i,\alpha}(P, t) = -\sum_{j=1}^N \frac{h_\alpha(Q_j)}{\alpha_j} g_j(P),$$

which is what we had to prove. □

One can also give an explicit formula for $s_{i,\alpha}(t, P)$. Since $\mathrm{Det}\,\Theta(x_i, t) = 0$, we can write a linear dependence relation

$$\Theta_{k1} = \sum_{j=2}^N \lambda_j^{(i)}(t) \Theta_{kj} \quad \text{for all } k,$$

and we see that the residue $s_{i,\alpha}(t, P)$ in equation (7.10) is given by ($\lambda_1^{(i)} = -1$)

(7.12)
$$s_{i,\alpha}(t, P) = \left\{ \frac{\sum_{k=1}^N \lambda_k^{(i)}(t) g_k(P)}{\prod_{j=1}^N (2\sigma_j e^{-4(\chi_j + 1)t - x_i - x_j}) \prod_{j \neq i} \sinh(x_i - x_j)} \right\} \mathrm{Det}\,\Theta_\alpha^{(i)}(t).$$

Here $\Theta_\alpha^{(i)}$ is obtained from $\Theta(x, t)$ by taking $x = x_i$ and replacing the first column by $h_\alpha(T_j)$. This equation must be compared with (2.22). In the rational case, we find a similar and simpler formula, including the same factor $\Theta_\alpha^{(i)}(t)$.

As in the elliptic case, we need the dual Baker–Akhiezer function ψ^+, and to this end we introduce the differential $d\Omega$ with poles of order 2 at the punctures P_j's such that $d\Omega = dw_j/w_j^2 + O(1/w_j)\,dw_j$ and vanishing on the $g + l - 1$ points γ_k. Let γ_k^+ be the $g + l - 1$ other zeros of $d\Omega$.

Let $h^{+,\alpha}(P)$ be the unique function with poles at the γ_k^+'s and such that $h^{+,\alpha}(P_j) = \delta_{\alpha j}$. In the trigonometric case, we introduce the function $g_j^+(P)$ with $g + l - 1$ poles γ_k^+, one pole at T_j with residue 1 (i.e., of the form $1/(k - \chi_j - 2)$), and l zeros at the P_j's. Then we define the dual Baker–Akhiezer function

$$\psi^{+,\alpha}(P, x, t)$$
(7.13)
$$= \left(h^{+,\alpha}(P) + \sum_{j=1}^{N} g_j^+(P) r_j^{+,\alpha}(x, t) \right) e^{-kx - k^2 t},$$

so that relations of the type (7.7) are satisfied with some coefficients σ_j^+. We choose σ_j^+ as

$$\sigma_j^+ = -\sigma_j \frac{d\Omega(T_j)}{d\Omega(Q_j)},$$

where the form $d\Omega$ is expressed on dk at Q_j and T_j. For this choice, the sum of the residues of $\psi^{+,\alpha}\psi_\beta\, d\Omega$ at Q_j and T_j vanishes. This condition ensures that the potential reconstructed from ψ^+ is the same as the one reconstructed from ψ.

Notice that the roles of Q_j and T_j are interchanged in the definitions of ψ and ψ^+.

A similar analysis holds in the rational case.

THEOREM 7.2. *The components of the Baker–Akhiezer function $\psi^+(x, t, P)$ are given by*

(7.14)
$$\psi^{+,\alpha}(P, x, t) = \frac{\mathrm{Det}\begin{pmatrix} h^{+,\alpha}(P) & g_j^+(P) \\ h^{+,\alpha}(Q_i) & \Theta_{ij}^+(x, t) \end{pmatrix}}{\mathrm{Det}(\Theta_{ij}^+(x, t))} e^{-kx - k^2 t},$$

where Θ^+ is the matrix with entries

(7.15)
$$\Theta_{ii}^+ = -\sigma_i^+ e^{2x + 4(\chi_i + 1)t} + g_i^+(Q_i), \qquad \Theta_{ij}^+ = g_j^+(Q_i).$$

In the rational case we define

(7.16)
$$\Theta_{ii}^+ = -x - 2\chi_i t - \sigma_i^+ + g_i^{(1)+}, \qquad \Theta_{ij}^+ = g_j^+(Q_i).$$

7.2. The potential.

THEOREM 7.3. *The vector Baker–Akhiezer function $\psi(x, t, P)$ is a solution of the equation $(\partial_t - \partial_x^2 + u(x, t))\psi = 0$, where the potential u is given by $u(x, t) = \sum \rho_i(t) V(x - x_i(t))$ and $\rho_i(t)$ is an $l \times l$ matrix of rank 1.*

PROOF. The usual arguments about the uniqueness of the Baker–Akhiezer function shows that $(\partial_t - \partial_x^2)\psi$ is of the form $-u(x, t)\psi$ with $u = 2\partial_x \xi_1^{\alpha j}(x, t)$. From equation (7.9) it is clear that $u(x, t)$ is of the form $\sum \rho_i(t) V(x - x_i(t))$. To compute ρ_i, let us expand around P_β,

$$g_i(P) = \frac{g_i^\beta}{k_\beta} + O\left(\frac{1}{k_\beta^2}\right), \qquad h_\alpha(P) = \delta_{\alpha\beta} + \frac{h_\alpha^\beta}{k_\beta} + O\left(\frac{1}{k_\beta^2}\right),$$

and take $s_{i,\alpha}(t, P)$ as given in (7.12). We find $\rho_{i,\alpha}^\beta = a_{i,\alpha} b_i^\beta$, where

$$a_{i,\alpha} = \frac{1}{Q_i(t)} \operatorname{Det} \Theta_\alpha^{(i)}(t),$$

$$b_i^\beta = -2Q_i(t) \frac{\sum_{k=1}^N \lambda_k^{(i)}(t) g_k^\beta}{\prod_{j=1}^N (2\sigma_j e^{-4(\chi_j+1)t - x_i - x_j}) \prod_{j \neq i} \sinh(x_i - x_j)}. \qquad \square$$

Alternatively one could use the dual Baker function ψ^+. It satisfies a Schrödinger equation with the same potential u as ψ. This is because the sum of the residues of the form $\psi^{+\alpha} \psi_\beta \Omega$ at the points Q_j and T_j vanishes, so that the same argument as in the elliptic case applies. This shows that Θ and Θ^+ have the same eigenvalues $-x_i(t)$ and gives alternative formulas for a_i^α and b_i^β, in particular,

$$b_i^\alpha = \frac{1}{Q_i^+(t)} \operatorname{Det} \Theta^{+,\alpha(i)}(t).$$

The normalizations $Q_i(t)$ and $Q_i^+(t)$ are as usual determined by the conditions $f_{ii} = 2$ and $\sum_\alpha b_i^\alpha = 1$.

This implies that $x_i(t)$, $a_i(t)$, $b_i(t)$ are the solutions of the trigonometric or rational model. Note that the curve is necessarily of the form of the spectral curve of the Calogero model.

7.3. Reconstruction formulas.

To construct the functions g_j, one can take advantage of the fact that we know on Γ the function $1/(k - \chi_j)$ which vanishes at the l punctures P_α and has a pole with residue 1 at Q_j. It has $l - 1$ other poles at some well-defined points $\delta_k^{(j)}$. The function $g_j(P)(k - \chi_j)$ has $g + l - 1$ poles γ_k and $l - 1$ zeros $\delta_k^{(j)}$. By the Riemann–Roch theorem, this function $F_j(P)$ is uniquely determined by these data and the normalization condition $F_j(Q_j) = 1$. One can give an expression in terms of theta functions as in (6.3). Then

$$g_j(P) = F_j(P)/(k - \chi_j).$$

In the standard Calogero–Moser model, we have $l = 1$, and $F_j = 1$.

Let us summarize the obtained results.

THEOREM 7.4. *Let Γ be a curve that is defined by the equation of the form (3.11) or (3.15) and $D = \gamma_1, \ldots, \gamma_{g+l-1}$ be a set of points in general position. Then the formulas*

$$\text{Det}\,\Theta(x_i(t), t) = 0, \tag{7.17}$$

$$a_{i,\alpha}(t) = \frac{1}{Q_i(t)} \text{Det}\,\Theta_\alpha^{(i)}, \qquad b_i^\alpha(t) = \frac{1}{Q_i^+(t)} \text{Det}\,\Theta^{+,\alpha(i)}, \tag{7.18}$$

where $Q_i(t)$ and $Q_i^+(t)$ are determined by the conditions $f_{ii} = 2$ and $\sum_\alpha b_i^\alpha = 1$, define the solutions of the system (1.4), (2.14), (2.15). *Here Θ is an $N \times N$ matrix with elements given in equations* (7.4) *in the trigonometric case and* (7.5) *in the rational case. Moreover $\Theta_\alpha^{(i)}$ is obtained from Θ by replacing its first column by $h_\alpha(T_j)$, $j = 1, \ldots, N$. The same is true for $\Theta^{+,\alpha(i)}$. Any solution of the system* (1.1) *may be obtained from these solutions taking into account the symmetries of the system.*

Appendix A

The Weierstrass σ-function of periods $2\omega_1, 2\omega_2$ is the entire function defined by

$$\sigma(z) = z \prod_{m,n \neq 0} \left(1 - \frac{z}{\omega_{mn}}\right) \exp\left[\frac{z}{\omega_{mn}} + \frac{1}{2}\left(\frac{z}{\omega_{mn}}\right)^2\right], \tag{7.19}$$

with $\omega_{mn} = 2m\omega_1 + 2n\omega_2$. The functions ζ and $\wp$ are

$$\zeta(z) = \sigma'(z)/\sigma(z), \qquad \wp(z) = -\zeta'(z). \tag{7.20}$$

The $\wp$-function is doubly periodic, and the σ-function and ζ-functions transform according to

$$\zeta(z + \omega_l) = \zeta(z) + \eta_l, \qquad \sigma(z + \omega_l) = -\sigma(z) e^{\eta_l(z + \omega_l/2)},$$

where $2(\eta_1\omega_2 - \eta_2\omega_1) = i\pi$. These functions have the symmetries

$$\sigma(-z) = -\sigma(z), \quad \zeta(-z) = -\zeta(z), \quad \wp(-z) = \wp(z). \tag{7.21}$$

Their behavior at the neighborhood of zero is

$$\sigma(z) = z + O(z^5), \quad \zeta(z) = z^{-1} + O(z^3), \quad \wp(z) = z^{-2} + O(z^2). \tag{7.22}$$

Setting

$$\Phi(x, z) = \frac{\sigma(z - x)}{\sigma(x)\sigma(z)} e^{\zeta(z)x} \tag{7.23}$$

it is easy to check that

$$\Phi(-x, z) = -\Phi(x, -z), \qquad \frac{d}{dx}\Phi(x, z) = \Phi(x, z)[\zeta(z + x) - \zeta(x)]. \tag{7.24}$$

The function $\Phi(x, z)$ is a double-periodic function of the variable z

(7.25) $$\Phi(x, z + 2\omega_l) = \Phi(x, z)$$

and has an expansion of the form

(7.26) $$\Phi(x, z) = (-z^{-1} + \zeta(x) + O(z)) e^{\zeta(z)x}$$

at the point $z = 0$. As a function of x, it has the following monodromy properties:

(7.27) $$\Phi(x + 2\omega_l, z) = \Phi(x, z) \exp 2(\zeta(z)\omega_l - \eta_l z)$$

and has a pole at the point $x = 0$, i.e.,

(7.28) $$\Phi(x, z) = x^{-1} + O(x).$$

Choosing the periods $\omega_1 = \infty$ and $\omega_2 = i\pi/2$, we obtain the hyperbolic case

(7.29) $$\sigma(z) \to \sinh(z) \exp\left(-\frac{z^2}{6}\right), \quad \zeta(z) \to \coth(z) - \frac{z}{3},$$
$$\wp(z) \to \frac{1}{\sinh^2(z)} + \frac{1}{3},$$

(7.30) $$\Phi(x, z) \to \frac{\sinh(z - x)}{\sinh(z) \sinh(x)} e^{x \coth z}.$$

In the rational limit, we have

$$\sigma(z) \to z, \quad \zeta(z) \to \frac{1}{z}, \quad \wp(z) \to \frac{1}{z^2}, \quad \Phi(x, z) \to \left(\frac{1}{x} - \frac{1}{z}\right) e^{x/z}.$$

Appendix B

Let us recall briefly some facts we need about Riemann's theta functions. First there is an embedding of the Riemann surface Γ into its Jacobian $J(\Gamma)$ by the Abel map.

Let a_i^0, b_i^0 be a basis of cycles on Γ with canonical matrix of intersections $a_i^0 \cdot a_j^0 = b_i^0 \cdot b_j^0 = 0$, $a_i^0 \cdot b_j^0 = \delta_{ij}$. In a standard way this basis defines a basis of normalized holomorphic differentials $\omega_j(P)$

(7.31) $$\oint_{a_j^0} \omega_i = \delta_{ij}.$$

The matrix of b-periods of these differentials

(7.32) $$B_{ij} = \oint_{b_i^0} \omega_j$$

defines the Riemann theta-function

(7.33) $$\theta(z_1, \ldots, z_g) = \sum_{m \in \mathbb{Z}^g} e^{2\pi i(m,z) + \pi i(Bm,m)}$$

on the torus $J(\Gamma)$, which is called the *Jacobian variety*

(7.34) $$J(\Gamma) = \mathbb{C}^g / \mathcal{B}.$$

The lattice $\mathcal{B}$ is generated by the basic vectors $e_i \in \mathbb{C}^g$ and by the vectors $B_j \in \mathbb{C}^g$ with coordinates B_{ij}.

The theta function has remarkable automorphy properties with respect to this lattice: for any $l \in \mathbb{Z}^g$ and $z \in \mathbb{C}^g$

(7.35) $$\theta(z+l) = \theta(z), \qquad \theta(z+Bl) = \exp[-i\pi(Bl,l) - 2i\pi(l,z)]\theta(z).$$

Let us choose a point $q_0 \in \Gamma$. Then the vector $A(P)$ with coordinates

(7.36) $$A_k(P) = \int_{q_0}^{P} \omega_k$$

defines the Abel map

(7.37) $$A: \Gamma \mapsto J(\Gamma)$$

which is an embedding of Γ into $J(\Gamma)$.

The fundamental theorem of Riemann expresses the intersection of the image of this embedding with the zero set of the theta function.

THEOREM. *Let* $Z = (Z_1, \ldots, Z_g) \in \mathbb{C}^g$ *be arbitrary. Either the function* $\theta(A(P) - Z)$ *vanishes identically for* $P \in \Gamma$ *or it has exactly g zeros* $P_1, \ldots, P_g$ *such that*

(7.38) $$A(P_1) + \cdots + A(P_g) = Z - \mathcal{K},$$

where $\mathcal{K}$ is the so-called vector of Riemann's constants, depending on the curve Γ and the point q_0, but independent of Z.

From this one can prove the Jacobi theorem, i.e., that any point in the Jacobian $J(\Gamma)$ is of the form $A(P_1), \ldots, A(P_g)$ for some divisor of degree g on Γ.

This allows to find a formula for a function that has g poles at points $\delta_1, \ldots, \delta_g$ and an additional pole at point Q^+. The dimension of the space of such functions is two. Of course it contains the constant. We choose the second basic function by the condition that it vanishes at some fixed point Q^-.

Let Z, Z^+, Z^-, Z^0 be vectors that are defined by the following formulas:

$$Z = \sum_{s=1}^{g} A(\delta_s) + \mathcal{K},$$

$$Z^+ = Z - A(\delta_1) + A(Q^+) = A(Q^+) + \sum_{s=2}^{g} A(\delta_s) + \mathcal{K},$$

$$Z^- = Z - A(\delta_1) + A(Q^-), \qquad Z^- + Z^0 = Z + Z^+.$$

Let us define the function

$$(7.39) \qquad f(P) = \frac{\theta(A(P) - Z^-)\theta(A(P) - Z^0)}{\theta(A(P) - Z)\theta(A(P) - Z^+)}.$$

From the Jacobi theorem it follows that the two factors in the denominator vanish at the points $\delta_1, \ldots, \delta_g$ and $Q^+, \delta_2, \ldots, \delta_g$ respectively. In a similar way, the two factors in the numerator vanish at $Q^+, \delta_2, \ldots, \delta_g$, and at g other points.

The zeros at $\delta_2, \ldots, \delta_g$ cancel between the numerator and the denominator, thereby leaving us with the correct divisor of zeros and poles. It remains to show that the function f is well defined on Γ. This is because the definition of Z^0 implies that the automorphy factors of the theta functions in equation (7.35) cancel between the numerator and the denominator when P describes b-cycles on Γ.

Acknowledgements

One of the authors (I. K.) would like to thank the Institute Henri Poincaré for the hospitality during the time when this work was done. This work was supported by RFFI grant 93-011-16087 and ISF grant MD800.

References

1. F. Calogero, *Exactly solvable one-dimensional many-body systems*, Lett. Nuovo Cimento **13** (1975), 411–415.
2. J. Moser, *Three integrable Hamiltonian systems connected with isospectral deformations*, Adv. Math. **16** (1975), 441–416.
3. A. M. Perelomov, *Completely integrable classical systems connected with semisimple Lie algebras*, Lett. Math. Phys. **1** (1977), 531–540.
4. I. M. Krichever, *Elliptic solutions of Kadomtsev–Petviashvili equation and integrable systems of particles*, Funktsional. Anal. i Prilozhen. **14** (1980), no. 4; English transl., Functional Anal. Appl. **14** (1980), no. 4, 282–290.
5. A. Gorski and N. Nekrasov, *Elliptic Calogero–Moser system from two-dimensional current algebra*, Preprint ITEP-NG/1-94, hepth 9401021.
6. J. Gibbons and T. Hermsen, *A generalization of the Calogero–Moser system*, Phys. D **11** (1984), 337.
7. D. Bernard, M. Gaudin, F. Hadane, and V. Pasquier, *Yang–Baxter equation in long-range interacting systems*, J. Phys. A **26** (1993), 5219.
8. J. Avan, O. Babelon, and E. Billey, *Exact Yangian symmetry in the classical Euler–Calogero–Moser model*, (hep-th 9312042), Phys. Lett. A **188** (1994), 263.
9. H. Airault, H. McKean, and J. Moser, *Rational and elliptic solutions of the KdV equation and related many-body problem*, Comm. Pure Appl. Math. **30** (1977), 95–125.
10. I. M. Krichever, *On rational solutions of Kadomtsev–Petviashvili equation and integrable systems of N particles on line*, Funktsional. Anal. i Prilozhen. **12** (1978), no. 1, 76–78; English transl. in Functional Anal. Appl. **12** (1978).
11. D. V. Choodnovsky and G. V. Choodnovsky, *Pole expansions of nonlinear partial differential equations*, Nuovo Cimento B **40** (1977), 339–350.
12. V. E. Zakharov and A. B. Shabat, *The integration scheme of nonlinear equations of mathematical physics with the help of the inverse scattering problem* I., Funktsional. Anal. i Prilozhen. **8** (1974), no. 3, 43–53; English transl. in Functional Anal. Appl. **8** (1974).
13. V. S. Druma, *On analytical solution of two-dimensional Korteweg–de Vries equation*, Pis′ ma Zh. Èksper. Teoret. Fiz. **19** (1973), no. 12, 219–225; English transl. in JETP Lett. **19** (1973).
14. I. M. Krichever, *Integration of nonlinear equations by algebraic-geometrical methods*, Funktsional. Anal. i Prilozhen. **11** (1977), no. 1, 15–31; English transl. in Functional Anal. Appl. **11** (1977).

15. J. Avan and M. Talon, *Classical R-matrix structure for the Calogero model*, Phys. Lett. B **303** (1993), 33.
16. J. Avan, O. Babelon, and M. Talon, *Construction of the classical R-matrices for the Toda and Calogero models*, Algebra i Analiz **6** (1994), no. 2, 67–89; English transl. in Leningrad Math. J. **6** (1995).

Translated by THE AUTHOR

Landau Institute for Theoretical Physics, Kosygina 2, Moscow, 117940, Russia

L.P.T.H.E. Université Paris VI (CNRS UA 280), Box 126, Tour 16, 1er étage, 4 place Jussieu, 75252 Paris Cedex 05, France

Symplectic and Poisson Geometry on Loop Spaces of Manifolds and Nonlinear Equations

OLEG MOKHOV

ABSTRACT. We consider some differential geometry classes of local and nonlocal Poisson and symplectic structures on loop spaces of smooth manifolds. These classes yield natural Hamiltonian or multi-Hamiltonian representations for some important nonlinear equations of mathematical physics and field theory, such as nonlinear sigma models with torsion, degenerate Lagrangian systems of field theory, systems of hydrodynamic type, N-component systems of Heisenberg magnet type, Monge–Ampère equations, the Krichever–Novikov equation and others. In particular, a complete classification of all nondegenerate Poisson bivectors $\omega^{ij}(x, u, u_x, u_{xx}, \ldots)$ depending on derivatives of the field variables $u^i(x)$ and the independent space variable x is obtained (u^i, $i = 1, \ldots, N$, are local coordinates on smooth manifold M). In other words, all Poisson brackets of the following form

$$\{u^i(x), u^j(y)\} = \omega^{ij}(x, u, u_x, u_{xx}, \ldots)\delta(x-y), \qquad \det(\omega^{ij}) \neq 0,$$

are explicitly described. In addition, we shall prove the integrability of a certain class of nonhomogeneous systems of hydrodynamic type and give a description of partial differential equations of associativity in 2D topological field theories (for solutions of special type of the Witten–Dijkgraaf–E. Verlinde–H. Verlinde (WDVV) system) as integrable nondiagonalizable weakly nonlinear systems of hydrodynamic type.

§1. Generalized Poisson bivectors on loop spaces of manifolds

Let an N-dimensional smooth manifold M with local coordinates $\{u^1, \ldots, u^N\}$ be given. A *classic Poisson bivector* $\omega^{ij}(u)$ on M is, by definition, a skew-symmetric $(2, 0)$-tensor on M ($\omega^{ij}(u) = -\omega^{ji}(u)$) that satisfies the following well-known relation (the Jacobi identity):

(1.1) $$\frac{\partial \omega^{ij}}{\partial u^s} \omega^{sk} + \frac{\partial \omega^{jk}}{\partial u^s} \omega^{si} + \frac{\partial \omega^{ki}}{\partial u^s} \omega^{sj} = 0.$$

Such a bivector defines a Poisson bracket on the space of smooth functions on M:

(1.2) $$\{f(u), g(u)\} = \omega^{ij}(u) \frac{\partial f}{\partial u^i} \frac{\partial g}{\partial u^j}$$

1991 *Mathematics Subject Classification.* Primary 53C15, 58F07; Secondary 58F05.

(or $\{u^i, u^j\} = \omega^{ij}(u)$), which is skew-symmetric
$$\{f(u), g(u)\} = -\{g(u), f(u)\}$$
and satisfies the Jacobi identity
$$\{f(u), \{g(u), h(u)\}\} + \{g(u), \{h(u), f(u)\}\} + \{h(u), \{f(u), g(u)\}\} = 0.$$

First of all, in this paper we study some natural infinite-dimensional generalizations of the Poisson bivectors and the corresponding Poisson brackets.

Let us consider the loop space of the manifold M, i.e., the space $L(M)$ of all smooth parametrized mappings $\gamma: S^1 \to M$, $\gamma(x) = \{u^i(x), x \in S^1\}$. The classical Poisson bivector $\omega^{ij}(u)$ gives the so-called *ultra-local Poisson bracket* on the loop space $L(M)$

$$(1.3) \qquad \{u^i(x), u^j(y)\} = \omega^{ij}(u(x))\delta(x-y),$$

or, in other words, we have a Poisson bracket on the space of functionals on the loop space $L(M)$

$$(1.4) \qquad \{F, G\} = \int_\gamma \frac{\delta F}{\delta u^i(x)} \omega^{ij}(u(x)) \frac{\delta G}{\delta u^j(x)} \, dx,$$

where F and G are arbitrary functionals on $L(M)$. In the infinite-dimensional case of $L(M)$, we can consider a natural generalization of the Poisson bracket (1.4). In fact, generally speaking, in this case a Poisson bivector ω^{ij} can depend also on derivatives of the fields $u^k(x)$: $\omega^{ij}(u, u_x, u_{xx}, \ldots)$.

PROBLEM 1. Describe all Poisson brackets on $L(M)$ of the form

$$(1.5) \quad \{F, G\} = \int_\gamma \frac{\delta F}{\delta u^i(x)} \omega^{ij}(u, u_x, u_{xx}, \ldots) \frac{\delta G}{\delta u^j(x)} \, dx, \qquad \det(\omega^{ij}) \neq 0.$$

REMARK 1.1. Note that in this paper we always consider only the tensors that depend on a finite number of derivatives of the fields $u^k(x)$.

DEFINITION 1.1. A bivector $\omega^{ij}(u, u_x, u_{xx}, \ldots)$ is called a *generalized Poisson bivector* on the loop space $L(M)$ if the tensor $\omega^{ij}(u, u_x, u_{xx}, \ldots)$ gives a Poisson bracket (1.5) on the space of functionals on $L(M)$.

THEOREM 1.1. *A nondegenerate tensor* $\omega^{ij}(u, u_x, u_{xx}, \ldots)$, $\det(\omega^{ij}) \neq 0$, *on the loop space* $L(M)$ *is a generalized Poisson bivector on* $L(M)$ *if and only if there is a closed 2-form* $\Omega_{ij}(u)$ *on* M ($d\Omega = 0$) *and a closed 3-form* $T_{ijk}(u)$ *on* M ($dT = 0$) *such that* $\det(T_{ijk}(u)u_x^k + \Omega_{ij}(u)) \neq 0$ *and*

$$(1.6) \qquad \omega^{ij}(T_{jks}(u)u_x^s + \Omega_{jk}(u)) = \delta_k^i.$$

REMARK 1.2. It follows from (1.6) that nondegenerate generalized Poisson bivectors ω^{ij} on $L(M)$ can depend only on the first derivatives of the field variables $u^i(x)$.

REMARK 1.3. If the manifold M is symplectic and the 2-form $\Omega_{ij}(u)$ is a symplectic form on M (i.e., $\det(\Omega_{ij}) \neq 0$), then for any closed 3-form $T_{ijk}(u)$ on M we always have $\det(T_{ijk}(u)u_x^k + \Omega_{ij}(u)) \neq 0$ and formula (1.6) gives explicitly a generalized Poisson bivector ω^{ij}. We also note here that for any closed 3-form T_{ijk} on M, $\det(T_{ijk}(u)u_x^k) \equiv 0$.

Theorem 1.1 describes translation invariant Poisson brackets. However, we can also consider Poisson bivectors $\omega^{ij}(x, u(x), u_x, u_{xx}, \ldots)$ depending explicitly on the independent space variable x.

THEOREM 1.2. *A tensor* $\omega^{ij}(x, u, u_x, u_{xx}, \dots)$, $\det(\omega^{ij}) \neq 0$, *gives a Poisson bracket*

$$\{F, G\} = \int_\gamma \frac{\delta F}{\delta u^i(x)} \omega^{ij}(x, u, u_x, u_{xx}, \dots) \frac{\delta G}{\delta u^j(x)} dx$$

on the space of functionals on $L(M)$ *if and only if there is an one-parameter family of 2-forms* $\Omega_{ij}(z, u)$ *on the manifold* M *and a closed 3-form* $R_{ijk}(u)$ *on* M *such that*

(1.7)
$$\det\left[\left(\int_0^x (d\Omega)_{ijk}\, dz + R_{ijk}(u)\right) u_x^k + \Omega_{ij}(x, u)\right] \neq 0,$$

$$\omega^{is}\left[\left(\int_0^x (d\Omega)_{sjk}(z, u(x))\, dz + R_{sjk}(u)\right) u_x^k + \Omega_{sj}(x, u(x))\right] = \delta^i_j,$$

where

$$(d\Omega)_{ijk}(z, u(x)) = \frac{\partial \Omega_{ij}(z, u(x))}{\partial u^k} + \frac{\partial \Omega_{ki}(z, u(x))}{\partial u^j} + \frac{\partial \Omega_{jk}(z, u(x))}{\partial u^i}$$

(*note that there are no closedness conditions for the 2-forms* Ω_{ij} *here in contrast with Theorem* 1.1).

REMARK 1.4. It is an interesting unsolved problem to classify all degenerate generalized Poisson bivectors $\omega^{ij}(x, u, u_x, u_{xx}, \dots)$ with respect to their dependence on derivatives of the fields $u^i(x)$.

In order to prove Theorems 1.1 and 1.2, we shall consider differential geometry objects which are inverse to nondegenerate Poisson bivectors ω^{ij}. It is well known that in the finite-dimensional case, we shall get a symplectic form $\omega_{ij}(u)$ on the manifold M: $\omega_{ij}(u)\omega^{jk}(u) = \delta_i^k$, $(d\omega)_{ijk} = 0$. In the next section we shall consider the corresponding generalizations of symplectic structures in the case of the infinite-dimensional loop space $L(M)$.

§2. Symplectic structures on loop spaces of manifolds

The tangent space of the loop space $L(M)$ at its point (a loop) γ consists of the all smooth vector fields ξ^i, $1 \leq i \leq N$, defined along the loop γ. We shall denote this space here by $T_\gamma L(M)$ (we note that $\xi^i(\gamma(x)) \in T_{\gamma(x)}M$ for each $x \in S^1$, where $T_{\gamma(x)}M$ is the tangent space of the manifold M at the point $\gamma(x)$).

Consider a skew-symmetric bilinear form $\omega(\xi, \eta)$ on the loop space $L(M)$,

(2.1) $$\omega(\xi, \eta) = \int_\gamma \xi^i A_{ij} \eta^j\, dx,$$

$\omega(\xi, \eta) = -\omega(\eta, \xi)$, $\xi = (\xi^1, \dots, \xi^N)$, $\eta = (\eta^1, \dots, \eta^N)$, $A = (A_{ij})$ is the skew-symmetric operator defined by the loop γ

$$A[\gamma]\colon T_\gamma L(M) \to T_\gamma^* L(M).$$

Introduce a differential of the 2-form $\omega(\xi, \eta)$ (which is a natural infinite-dimensional generalization of the usual differential in the Lie algebra of vector fields on a manifold):

(2.2) $$(d\omega)(\xi, \eta, \zeta) = \sum_{(\xi, \eta, \zeta)} \left\{ \int_\gamma \xi^i \frac{\delta \omega(\eta, \zeta)}{\delta u^i}\, dx + \omega(\xi, [\eta, \zeta]) \right\},$$

where
$$[\eta, \zeta]^j = \eta^i_{(k)} \frac{\partial \zeta^j}{\partial u^i_{(k)}} - \zeta^i_{(k)} \frac{\partial \eta^j}{\partial u^i_{(k)}}$$

is the commutator of the vector fields η and ζ, $f_{(k)} = d^k f/dx^k$, and the sign $\sum_{(\xi, \eta, \zeta)}$ means that the sum is taken over all cyclic permutations of the elements (ξ, η, ζ).

If $d\omega = 0$, i.e., the 2-form (2.1) is closed, then the operator A_{ij} is said to be a *symplectic operator* (ω is called a *presymplectic form* or *presymplectic structure*).

REMARK 2.1. In the infinite-dimensional case, a presymplectic structure is often said to be *symplectic* even if the corresponding 2-form ω is degenerate on $L(M)$. We shall adhere to this terminology here (see [30] for references on symplectic operators, the general theory of local Hamiltonian and symplectic structures, and their connections with the integrability of nonlinear partial differential equations).

LEMMA 2.1. *A nondegenerate tensor $\omega^{ij}(x, u, u_x, u_{xx}, \ldots)$, $\det \omega^{ij} \neq 0$, is a generalized Poisson bivector on $L(M)$ if and only if the tensor $\omega_{ij}(x, u, u_x, u_{xx}, \ldots)$, where $\omega^{is}\omega_{sj} = \delta^i_j$, is a symplectic operator.*

It was shown by author in [13] that it is sufficient to verify the Jacobi identity only for all linear functionals of the form

$$F = \int_\gamma f_i(x) u^i(x)\, dx,$$

where $f_i(x)$, $i = 1, \ldots, N$, are arbitrary functions. Considering vector fields ξ of the form $\xi^i = \omega^{ij} f_j(x)$, it is easy to check that the condition $(d\omega)(\xi, \eta, \zeta) = 0$ for a symplectic structure $\omega_{ij}(x, u, u_x, u_{xx}, \ldots)$ is equivalent to the Jacobi identity on linear functionals for the bracket

$$\{u^i(x), u^j(y)\} = \omega^{ij}(x, u, u_x, u_{xx}, \ldots)\delta(x - y).$$

Now consider the symplectic operators $\omega_{ij}(u, u_x, u_{xx}, \ldots)$.

LEMMA 2.2. *An operator $\omega_{ij}(u, u_x, u_{xx}, \ldots)$ is symplectic if and only if*
(1) $\omega_{ij} = -\omega_{ji}$ *(skew-symmetry);*
(2) *for any vector-valued functions $\eta(x) = (\eta^1(x), \ldots, \eta^N(x))$, $\zeta(x) = (\zeta^1(x), \ldots, \zeta^N(x))$ we have the following identity:*

$$(2.3) \quad (-1)^k \left(\frac{d}{dx}\right)^k \left(\frac{\partial \omega_{ij}}{\partial u^s_{(k)}} \eta^i(x)\zeta^j(x)\right) + \frac{\partial \omega_{si}}{\partial u^j_{(k)}} \eta^i(x)\zeta^j_{(k)}(x) - \frac{\partial \omega_{sj}}{\partial u^i_{(k)}} \eta^i_{(k)}(x)\zeta^j(x) \equiv 0;$$

(the closedness condition for ω_{ij}).

Now consider $R = \max\{m : \text{there exist } i, j, s \text{ such that } \partial \omega_{ij}/\partial u^s_{(m)} \neq 0\}$.

If $R \geq 2$, then the coefficient at the term $\eta_x^i(x)\zeta_{(R-1)}^j(x)$ in (2.3) is equal to $(-1)^R R \partial \omega_{ij}/\partial u_{(R)}^s$, i.e., $R \leq 1$ and the form ω_{ij} depends only on $u^k(x)$ and the first derivatives $u_x^k(x)$.

Now it is easy to show from the identity (2.3) that ω_{ij} must be quasilinear with respect to the first derivatives $u_x^k(x)$, i.e.,

$$\omega_{ij}(u, u_x) = T_{ijk}(u)u_x^k + \Omega_{ij}(u). \tag{2.4}$$

Note that all symplectic operators of this type were described by the author in [8], where differential geometry homogeneous symplectic operators of the first order

$$A_{ij} = g_{ij}\frac{d}{dx} + b_{ijk}(u)u_x^k$$

were studied in detail (allowing the degeneration of g_{ij}, we get a complete description of quasilinear symplectic structures $\omega_{ij}(u, u_x)$).

It turns out that the symplectic operators ω_{ij} of the form (2.4) give Hamiltonian (or symplectic) structures of some special Lagrangian systems in classical field theory (a certain degeneration of nonlinear sigma models).

THEOREM 2.1. *Lagrangian systems $\delta S/\delta u = 0$ given by the action*

$$S = \int (a_{ij}(u)u_x^i u_t^j + b_i(u)u_t^i + U(x, t, u, u_x, u_{xx}, \dots))\, dx\, dt, \tag{2.5}$$

where $b_i(u)$, $U(x, t, u, u_x, u_{xx}, \dots)$ are arbitrary functions and skew-symmetric functions $a_{ij}(u)$ ($a_{ij} = -a_{ji}$), are Hamiltonian systems with respect to the corresponding symplectic structures of the form (2.4).

Conversely, any Hamiltonian system of the form

$$(T_{ijk}(u)u_x^k + \Omega_{ij}(u))u_t^j = \frac{\delta U}{\delta u^i}$$

is a Lagrangian system with respect to the action of the type (2.5).

In the next section we shall consider the symplectic geometry corresponding to the general nonlinear sigma models with torsion.

§3. Nonlinear sigma models with torsion and symplectic geometry on loop spaces of manifolds

Consider two-dimensional systems generated by general nonlinear sigma model actions of the form

$$S = \int \left(\frac{1}{2}r_{ij}(u)u_x^i u_t^j + U(u)\right) dx\, dt, \tag{3.1}$$

where $r_{ij}(u)$ is an arbitrary tensor and $U(u)$ is an arbitrary function on M. As it was shown by the author in [8], the corresponding Lagrangian system $\delta S/\delta u = 0$ always has the symplectic representation

$$A_{ij}u_t^j = \frac{\delta U}{\delta u}, \tag{3.2}$$

where

$$A_{ij} = \frac{1}{2}(r_{ij} + r_{ji})\frac{d}{dx} + \frac{1}{2}\left(\frac{\partial r_{ki}}{\partial u^j} + \frac{\partial r_{ij}}{\partial u^k} - \frac{\partial r_{kj}}{\partial u^i}\right)u_x^k \qquad (3.3)$$

is a symplectic operator.

If $\det(r_{ij} + r_{ji}) \neq 0$ (in this case we have a Riemannian or pseudo-Riemannian manifold M with the metric $g_{ij} = \frac{1}{2}(r_{ij} + r_{ji})$), then the symplectic operator (3.3) gives a symplectic form $\omega(\xi, \eta)$ on $L(M)$ that can be expressed as

$$\omega(\xi, \eta) = \int_\gamma \langle \xi, \nabla_{\dot\gamma}\eta \rangle, \qquad (3.4)$$

where $\langle \xi, \eta \rangle = \frac{1}{2}(r_{ij} + r_{ji})\xi^i \eta^j$ and $\nabla_{\dot\gamma}$ is the covariant derivation along γ generated by some differential geometry connection $\Gamma^i_{jk}(u)$ on M. Let us describe all symplectic forms of the form (3.4) on $L(M)$.

THEOREM 3.1 [8]. *Let a Riemannian or pseudo-Riemannian manifold* (M, g_{ij}) *be given. A differential geometry connection* $\Gamma^i_{jk}(u)$ *on* M *defines a symplectic form* (3.4), *where* $\langle \xi, \eta \rangle = g_{ij}\xi^i\eta^j$, *if and only if*

(1) *the connection* $\Gamma^i_{jk}(u)$ *is compatible with the metric* g_{ij}, *i.e.*,

$$\nabla_k g_{ij} \equiv \frac{\partial g_{ij}}{\partial u^k} - g_{is}\Gamma^s_{jk} - g_{js}\Gamma^s_{ik} = 0;$$

(2) *the torsion*

$$T_{ijk} = g_{is}T^s_{jk}, \qquad T^i_{jk} \equiv \Gamma^i_{jk} - \Gamma^i_{kj}$$

is a closed 3-form on M.

Any closed 3-form on the (pseudo-)Riemannian manifold (M, g_{ij}) gives the only differential geometry connection $\Gamma^i_{jk}(u)$ compatible with the metric $g_{ij}(u)$, with torsion tensor determined explicitly by this closed 3-form. Correspondingly, any metric $g_{ij}(u)$ and any closed 3-form on M generates the symplectic form (3.4) on $L(M)$.

The corresponding symplectic operator has the form

$$M_{ij} = g_{ij}(u)\frac{d}{dx} + g_{is}(u)\Gamma^s_{jk}(u)u_x^k, \qquad (3.5)$$

where locally we have

$$\Gamma^i_{jk}(u) = \frac{1}{2}g^{is}(u)\left(\frac{\partial g_{sk}}{\partial u^j} + \frac{\partial g_{js}}{\partial u^k} - \frac{\partial g_{jk}}{\partial u^s} + T_{sjk}(u)\right), \qquad (3.6)$$

$$T_{ijk}(u) = \frac{1}{2}\left(\frac{\partial f_{ij}}{\partial u^k} + \frac{\partial f_{jk}}{\partial u^i} + \frac{\partial f_{ki}}{\partial u^j}\right), \qquad (3.7)$$

$$f_{ij}(u) = -f_{ji}(u). \qquad (3.8)$$

COROLLARY. *Let (M, g_{ij}) be a Riemannian or pseudo-Riemannian manifold. Then closed 2-forms of the form (3.4) on $L(M)$ are in one-to-one correspondence with closed 3-forms on M.*

An elementary conclusion is that on any two-dimensional Riemannian (or pseudo-Riemannian) manifold (M, g_{ij}) there is a single symplectic 2-form (3.4) on the loop space $L(M)$, and correspondingly a single symplectic operator (3.5), generated by the Levi–Civita connection. As for N-dimensional Riemannian (or pseudo-Riemannian) manifolds, the conclusion is that there always exist an symplectic 2-form (3.4), and a symplectic operator (3.5), generated by the Levi–Civita connection.

It must be noted that the symplectic form (3.4) is always degenerate (i.e., in fact, it is only a presymplectic form on $L(M)$) and the null space of $\omega(\xi, \eta)$ consists of vector fields parallel along γ. In particular, the velocity vector field $\{u_x^i\}$ belongs to the null space of $\omega(\xi, \eta)$ if γ is a geodesic loop. Another subject is the interrelation between the symplectic forms under consideration and classical finite-dimensional symplectic structures on manifolds of geodesics. Recall that for N-dimensional Riemannian manifolds (M, g_{ij}) whose geodesics are periodic and of equal length there exists a $(2N - 2)$-dimensional symplectic manifold CM of geodesics. The symplectic structure on CM is determined by the curvature form of the S^1-connection in the principle bundle UM of unit tangent vectors of (M, g_{ij}), where the S^1-connection is generated by the canonical 1-form on T^*M (Reeb theorem, see [33–36]). The tangent space $T_\gamma CM$ at the point γ is isomorphic to the space of normal Jacobian fields along the geodesic γ.

THEOREM 3.2 [8]. *The restriction of the symplectic form (3.4) generated on $L(M)$ by the Levi–Civita connection to the finite-dimensional subspace of normal Jacobian fields along the geodesics γ coincides with the symplectic Reeb form, which is a closed nondegenerate 2-form on CM defining a symplectic structure.*

THEOREM 3.3 [8]. *Formulas (3.5)–(3.8) give a complete description of all symplectic operators of the following form*

$$(3.9) \qquad A_{ij} = a_{ij}(u) \frac{d}{dx} + b_{ijk}(u) u_x^k, \qquad \det a_{ij}(u) \neq 0$$

(*homogeneous differential operators of the first order with nondegenerate leading coefficient*).

LEMMA 3.1. *An operator of the form (3.9) is symplectic if and only if the following conditions hold*:

$$(3.10) \qquad a_{ij} = a_{ji},$$

$$(3.11) \qquad \frac{\partial a_{ij}}{\partial u^k} = b_{ijk} + b_{jik},$$

$$(3.12) \qquad \frac{\partial a_{ij}}{\partial u^k} = b_{ikj} + b_{jki},$$

$$(3.13) \qquad \frac{\partial b_{jmk}}{\partial u^i} - \frac{\partial b_{jmi}}{\partial u^k} + \frac{\partial b_{ijk}}{\partial u^m} - \frac{\partial b_{imk}}{\partial u^j} = 0.$$

REMARK 3.1. Hamiltonian operators of the form (3.9) were classified by Dubrovin and Novikov [2] (see also [1, 4]). In this case, the leading coefficient a^{ij} must be a flat metric on the manifold M and $b^{ij}_k(u)$ is generated by this flat metric. These Hamiltonian operators play a very important role in the theory of integrability of diagonalizable systems of hydrodynamic type [5, 6] (see also [1, 4]).

Thus, any nonlinear sigma model (3.1) has a natural Hamiltonian (symplectic) representation (3.2) given by the symplectic operator (3.3). If a nonlinear sigma model system also has another Hamiltonian representation compatible with (3.2), then this nonlinear sigma model system will be integrable according to the Lenart–Magri scheme ([37]). The important problem of finding Hamiltonian operators compatible with the symplectic operator (3.3) was formulated by the author in 1989 (see [8, 11, 38–40]). For some known two-component integrable nonlinear sigma models without torsion (r_{ij} is a symmetric tensor in (3.1)) such second Hamiltonian representations compatible with (3.2), (3.3) were found in [41].

EXAMPLE 3.1 [41].

$$(3.14) \qquad S = \int \left(\frac{1}{2} \frac{u^1_x u^2_t + u^2_x u^1_t}{u^1 u^2 + c} + k u^1 u^2 \right) dx\, dt,$$

where c and k are arbitrary constants. In this case

$$(r_{ij}) = \frac{1}{u^1 u^2 + c} \begin{pmatrix} 0 & 1 \\ 1 & 0 \end{pmatrix}, \qquad U(u) = k u^1 u^2.$$

The first symplectic operator (3.3) has the form

$$(3.15) \qquad A = \frac{1}{u^1 u^2 + c} \begin{pmatrix} 0 & D \\ D & 0 \end{pmatrix} - \frac{1}{(u^1 u^2 + c)^2} \begin{pmatrix} 0 & u^1 u^2_x \\ u^2 u^1_x & 0 \end{pmatrix}.$$

The second Hamiltonian operator B for the action (3.14) is nonlocal and compatible with (3.15):

$$(3.16)$$
$$B = \begin{pmatrix} 0 & -(u^1 u^2 + c) \\ (u^1 u^2 + c) & 0 \end{pmatrix} + \begin{pmatrix} u^1_x D^{-1} \circ u^1 & -u^1_x D^{-1} \circ u^2 \\ u^2_x D^{-1} \circ u^1 & -u^2_x D^{-1} \circ u^2 \end{pmatrix}$$
$$+ \begin{pmatrix} u^1 D^{-1} \circ u^1_x & u^1 D^{-1} \circ u^2_x \\ -u^2 D^{-1} \circ u^1_x & -u^2 D^{-1} \circ u^2_x \end{pmatrix}.$$

The recursion operator $L^i_j = B^{is} A_{sj}$ for this integrable nonlinear sigma model (3.14) has the following form

$$L = \begin{pmatrix} -D & 0 \\ 0 & D \end{pmatrix} + \frac{1}{u^1 u^2 + c} \begin{pmatrix} u^1 u^2_x & 2 u^1 u^1_x \\ -2 u^2 u^2_x & -u^2 u^1_x \end{pmatrix}$$

$$(3.17) \quad + \begin{pmatrix} u^1_x D^{-1} \circ \frac{c u^2_x}{(u^1 u^2 + c)^2} & u^1_x D^{-1} \circ \frac{-c u^1_x}{(u^1 u^2 + c)^2} \\ u^2_x D^{-1} \circ \frac{c u^2_x}{(u^1 u^2 + c)^2} & u^2_x D^{-1} \circ \frac{-c u^1_x}{(u^1 u^2 + c)^2} \end{pmatrix}$$

$$+ \begin{pmatrix} u^1 D^{-1} \circ \left(\frac{u^1 (u^2_x)^2}{(u^1 u^2 + c)^2} - \frac{u^2_{xx}}{u^1 u^2 + c} \right) & u^1 D^{-1} \circ \left(\frac{u^2 (u^1_x)^2}{(u^1 u^2 + c)^2} - \frac{u^1_{xx}}{u^1 u^2 + c} \right) \\ -u^2 D^{-1} \circ \left(\frac{u^1 (u^2_x)^2}{(u^1 u^2 + c)^2} - \frac{u^2_{xx}}{u^1 u^2 + c} \right) & -u^2 D^{-1} \circ \left(\frac{u^2 (u^1_x)^2}{(u^1 u^2 + c)^2} - \frac{u^1_{xx}}{u^1 u^2 + c} \right) \end{pmatrix}.$$

EXAMPLE 3.2 [41].

$$(3.18) \qquad S = \int \left(\frac{1}{2} \frac{u_x^1 u_t^2 + u_x^2 u_t^1}{u^1 + u^2} + k(u^1 + u^2) \right) dx \, dt,$$

where k is an arbitrary constant. In this case:

$$(r_{ij}) = \frac{1}{u^1 + u^2} \begin{pmatrix} 0 & 1 \\ 1 & 0 \end{pmatrix}, \qquad U(u) = k(u^1 + u^2).$$

The first symplectic operator (3.3) then has the following form:

$$(3.19) \qquad A = \frac{1}{u^1 + u^2} \begin{pmatrix} 0 & D \\ D & 0 \end{pmatrix} - \frac{1}{(u^1 + u^2)^2} \begin{pmatrix} 0 & u_x^2 \\ u_x^1 & 0 \end{pmatrix}.$$

The second Hamiltonian operator B for the action (3.18) is also nonlocal and compatible with (3.19):

$$(3.20) \qquad B = \begin{pmatrix} 0 & -(u^1 + u^2) \\ (u^1 + u^2) & 0 \end{pmatrix} + \begin{pmatrix} u_x^1 D^{-1} & -u_x^1 D^{-1} \\ u_x^2 D^{-1} & -u_x^2 D^{-1} \end{pmatrix}$$
$$+ \begin{pmatrix} D^{-1} \circ u_x^1 & D^{-1} \circ u_x^2 \\ -D^{-1} \circ u_x^1 & -D^{-1} \circ u_x^2 \end{pmatrix}.$$

The recursion operator $L_j^i = B^{is} A_{sj}$ has the form
$$(3.21)$$
$$L = \begin{pmatrix} -D & 0 \\ 0 & D \end{pmatrix} + \frac{1}{u^1 + u^2} \begin{pmatrix} u_x^2 & 2u_x^1 \\ -2u_x^2 & -u_x^1 \end{pmatrix}$$
$$+ \begin{pmatrix} -u_x^1 D^{-1} \circ \frac{u_x^2}{(u^1 + u^2)^2} & u_x^1 D^{-1} \circ \frac{u_x^1}{(u^1 + u^2)^2} \\ -u_x^2 D^{-1} \circ \frac{u_x^2}{(u^1 + u^2)^2} & u_x^2 D^{-1} \circ \frac{u_x^1}{(u^1 + u^2)^2} \end{pmatrix}$$
$$+ \begin{pmatrix} D^{-1} \circ \left(\frac{(u_x^2)^2}{(u^1 + u^2)^2} - \frac{u_{xx}^2}{u^1 + u^2} \right) & D^{-1} \circ \left(\frac{(u_x^1)^2}{(u^1 + u^2)^2} - \frac{u_{xx}^1}{u^1 + u^2} \right) \\ D^{-1} \circ \left(-\frac{(u_x^2)^2}{(u^1 + u^2)^2} + \frac{u_{xx}^2}{u^1 + u^2} \right) & D^{-1} \circ \left(-\frac{(u_x^1)^2}{(u^1 + u^2)^2} + \frac{u_{xx}^1}{u^1 + u^2} \right) \end{pmatrix}.$$

§4. Homogeneous symplectic forms on loop spaces of manifolds

Consider the matrix entries of the operator A of the form

$$(4.1) \qquad A_{ij}^{[m]} = a_{ij}^{[m]}(u) \frac{d^m}{dx^m} + b_{ijk}^{[m]}(u) u_x^k \frac{d^{m-1}}{dx^{m-1}}$$
$$+ (c_{ijkl}^{[m]}(u) u_x^k u_x^l + d_{ijk}^{[m]}(u) u_{xx}^k) \frac{d^{m-2}}{dx^{m-2}} + \dots,$$

where all the terms in (4.1) are homogeneous of degree m with respect to the natural grading

$$\deg(fg) = \deg f + \deg g,$$
$$\deg f(u(x)) = \deg u(x) = 0,$$
$$\deg \frac{d^k u}{dx^k} = \deg \frac{d^k}{dx^k} = k.$$

Symplectic operators of the form (4.1) and the corresponding symplectic forms on the loop space $L(M)$

(4.2) $$\omega(\xi, \eta) = \int_\gamma \xi^i (A_{ij}^{[m]} \eta^j) \, dx$$

were introduced and studied by the author in [9, 38, 39].

It can be proved easily that for the symplectic form (4.2) the coefficients of the operator (4.1) are transformed under changes of coordinates $u^i = u^i(v)$ on the N-dimensional manifold M as differential geometry objects. Under the assumption of nondegeneracy of the higher coefficient $a_{ij}^{[m]}(u)$, the conditions reflecting the symplecticity of the operator (4.1) are in their essence certain natural differential geometry type restrictions imposed on connections defined on the manifold M equipped with the metric $a_{ij}^{[m]}$ (symmetric for any odd m, $m = 2k + 1$, or skew-symmetric for any even m, $m = 2k$) and the coefficients can be expressed in an invariant form via the metric, the curvature and the torsion of the connections. This allows us to obtain an invariant geometric description of the symplectic operators (4.1).

DEFINITION 4.1. Symplectic operators of the form (4.1) are called *symplectic operators of differential geometry type*.

REMARK 4.1. Poisson structures of the form (4.1) were introduced by Dubrovin and Novikov [3] as a natural generalization of Poisson structures of hydrodynamic type ($m = 1$). The case $m = 2$ for Poisson structures of the form (4.1) was studied in [44–46].

For $m = 0$ the symplecticity conditions are

$$a_{ij}^{[0]}(u) = -a_{ji}^{[0]}(u), \qquad \sum_{(i,j,k)} \frac{\partial a_{ij}^{[0]}}{\partial u^k} = 0,$$

where the sum is taken with respect to all cyclic permutations of i, j, k. This means that

$$\Omega_0 = a_{ij}^{[0]}(u) \, du^i \wedge du^j$$

is an arbitrary closed 2-form on M. If $\det a_{ij}^{[0]} \neq 0$, we have the symplectic structure Ω_0 on M and the symplectic structure

$$\omega(\xi, \eta) = \int_\gamma \Omega_0(\xi, \eta) \, dx$$

on the loop space $L(M)$.

The case $m = 1$ corresponds exactly to the symplectic forms (3.4). It is also possible to get a complete description of the matrix symplectic operators (4.1) for $m = 2$ [9]. In this case we deal with the operators of the form

$$(4.3) \quad A_{ij}^{[2]} = a_{ij}^{[2]}(u) \frac{d^2}{dx^2} + b_{ijk}^{[2]}(u) u_x^k \frac{d}{dx} + c_{ijkl}^{[2]}(u) u_x^k u_x^l + d_{ijk}^{[2]}(u) u_{xx}^k.$$

LEMMA 4.1. *Operator* (4.3) *is symplectic if and only if the following conditions are valid*:

$$(4.4) \quad a_{ij}^{[2]} = -a_{ji}^{[2]},$$

$$(4.5) \quad \frac{\partial a_{ij}^{[2]}}{\partial u^k} = \frac{1}{2}(b_{ijk}^{[2]} - b_{jik}^{[2]}),$$

$$(4.6) \quad d_{ijk}^{[2]} = \frac{1}{2}(b_{ijk}^{[2]} + b_{kji}^{[2]} - b_{kij}^{[2]}),$$

$$(4.7) \quad c_{ijkl}^{[2]} + c_{ijlk}^{[2]} = \frac{\partial d_{ijk}^{[2]}}{\partial u^l} + \frac{\partial d_{ijl}^{[2]}}{\partial u^k} + \frac{\partial d_{lik}^{[2]}}{\partial u^j} - \frac{\partial d_{ljk}^{[2]}}{\partial u^i}.$$

THEOREM 4.1 [9]. *Under the assumption* $\det a_{ij}^{[2]}(u) \neq 0$, *symplectic operators of the form* (4.3) *exist if and only if* M *is an even-dimensional manifold with an almost symplectic structure*.

Recall that a manifold with an almost symplectic structure is an even-dimensional M endowed with a nondegenerate skew-symmetric metric $g_{ij}(u)$.

Let $a_{ij}^{[2]}(u) = g_{ij}(u)$. Consider an arbitrary symplectic operator of the form (4.2) and introduce the coefficients $\Gamma_{jk}^i(u)$ by the relation

$$b_{ijk}^{[2]}(u) = 2 g_{is}(u) \Gamma_{jk}^s(u).$$

LEMMA 4.2 [9]. *The coefficients* $\Gamma_{jk}^i(u)$ *define a symplectic connection on* (M, g_{ij}), *i.e., a differential geometry connection compatible with the almost symplectic structure of the manifold*:

$$\nabla_k g_{ij} \equiv \frac{\partial g_{ij}}{\partial u^k} - g_{is} \Gamma_{jk}^s - g_{js} \Gamma_{ik}^s = 0.$$

It turns out that any symplectic connection $\Gamma_{jk}^i(u)$ on a manifold with an almost symplectic structure (M, g_{ij}) generates a unique symplectic operator (4.3).

THEOREM 4.2 [9]. *If* (M, g_{ij}) *is a manifold with an almost symplectic structure, then there exists a one-to-one correspondence between the symplectic connections* Γ_{jk}^i *on* (M, g_{ij}) *and closed 2-forms* (4.2), $m = 2$, *on the space* $L(M)$ *of smooth loops on* M. *This correspondence is expressed by the formula*

$$(4.8) \quad \omega(\xi, \eta) = \int_{S^1} \left\{ \langle \xi, \nabla_v^2 \eta \rangle + \frac{1}{2} \langle v, R(\xi, \eta) v \rangle + \left\langle v, \sum_{(\eta, \xi, v)} R(\eta, \xi) v \right\rangle \right. $$
$$\left. + \langle T(v, \xi), T(v, \eta) \rangle + \langle T(\xi, \eta), \nabla_v v \rangle \right\} dx,$$

where $\langle \xi, \eta \rangle = g_{ij}\xi^i\eta^j$, the sum $\sum_{(\eta,\xi,v)}$ is taken over all cyclic permutations of the elements (η, ξ, v), v is the velocity vector of the loop $\gamma(x)$, i.e., $v^i = u^i_x$, ∇_v is the covariant derivation along γ, $[T(\xi, \eta)]^i = T^i_{kl}\xi^k\eta^l$, $T^i_{jk} \equiv \Gamma^i_{jk} - \Gamma^i_{kj}$ is the torsion tensor of the connection and $[R(\xi, \eta)\zeta]^i = R^i_{jkl}\xi^k\eta^l\zeta^j$,

$$R^i_{jkl} = -\frac{\partial \Gamma^i_{jl}}{\partial u^k} + \frac{\partial \Gamma^i_{jk}}{\partial u^l} - \Gamma^i_{pk}\Gamma^p_{jl} + \Gamma^i_{pl}\Gamma^p_{jk}$$

is the curvature tensor of the connection.

The formula (4.8) gives the general expression for closed 2-forms (4.2), $m = 2$, and also describes all the symplectic operators of the form (4.3).

On symplectic manifolds there is a special class of symmetric symplectic connections satisfying the condition $T^i_{jk} = 0$. Such connections exist on (M, g_{ij}) if and only if the manifold (M, g_{ij}) is indeed symplectic, i.e., $(dg)_{ijk} = 0$.

COROLLARY 4.1 [9]. *If a symplectic connection $\Gamma^i_{jk}(u)$ is symmetric, then the closed 2-form (4.8) can be presented in the form*

$$(4.9) \qquad \omega(\xi, \eta) = \int_{S^1} \left\{ \langle \xi, \nabla^2_v \eta \rangle + \frac{1}{2}\langle v, R(\xi, \eta)v \rangle \right\} dx.$$

This formula describes the most natural class of homogeneous symplectic forms of the second order on loop spaces of symplectic manifolds.

It is worth noting that the class of symplectic connections on manifolds with almost symplectic structures is rich enough. Indeed, any such manifold has infinitely many distinct symplectic connections and correspondingly infinitely many distinct symplectic forms (4.8) on its loop space. Similarly, any symplectic manifold has infinitely many symmetric symplectic connections and the corresponding symplectic forms (4.9) on the loop space.

§5. Lagrangian systems and symplectic structures on loop spaces

In this section we consider two-dimensional Lagrangian systems generated by actions of the form ([14, 12])

$$(5.1) \qquad S = \int (g_i(x, u, u_x, \ldots)u^i_t + h(x, u, u_x, \ldots))\, dx\, dt,$$

where g_i, h are arbitrary functions depending on the independent spatial variable x, the fields $u^i(x, t)$, and their derivatives $u^i_{(n)} = \partial^n u^i / \partial x^n$ with respect to x.

The corresponding Lagrangian system $\delta S / \delta u^i = 0$ has the form

$$(5.2) \qquad M_{ij} u^j_t = \frac{\delta H}{\delta u^i},$$

where

$$(5.3) \qquad M_{ij} = \frac{\partial g_i}{\partial u^j_{(n)}} \frac{d^n}{dx^n} - (-1)^n \frac{d^n}{dx^n} \circ \frac{\partial g_j}{\partial u^i_{(n)}}$$

and

$$H = \int h\, dx.$$

LEMMA 5.1 [12, 14]. *The differential operator M_{ij} is symplectic and the bilinear form*

(5.4) $$\omega(\xi, \eta) = \int_{S^1} \xi^i M_{ij} \eta^j \, dx,$$

where $\xi, \eta \in T_\gamma L(M)$, is a symplectic form on the loop space $L(M)$.

In other words, the inverse operator $K^{ij} = (M^{-1})^{ij}$ such that $K^{ij} M_{jk} = \delta_k^i$ always gives a nonlocal Poisson bracket

(5.5) $$\{u^i(x), u^j(y)\} = K^{ij}[u(x)]\delta(x - y)$$

and hence we always have the following nonlocal Hamiltonian representation for Lagrangian systems generated by the actions (5.1):

(5.6) $$u_t^i = K^{ij} \frac{\delta H}{\delta u^j} \equiv \{u^i(x), H\}.$$

The corresponding symplectic representation for these Lagrangian systems has the form ([14])

(5.7) $$\omega(\delta u, u_t) = \delta H,$$

where the relation (5.7) is valid for arbitrary variations $\delta u^i(x)$ of the fields $u^i(x)$, H is a functional on the loop space $L(M)$.

Some special actions of the form (5.1) generate important nonlinear equations of mathematical physics and field theory such as that in nonlinear sigma models (see §3 above), Monge–Ampère equations, some systems of hydrodynamic type, the Krichever–Novikov equation and so on; the corresponding symplectic representations are very useful and effective for investigation of integrability and the procedures of averaging. Of course, it is easy to generalize this construction for arbitrary degenerate Lagrangian quasilinear with respect to higher t-derivatives of the fields $u^i(x, t)$.

Let us consider some examples.

EXAMPLE 5.1. The Korteweg–de Vries equation (KdV).
Consider the following simple action of the type (5.1):

$$S = \int \left(\frac{1}{2} u_x u_t - \frac{1}{2} u_{xx}^2 - u_x^3 \right) dx \, dt.$$

The symplectic representation (5.2)–(5.3) for the corresponding Lagrangian equation $\delta S/\delta u = 0$ has the form

$$\frac{d}{dx}(u_t) = -\frac{\delta H}{\delta u}, \qquad H = \int \left(u_x^3 + \frac{1}{2} u_{xx}^2 \right) dx.$$

After the field transformation $v(x) = u_x(x)$, we obtain the usual Hamiltonian representation of the KdV equation:

$$v_t = \frac{d}{dx} \frac{\delta H}{\delta v}, \qquad H = \int \left(v^3 + \frac{1}{2} v_x^2 \right) dx, \qquad v_t = 6vv_x - v_{xxx}.$$

REMARK 5.1. Note that if $v^i(x) = u_x^i(x)$, then for any functional H the relation

$$\frac{\delta H}{\delta u^i} = -\frac{d}{dx} \frac{\delta H}{\delta v^i}$$

is valid for the variation derivatives.

EXAMPLE 5.2 [10]. The Krichever–Novikov equation (KN).
Consider the special action of the form (5.1)

$$(5.8) \qquad S = \int \left(\frac{1}{2} \frac{u_t}{u_x} - \frac{1}{2} \frac{u_{xx}^2}{u_x^2} - \frac{1}{3} \frac{R(u)}{u_x^2} \right) dx\, dt,$$

where $R(u) = a_3 u^3 + a_2 u^2 + a_1 u + a_0$ is an arbitrary polynomial of degree three, $a_i = \text{const}$, $i = 0, 1, 2, 3$.

In this case the Lagrangian equation $\delta S/\delta u = 0$ has the symplectic representation (5.2)–(5.3) with the symplectic operator

$$(5.9) \qquad M = -\frac{1}{u_x} \frac{d}{dx} \circ \frac{1}{u_x},$$

$$(5.10) \qquad \left(\frac{1}{u_x} \frac{d}{dx} \frac{1}{u_x} \right) u_t = \frac{\delta H}{\delta u},$$

$$(5.11) \qquad H = \int \left(\frac{1}{2} \frac{u_{xx}^2}{u_x^2} + \frac{1}{3} \frac{R(u)}{u_x^2} \right) dx.$$

The formulas (5.10)–(5.11) give the well-known symplectic (and Hamiltonian) representation ([20]) for the Krichever–Novikov equation ([21])

$$(5.12) \qquad u_t = u_{xxx} - \frac{3}{2} \frac{u_{xx}^2}{u_x} + \frac{R(u)}{u_x},$$

which is a completely integrable equation with an infinite family of conservation laws, a representation of zero curvature with parameter on an elliptic curve, and plays an important role in the contemporary theory of integrable nonlinear systems of mathematical physics.

We recall that the equation (5.12) first appeared in [21] (in slightly different form, see also [22, 23]) as the deformation equation (by virtue of the Kadomtsev–Petviashvili (KP) equation) for the Tyurin parameters corresponding to the general solution of the commutation equation for ordinary differential operators with the spectral curve of genus $g = 1$ and fixed rank $l = 2$ of the vector bundle of common eigenfunctions over the spectral curve. (Note that the corresponding integrable Boussinesq type system of deformations of the Tyurin parameters by virtue of the KP equation for the general solution of the commutation equation of ordinary differential operators of genus $g = 1$ and rank $l = 3$ was obtained in the author's papers [24, 25, 13]). Correspondingly, any solution of the KN equation determines a solution of KP by an explicit formula [21, 22]. It is shown in [26, 27] that among all nonlinear equations (up to local changes of the field variable $u = u(v)$) of Korteweg–de Vries type

$$u_t = u_{xxx} + f(u, u_x, u_{xx})$$

having infinite series of local conservation laws, only the KN equation (in the generic case, i.e., when the polynomial $R(u)$ has no multiple roots) cannot be reduced to the KdV equation by differential substitutions of the form

$$u = F(v, v_x, \ldots, v_{(m)}).$$

However, the question of a possible connection between the KN and KdV equations via more complicated transformations is still unsolved.

The KN equation appeared as the first nonformal example of a system whose Hamiltonian structure is defined by a nontrivial differential symplectic operator. The action (5.8) gives a Lagrangian representation for the KN equation ([10]) and explains the natural origin of the symplectic operator (5.9).

The KN equation (5.12) is related to the Hamiltonian equation ([10])

$$v_t = -\left(\frac{d}{dz}\right)^{-1} \frac{\delta H}{\delta v}, \tag{5.13}$$

$$H = -\int \left[\frac{1}{2} \frac{v_{zz}^2}{v_z^3} + \frac{1}{3} R(z) v_z^3\right] dz, \tag{5.14}$$

by the point transformation of hodograph type

$$\begin{cases} x = v, \\ u = z. \end{cases} \tag{5.15}$$

One can assume that $v(z)$ is a new scalar field which is the inverse of the field $u(x)$, i.e., $u(v(z)) = z$ and $v(u(x)) = x$.

Introducing the new field $w(z) = v_z(z)$, we obtain a canonical Hamiltonian representation of the Krichever–Novikov equation (5.12) ([10]):

$$w_t = \frac{d}{dz} \frac{\delta H}{\delta w}, \tag{5.16}$$

$$H = -\int \left[\frac{1}{2} \frac{w_z^2}{w^3} + \frac{1}{3} R(z) w^3\right] dz. \tag{5.17}$$

Note that the Hamiltonian H is not invariant with respect to translations in z.

As it was shown by author in [10], the KN equation is also simply related with the canonical Hamiltonian system

$$\begin{cases} q_t = \delta F/\delta p, \\ p_t = -\delta F/\delta q, \end{cases} \tag{5.18}$$

where the Hamiltonian $F[p, q]$ in the conjugate variables $p(z), q(z)$ has the form

$$F[p, q] = \int \left(p(z)\left(\frac{q_{zzz}}{q_z^3} - \frac{3}{2} \frac{q_{zz}^2}{q_z^4} - R(z) q_z^2\right) + \frac{1}{3} R(z) q_z^3 - \frac{3}{2} \frac{q_{zz}^2}{q_z^3}\right) dz. \tag{5.19}$$

In [28–30] symplectic structures of the degenerate KN equation ($R(u) \equiv 0$) are considered (as it is known [27], in this case the KN equation reduces to the KdV equation by a differential substitution). The degenerate KN equation has the form

$$u_t = u_x S[u], \tag{5.20}$$

where

$$S[u] = \frac{u_{xxx}}{u_x} - \frac{3}{2} \frac{u_{xx}^2}{u_x^2}$$

is the Schwarzian of u, and has a compatible pair of differential symplectic operators found by Dorfman ([28–30, 12]) and a number of other interesting properties studied by various authors (see [31], for example). For the multicomponent analog of the degenerate KN equation (5.20) a multisymplectic representation was found by Fordy and Antonowicz [32].

It is interesting that as it was discovered by author and Ferapontov ([15], see also [16, 17] and §§7, 9 in this paper) the symplectic operator (5.9) and its multicomponent generalization

$$(5.21) \qquad M_{ij} = \frac{1}{K} \frac{1}{u_x^i} \frac{d}{dx} \circ \frac{1}{u_x^j},$$

$K = $ const, arises naturally in the Dubrovin–Novikov theory of Hamiltonian systems of hydrodynamic type developed in [1–4]. The Hamiltonian structure of these systems introduced and studied by Dubrovin and Novikov in [2] is generated by metrics of zero curvature (all these Hamiltonian systems of hydrodynamic type have a natural Lagrangian representation of the type considered above (5.1)–(5.3) and we shall consider them in §7). It was shown in [15] that with the help of the operator (5.21) one can adequately extend the theory of Hamiltonian systems of hydrodynamic type to metrics of arbitrary constant Riemannian curvature K. In particular, such Hamiltonian diagonalizable systems of hydrodynamic type will be integrable (they are semi-Hamiltonian in the sense of Tsarev [5, 6]). All systems of hydrodynamic type derived by averaging well-known integrable equations of soliton theory (KdV, Nonlinear Schrödinger, Sine-Gordon, and so on) have Hamiltonian structures generated by metrics of nonzero constant Riemannian curvature (along with structures of Dubrovin–Novikov type) (see [19]).

§6. Symplectic and Poisson structures of the Monge–Ampère equations

EXAMPLE 6.1. The Monge–Ampère equations.
Consider the following action of type (5.1) [50]:

$$(6.1) \qquad S = \int \left[\frac{1}{2} u_x^2 q_t - u_x q_x u_t - \frac{1}{2} q^2 u_{xx} + \Phi(x, t, u, u_x) \right] dx\, dt,$$

where $\Phi(x, t, u, u_x)$ is an arbitrary function.

The corresponding Lagrangian system $\delta S/\delta u^i = 0$ has natural symplectic representation (5.2)–(5.3) of the form

$$(6.2) \qquad M \begin{pmatrix} u \\ q \end{pmatrix}_t = \begin{pmatrix} \delta F/\delta u \\ \delta F/\delta q \end{pmatrix},$$

where

$$(6.3) \qquad M = \begin{pmatrix} q_x\, d/dx + d/dx \circ q_x & -u_{xx} \\ u_{xx} & 0 \end{pmatrix}$$

is a symplectic operator,

$$(6.4) \qquad F = \int \left[\frac{1}{2} q^2 u_{xx} - \Phi(x, t, u, u_x) \right] dx.$$

The symplectic system (6.2)–(6.4) is equivalent to the Monge–Ampère equation ([50])

(6.5) $$u_{xx}u_{tt} - (u_{xt})^2 = \delta\Phi/\delta u .$$

In the case of the Monge–Ampère equations, the Poisson bracket (5.5) is also local and the corresponding Hamiltonian operator $K^{ij} = (M^{-1})^{ij}$ has the form ([50])

(6.6) $$K = \begin{pmatrix} 0 & 1/u_{xx} \\ -1/u_{xx} & (q_x/u_{xx}^2)\,d/dx + d/dx \circ q_x/u_{xx}^2 \end{pmatrix}.$$

And what is more, this fact is valid for all local Hamiltonian operators of the form

(6.7) $$K = \begin{pmatrix} 0 & B \\ -B & A\,d/dx + A \circ d/dx \end{pmatrix},$$

where B and A are some functions of x, u, u_x, u_{xx}, ..., q, q_x, q_{xx}, ... ($B \neq 0$), which were considered by Nutku and Sarıoğlu in [50] in connection with Poisson structures of the Monge–Ampère equations. Namely, for any local Hamiltonian differential operator of the form (6.7) the corresponding inverse symplectic operator M is also local differential operator and has the form

(6.8) $$M = \begin{pmatrix} (A/B^2)\,d/dx + d/dx \circ A/B^2 & -1/B \\ 1/B & 0 \end{pmatrix}.$$

THEOREM 6.1. *The operator*

(6.9) $$M = \begin{pmatrix} [\frac{\partial g}{\partial u_x} - (\frac{\partial g}{\partial u_{xx}})_x]\frac{d}{dx} + \frac{d}{dx} \circ [\frac{\partial g}{\partial u_x} - (\frac{\partial g}{\partial u_{xx}})_x] & \frac{\partial g}{\partial q} \\ -\frac{\partial g}{\partial q} & 0 \end{pmatrix}$$

is a symplectic operator of the form (6.8) *for arbitrary function* $g(x, t, u, u_x, u_{xx}, q)$. *The corresponding local Hamiltonian operator K has the form* (6.7), *where*

(6.10) $$B = -\frac{1}{\partial g/\partial q}, \qquad A = \frac{\partial g/\partial u_x - (\partial g/\partial u_{xx})_x}{(\partial g/\partial q)^2} .$$

§7. Symplectic and nonlocal Poisson structures of homogeneous and nonhomogeneous systems of hydrodynamic type

EXAMPLE 7.1. Systems of hydrodynamic type.
Consider the following action of type (5.1):

(7.1) $$S = \int \left(\frac{1}{2} g_{ij} u_x^i u_t^j - h(u_x) \right) dx\,dt ,$$

where (g_{ij}) is a constant nondegenerate symmetric tensor, $h(v)$ is an arbitrary function.

The corresponding Lagrangian system $\delta S/\delta u^i = 0$ has the symplectic representation (5.2)–(5.3) of the form

$$(7.2) \qquad g_{ij} \frac{d}{dx}(u_t^j) = -\frac{\delta H}{\delta u^i}, \qquad H = \int h(u_x)\, dx.$$

After the field transformation $v^i(x) = u_x^i(x)$ (see Remark 5.1) we obtain a Hamiltonian system of hydrodynamic type in flat coordinates

$$(7.3) \qquad v_t^i = g^{ij} \frac{d}{dx} \frac{\delta H}{\delta v^j}, \qquad H = \int h(v)\, dx.$$

In other words, the action (7.1) gives the general class of Hamiltonian systems of hydrodynamic type (written in flat coordinates $(v^1, \ldots, v^N)$) introduced and studied by Dubrovin and Novikov [1–4].

We recall very briefly the basic notions and results on Poisson structures of hydrodynamic type that we shall need.

Consider the one-dimensional systems of hydrodynamic type, i.e., in other words, the one-dimensional evolution quasilinear systems of the first order

$$(7.4) \qquad u_t^i = v_j^i(u) u_x^j,$$

where $v_j^i(u)$ is an arbitrary $N \times N$ matrix function of $u = (u^1, \ldots, u^N)$, $u^i = u^i(x, t)$, $i = 1, \ldots, N$.

The Hamiltonian systems of hydrodynamic type considered by Dubrovin and Novikov in [1–4] have the form

$$(7.5) \qquad u_t^i = \{u^i, H\},$$

where H is a functional of hydrodynamic type, i.e.,

$$(7.6) \qquad H = \int h(u)\, dx,$$

and the Poisson bracket has the form

$$(7.7) \qquad \{u^i(x), u^j(y)\} = g^{ij}(u(x))\delta_x(x-y) + b_k^{ij}(u(x))u_x^k \delta(x-y)$$

(the Poisson bracket of Dubrovin–Novikov type [2]). It was shown in [2] that if $\det[g^{ij}(u)] \neq 0$ then the expression (7.7) gives a Poisson bracket if and only if

(1) $g^{ij}(u)$ is a metric of zero Riemannian curvature (i.e., simply a flat metric);
(2) $b_k^{ij}(u) = -g^{is}(u)\Gamma_{sk}^j(u)$, where $\Gamma_{sk}^j(u)$ are the coefficients of the differential geometry connection generated by the metric g^{ij}, i.e., the only symmetric connection compatible with the metric (the Levi–Civita connection).

Thus, there always exist local variables $v^i = v^i(u)$ in which the Poisson bracket (7.7) is simply constant:

$$\{v^i(x), v^j(y)\} = \varepsilon^i \delta^{ij} \delta_x(x-y), \tag{7.8}$$

where $\varepsilon^i = \pm 1$, $i = 1, \ldots, N$. Correspondingly, it is easy to give a Lagrangian description (7.1) of these systems. The Hamiltonian systems of hydrodynamic type have the form

$$u_t^i = [\nabla^i \nabla_j h(u)] u_x^j, \tag{7.9}$$

where ∇ is the covariant derivative generated by a zero curvature metric. As it was shown in [5, 6], if the Hamiltonian system of hydrodynamic type (7.9) has Riemann invariants (i.e., the matrix $v_j^i(u) = \nabla^i \nabla_j h(u)$ is diagonalizable), then it is integrable.

Multidimensional Poisson structures of hydrodynamic type were introduced and studied in [3, 7].

In [15] (see also [16, 17]) a nonlocal generalization of Hamiltonian theory of the systems of hydrodynamic type (7.4) connected with the nonlocal Poisson brackets of the form

$$\{u^i(x), u^j(y)\} = g^{ij}(u(x))\delta_x(x-y) + b_k^{ij}(u(x))u_x^k \delta(x-y) \\ + K u_x^i (d/dx)^{-1} u_x^j \delta(x-y) \tag{7.10}$$

was suggested.

It is easy to show that for any Hamiltonian functional H of hydrodynamic type ($H = \int h(u)\,dx$) the Poisson bracket of the form (7.10) also always generates a system of hydrodynamic type (7.4).

REMARK 7.1 [16]. The expression (7.10) is the most general form of Poisson bracket with the property of generating a system of hydrodynamic type (7.4) for any Hamiltonian H of hydrodynamic type (7.6).

THEOREM 7.1 [15]. *If* $\det[g^{ij}(u)] \neq 0$, *then expression* (7.10) *gives a Poisson bracket if and only if*
  (1) $g^{ij}(u)$ *is a metric of constant Riemannian curvature* K;
  (2) $b_k^{ij}(u) = -g^{is}\Gamma_{sk}^j(u)$, *where* $\Gamma_{sk}^j(u)$ *are the coefficients of the connection generated by the metric* g^{ij} (*the Levi–Civita connection*).

The corresponding Hamiltonian systems of hydrodynamic type have the form

$$u_t^i = [\nabla^i \nabla_j h(u) + K\delta_j^i h(u)] u_x^j, \tag{7.11}$$

where ∇ is the covariant derivative generated by a metric of constant Riemannian curvature K.

If $\det[g^{ij}(u)] = 0$, then the description of the nonlocal Poisson brackets (7.10) is much more complicated [16].

Note that the Poisson brackets (7.7) and (7.10) define natural differential geometry Poisson brackets on loop spaces of arbitrary flat and constant Riemannian curvature manifolds respectively, just like symplectic structures of differential geometry type do on loop spaces of arbitrary Riemannian manifolds [8, 9].

Now consider nonhomogeneous systems of hydrodynamic type

$$(7.12) \qquad u_t^i = v_j^i(u) u_x^j + f^i(u).$$

Local Poisson structures for the systems (7.12) studied by Dubrovin and Novikov [3, 4] also have a natural nonlocal generalization. We consider here nonlocal Hamiltonian nonhomogeneous systems of hydrodynamic type

$$(7.13) \qquad u_t^i = \{u^i, H\},$$

where H is a functional of hydrodynamic type (7.6), and the nonlocal nonhomogeneous Poisson bracket of hydrodynamic type has the form

$$(7.14) \quad \begin{aligned}\{u^i(x), u^j(y)\} &= g^{ij}(u(x))\delta_x(x-y) + b_k^{ij}(u(x))u_x^k\delta(x-y) \\ &+ Ku_x^i(d/dx)^{-1}u_x^j\delta(x-y) + \omega^{ij}(u(x))\delta(x-y).\end{aligned}$$

It is easy to show that the expression (7.14) is a Poisson bracket if and only if it is a sum of two compatible Poisson brackets (7.10) and (1.3).

THEOREM 7.2 [52]. *The Poisson brackets* (7.10) *and* (1.3) *are compatible if and only if the bivector* $\omega^{ij}(u)$ *is a Killing bivector on the manifold* (M, g^{ij}) *of constant Riemannian curvature* K, *i.e.*,

$$(7.15) \qquad \nabla^i \omega^{jk} + \nabla^j \omega^{ik} = 0,$$

where ∇ *is the covariant derivative generated by the metric* $g^{ij}(u)$ *of constant Riemannian curvature* K.

§8. On some integrable nonhomogeneous systems of hydrodynamic type

In this section we consider a special class of nonhomogeneous systems of hydrodynamic type with quadratic nonlinearity ([49, 48]):

$$(8.1) \qquad u_t^i = a^i u_x^i + \sum_{k,j} b_{jk}^i u^j u^k + \sum_k c_k^i u^k,$$

where summation over repeating indices is not assumed, a^i, b_{jk}^i, and c_k^i are constant tensors, $i, j, k = 1, \ldots, N$.

There are a number of well-known integrable systems among (8.1).

EXAMPLE 8.1. The N-wave equation.

For example, an integrable real-valued exact resonance system of parametric interaction of three wave packets in nonlinear optics ($N = 3$) is a special case of (8.1):

$$(8.2) \qquad \begin{cases} u_t^1 = a^1 u_x^1 - \varepsilon u^2 u^3, \\ u_t^2 = a^2 u_x^2 + \varepsilon u^1 u^3, \\ u_t^3 = a^3 u_x^3 + \varepsilon u^1 u^2, \end{cases}$$

where a^i, ε are some constants.

EXAMPLE 8.2. The KdV equation.

Consider the KdV equation (see Example 5.1) as an evolution system with respect to x:

(8.3)
$$\begin{cases} u^1_x = u^2, \\ u^2_x = u^3, \\ u^3_x = -u^1_t + 6u^1 u^2. \end{cases}$$

It was shown in [47] that the KdV system (8.3) is Hamiltonian with respect to some nonhomogeneous Poisson structures of hydrodynamic type (7.14), which are in fact induced by well-known Magri and Gardner brackets for KdV. After the local quadratic unimodular change of field variables [48]

(8.4)
$$\begin{cases} u^1 = (w^1 - w^3)/\sqrt{2}, \\ u^2 = w^2, \\ u^3 = (w^1 + w^3)/\sqrt{2} + (w^1 - w^3)^2, \end{cases}$$

we obtain a Hamiltonian representation for the KdV system (8.3) ([48]) generated by the simplest infinite dimensional Kac–Moody Lie algebra $\widehat{sl}(2)$ and some quadratic Hamiltonian H:

(8.5) $$w^i_x = M^{ij}(t) \frac{\delta H}{\delta w^j},$$

$$H = -\int [(w^1 - w^3)^2 - \sqrt{2}(w^1 + w^3)] dt,$$

(8.6)
$$(M^{ij}(t)) = \begin{pmatrix} 1 & 0 & 0 \\ 0 & -1 & 0 \\ 0 & 0 & -1 \end{pmatrix} \frac{d}{dt} + \begin{pmatrix} 0 & -2w^3(t) & 2w^2(t) \\ 2w^3(t) & 0 & 2w^1(t) \\ -2w^2(t) & -2w^1(t) & 0 \end{pmatrix},$$

where M is the Hamiltonian operator given by $\widehat{sl}(2)$ [48].

The second nonhomogeneous Poisson structure of hydrodynamic type for the KdV system (8.3) is compatible with (8.6) and has the following form [48]:

(8.7)
$$(L^{ij}(t)) = \frac{1}{2} \begin{pmatrix} 1 & 0 & 1 \\ 0 & 0 & 0 \\ 1 & 0 & 1 \end{pmatrix} \frac{d}{dt} + (w^1 - w^3) \begin{pmatrix} 0 & 1 & 0 \\ -1 & 0 & -1 \\ 0 & 1 & 0 \end{pmatrix}$$
$$+ \frac{1}{\sqrt{2}} \begin{pmatrix} 0 & 1 & 0 \\ -1 & 0 & 1 \\ 0 & -1 & 0 \end{pmatrix};$$

here the metric $g^{ij}(w)$ in (8.7) is degenerate (see (7.14)).

Correspondingly, the second Hamiltonian representation for the KdV system (8.3) has the form

$$w^i_x = L^{ij}(t) \frac{\delta H}{\delta w^j},$$

where H is the quadratic Hamiltonian

$$H = -\frac{1}{2} \int [(w^1)^2 - (w^2)^2 - (w^3)^2] \, dt \,.$$

The second Poisson structure (8.7) for the KdV system (8.3) is generated by the three-dimensional nilpotent non-Abelian Lie algebra $\mathcal{G}_0$ (it is a Lie algebra of type II according to the Bianchi classification of three-dimensional Lie algebras) and 2-cocycle on its loop algebra. Note that any nilpotent non-Abelian Lie algebra contains a subalgebra isomorphic to $\mathcal{G}_0$.

Consider the special system of type (8.1):

(8.8) $$u_t^i = a^i u_x^i + u^i \sum_k (a^i - a^k) u^k \,,$$

where $a^i \neq a^j$, if $i \neq j$, $i, j = 1, \ldots, N$.

THEOREM 8.1. *The system* (8.8) *is integrable and equivalent to some integrable homogeneous diagonal Hamiltonian system of hydrodynamic type by a combination of a reciprocal transformation and some changes of the fields and independent variables.*

First of all, let us introduce new field variables $w^i(x)$ by the relation $u^i = e^{w^i}$. Then our system (8.8) takes the following form:

(8.9) $$w_t^i = a^i w_x^i + \sum_k (a^i - a^k) e^{w^k} \,.$$

Consider the so-called reciprocal transformations of the system (8.9), or, in other words, transformations of the independent variables x and t corresponding to a solution $w(x, t)$ of the system (see also [42, 51]):

(8.10) $$\begin{aligned} dx' &= \varphi_1(x, t, w) \, dx - \psi_1(x, t, w) \, dt \,, \\ dt' &= \varphi_2(x, t, w) \, dx - \psi_2(x, t, w) \, dt \,, \end{aligned}$$

where
(8.11)
$$\frac{\partial \varphi_i(x, t, w)}{\partial t} + \frac{\partial \psi_i(x, t, w)}{\partial x} = 0 \,, \quad i = 1, 2, \quad \Delta = \varphi_1 \psi_2 - \varphi_2 \psi_1 \neq 0 \,.$$

Here $w^i(x, t)$ is an arbitrary solution of the system (8.9), φ_i, ψ_i, $i = 1, 2$, are generated by some conservation laws (8.11) of the system (8.9).

Let us consider the following two conservation laws of the system (8.9):
1) $\varphi_1 = -\sum_k e^{w^k}$, $\psi_1 = \sum_k a^k e^{w^k}$,
2) $\varphi_2 = 0$, $\psi_2 = 1$.
Using the relations

$$w_x^i = w_{x'}^i \varphi_1 + w_{t'}^i \varphi_2 \,, \qquad w_t^i = -w_{x'}^i \psi_1 - w_{t'}^i \psi_2$$

after the corresponding reciprocal transformation (8.10)–(8.11), we obtain the following nonhomogeneous system of hydrodynamic type:

$$(8.12) \qquad w^i_{t'} = \left(\sum_k (a^i - a^k) e^{w^k}\right)(w^i_{x'} - 1).$$

After the transformations

$$v^i = w^i - x', \qquad x'' = -e^{-x'},$$

we obtain a diagonal weakly nonlinear semi-Hamiltonian homogeneous system of hydrodynamic type which is integrable by the Tsarev theorem [5, 6]:

$$(8.13) \qquad v^i_{t'} = \left(\sum_k (a^i - a^k) e^{v^k}\right) v^i_{x''}.$$

§9. Killing–Poisson bivectors on Riemannian manifolds and integrable bi-Hamiltonian hierarchies of Heisenberg magnet type

The Killing–Poisson bivector $\omega^{ij}(u)$ on an N-dimensional Riemannian manifold (M, g_{ij}) is (see [53]) a skew-symmetric tensor on M ($\omega^{ij} = -\omega^{ji}$) which satisfies well-known relations (1.1) (the Jacobi identity for the Poisson bivector $\omega^{ij}(u)$) and (7.15) (the identity for the Killing bivector $\omega^{ij}(u)$ on the Riemannian manifold (M, g_{ij}) with the Levi–Civita connection ∇ generated by the metric g_{ij}). It was shown in §7 (see also [52]) that Killing–Poisson bivectors on the manifolds of constant Riemannian curvature K define natural nonlocal (if $K \neq 0$) Poisson structures for nonhomogeneous systems of hydrodynamic type. Here we give a complete description of the Killing–Poisson bivectors on the manifolds of constant Riemannian curvature in terms of Lie algebras with invariant scalar products and show that the compatible pairs of Poisson structures given by these bivectors also generate integrable bi-Hamiltonian hierarchies of Heisenberg magnet type [52].

It is easy to describe all Killing–Poisson bivectors on flat spaces or, in other words, spaces of zero Riemannian curvature. In this case we shall consider flat coordinates $(v^1, \ldots, v^N)$ in which the metric g_{ij} is constant. A tensor $\omega^{ij}(v)$ is a Killing–Poisson bivector on a flat space (M, g_{ij}) if and only if in flat coordinates $(v^1, \ldots, v^N)$ the following conditions are valid:

$$(9.1) \qquad \omega^{ij}(v) = c^{ij}_k v^k + d^{ij},$$

where c^{ij}_k, d^{ij} are constants such that

(1) c^{ij}_k are the structural constants of a Lie algebra with invariant scalar product $\langle \, , \, \rangle$ given by the metric g_{ij}: $\langle \operatorname{ad} X(Y), Z \rangle = -\langle Y, \operatorname{ad} X(Z) \rangle$;

(2) d^{ij} is a 2-cocycle on this Lie algebra:

$$d^{ij} = -d^{ji}, \qquad \sum_{(i,j,k)} c^{ij}_s d^{sk} = 0,$$

where the sum is taken with respect to all cyclic permutations of the indices i, j, k.

For example, any semi-simple Lie algebra gives a Killing–Poisson bivector on flat space (we must take the Killing metric on the Lie algebra as the metric for the corresponding flat space). This description coincides with the classification of local nonhomogeneous Poisson brackets of hydrodynamic type [3, 4].

Let us consider an arbitrary metric g^{ij} of constant Riemannian curvature K in canonical variables $u^1, \ldots, u^N$:

$$(g^{ij}(u)) = (\lambda(u))^2 \begin{pmatrix} \varepsilon_1 & \cdots & 0 \\ \vdots & \ddots & \vdots \\ 0 & \cdots & \varepsilon_N \end{pmatrix}, \quad \varepsilon_i = \pm 1, \quad \lambda(u) = 1 + \frac{K}{4} \sum_i \varepsilon_i (u^i)^2.$$

THEOREM 9.1. *For $N = 2$ a tensor $\omega^{ij}(u)$ is a Killing–Poisson bivector on some space (M, g_{ij}) of constant Riemannian curvature K if and only if in canonical variables (u^1, u^2)*

(9.2) $$(\omega^{ij}(u)) = c\lambda^2(u) \begin{pmatrix} 0 & 1 \\ -1 & 0 \end{pmatrix},$$

where c is an arbitrary constant.

As it was shown in [52] (see also §7) any Killing–Poisson bivector $\omega^{ij}(u)$ on the space (M, g_{ij}) of constant Riemannian curvature K define a pair of compatible Hamiltonian operators

(9.3) $$M_1^{ij} = g^{ij}(u) \frac{d}{dx} - g^{is}(u) \Gamma^j_{sk}(u) u_x^k + K u_x^i \left(\frac{d}{dx}\right)^{-1} u_x^j,$$

(9.4) $$M_2^{ij} = \omega^{ij}(u),$$

which generate the integrable hierarchy of the generalized N-component Heisenberg magnet equations

$$S_t = [S, S_{xx}], \quad S^2 = 1,$$

where S is an $(N+1)$-vector and $[\ ,\ ]$ is the commutator in an $(N+1)$-dimensional Lie algebra equipped with an invariant inner product. The classical Heisenberg magnet corresponds to the simplest case of the two-dimensional sphere ($N = 2$). For the two-dimensional sphere $S^{1^2} + S^{2^2} + S^{3^2} = 1$ in the stereographic projection coordinates

$$S^1 = u^1/P, \quad S^2 = u^2/P, \quad S^3 = (P-1)/P,$$

where $P = (u^{1^2} + u^{2^2} + 1)/2$, the metric is expressed as

$$(g_{ij}) = \frac{1}{P^2} \begin{pmatrix} 1 & 0 \\ 0 & 1 \end{pmatrix}.$$

The corresponding nonlocal Hamiltonian operator (9.3) generated by the metric has the form

(9.5) $$M_1 = P^2 \begin{pmatrix} d & 0 \\ 0 & d \end{pmatrix} + P \begin{pmatrix} u^1 u_x^1 + u^2 u_x^2 & u^1 u_x^2 - u^2 u_x^1 \\ u^2 u_x^1 - u^1 u_x^2 & u^1 u_x^1 + u^2 u_x^2 \end{pmatrix} + \begin{pmatrix} u_x^1 d^{-1} \circ u_x^1 & u_x^1 d^{-1} \circ u_x^2 \\ u_x^2 d^{-1} \circ u_x^1 & u_x^2 d^{-1} \circ u_x^2 \end{pmatrix}.$$

By Theorem 9.1, the unique (up to a constant factor) Killing–Poisson bivector on the two-dimensional sphere in these coordinates has the form

$$(9.6) \qquad M_2 = (\omega^{ij}(u)) = \begin{pmatrix} 0 & -P^2 \\ P^2 & 0 \end{pmatrix}.$$

Consider the recursion operator $R = M_1(M_2)^{-1}$ corresponding to the compatible Hamiltonian pair (9.5), (9.6) and apply it to the simplest translation flow

$$\begin{pmatrix} u^1 \\ u^2 \end{pmatrix}_t = \begin{pmatrix} u^1 \\ u^2 \end{pmatrix}_x.$$

We obtain a new system

$$\begin{pmatrix} u^1 \\ u^2 \end{pmatrix}_t = R \begin{pmatrix} u^1 \\ u^2 \end{pmatrix}_x,$$

or

$$(9.7) \qquad \begin{cases} u_t^1 = u_{xx}^2 + (u^2(u_x^1)^2 - 2u^1 u_x^1 u_x^2 - u^2(u_x^2)^2)/P, \\ u_t^2 = -u_{xx}^1 - (u^1(u_x^2)^2 - 2u^2 u_x^1 u_x^2 - u^1(u_x^1)^2)/P, \end{cases}$$

which coincides with the classical Heisenberg magnet equations

$$\vec{S}_t = \vec{S} \times \vec{S}_{xx}, \qquad \vec{S}^2 = 1.$$

THEOREM 9.2 [52]. *The compatible Hamiltonian pair (9.5), (9.6) generates the hierarchy of the Heisenberg magnet equations.*

Explicit bi-Hamiltonian representation of the Heisenberg magnet equations (9.7) has the form

$$(9.8) \qquad \begin{pmatrix} u^1 \\ u^2 \end{pmatrix}_t = M_1 \begin{pmatrix} \delta G/\delta u^1 \\ \delta G/\delta u^2 \end{pmatrix} = M_2 \begin{pmatrix} \delta H/\delta u^1 \\ \delta H/\delta u^2 \end{pmatrix}$$

with Hamiltonians

$$(9.9) \qquad G = \int \frac{u^2 u_x^1 - u^1 u_x^2}{(2P-1)P} \, dx, \qquad H = \frac{1}{2} \int \frac{(u_x^1)^2 + (u_x^2)^2}{P^2} \, dx.$$

Let us give a complete description of the Killing–Poisson bivectors on the N-dimensional sphere [52]

$$(9.10) \qquad \sum_{k=1}^{N+1} (S^k)^2 = 1.$$

Let c_k^{ij} be the structural constants of an $(N+1)$-dimensional Lie algebra with invariant scalar (Euclidean) product; we assume that $c_k^{ij} + c_i^{kj} = 0$. Consider the Lie-Poisson bivector $\Omega^{ij} = c_k^{ij} S^k$ and restrict it on the sphere (9.10). After this restriction, we obtain a Killing–Poisson bivector on the N-dimensional sphere. The converse is also true.

THEOREM 9.3 [52]. *Each Killing–Poisson bivectors on N-dimensional sphere can be obtained by the above construction from a Lie algebra with an invariant scalar product.*

In coordinates $(u^1, \ldots, u^N)$ of the stereographic projection
$$\tag{9.11} S^1 = \frac{u^1}{P}, \quad \ldots, \quad S^N = \frac{u^N}{P}, \quad S^{N+1} = \frac{P-1}{P}, \quad P = \frac{1}{2}\left(\sum_{s=1}^{N}(u^s)^2 + 1\right)$$

we have

$$\tag{9.12} \omega^{ij} = \Omega|_{S^N} = P\sum_{s=1}^{N}(c_{N+1}^{si}u^s u^j - c_{N+1}^{sj}u^s u^i + c_s^{ij}u^s) + c_{N+1}^{ij}(P-1)P.$$

For example, the Killing–Poisson bivector (9.6) on the two-dimensional sphere is a result of restriction of the Lie–Poisson bivector

$$\Omega = \begin{pmatrix} 0 & S^3 & -S^2 \\ -S^3 & 0 & S^1 \\ S^2 & -S^1 & 0 \end{pmatrix}.$$

In order to obtain a complete description of the Killing–Poisson bivectors on arbitrary spaces of constant Riemannian curvature, it is necessary to consider Lie algebras equipped with arbitrary invariant scalar products (of an arbitrary signature) and apply the same construction. The corresponding Killing–Poisson bivector gives a compatible Hamiltonian pair (9.3), (9.4) which generates a bi-Hamiltonian integrable N-component system of the Heisenberg magnet type.

§10. Nonlinear partial differential equations of associativity in 2D topological field theories and nondiagonalizable integrable systems of hydrodynamic type

In this section we consider the so-called nonlinear partial differential equations of associativity in 2D topological field theories (see [54–57]) and give their description as integrable nondiagonalizable weakly nonlinear systems of hydrodynamic type. For systems of this type, the corresponding general differential geometry integrability theory connected with Poisson structures of hydrodynamic type can be developed.

We recall very briefly, following Dubrovin [54], the basic mathematical concepts connected with the Witten–Dijkgraaf–E. Verlinde–H. Verlinde (WDVV) system that first appeared in two-dimensional topological field theories [56, 57], and its relations to the Dubrovin type equations of associativity.

Consider a function $F(t)$, $t = (t^1, \ldots, t^N)$, such that the following three conditions are satisfied for its third derivatives

$$c_{\alpha\beta\gamma}(t) = \frac{\partial^3 F(t)}{\partial t^\alpha \partial t^\beta \partial t^\gamma} :$$

(1) *normalization*, i.e., $\eta_{\alpha\beta} = c_{1\alpha\beta}(t)$ is a constant nondegenerate matrix;

(2) *associativity*, i.e., the functions $c_{\alpha\beta}^{\gamma}(t) = \eta^{\gamma\varepsilon} c_{\varepsilon\alpha\beta}(t)$ for any t define the structure of an associative algebra A_t in the N-dimensional space with a basis $e_1, \ldots, e_N$:

$$e_\alpha \cdot e_\beta = c_{\alpha\beta}^{\gamma}(t) e_\gamma;$$

(3) $F(t)$ is a quasihomogeneous function of its variables:

$$F(c^{d_1} t^1, \ldots, c^{d_N} t^N) = c^{d_F} F(t^1, \ldots, t^N)$$

for any nonzero c and for some numbers $d_1, \ldots, d_N, d_F$.

The resulting system of equations for $F(t)$ is called the *Witten–Dijkgraaf–E. Verlinde–H. Verlinde* (WDVV) system [56, 57] (see also [54–55]). It was shown by Dubrovin [54] that solutions of the WDVV system can be reduced by a linear change of coordinates to two special types:

(1) in the physically most important case,

$$(10.1) \quad F(t) = \frac{1}{2}(t^1)^2 t^N + \frac{1}{2} t^1 \sum_{\alpha=2}^{N-1} t^\alpha t^{N-\alpha+1} + f(t^2, \ldots, t^N),$$

for some function $f(t^2, \ldots, t^N)$;

(2) in a certain special case,

$$(10.2) \quad F(t) = \frac{c}{6}(t^1)^3 + \frac{1}{2} t^1 \sum_{\alpha=1}^{N-1} t^\alpha t^{N-\alpha+1} + f(t^2, \ldots, t^N)$$

for a nonzero constant c.

If $N = 3$ (first nontrivial case for the condition of associativity in the algebra A_t), then for solutions of the first type (10.1) of the WDVV system, the associativity condition in the algebra A_t gives the following nonlinear partial differential equation for the function $f(x, y)$ [54]:

$$(10.3) \quad f_{xxy}^2 = f_{yyy} + f_{xxx} f_{xyy}.$$

Let us introduce new variables a, b, c, d such that

$$(10.4) \quad a = f_{xxx}, \quad b = f_{xxy}, \quad c = f_{xyy}, \quad d = f_{yyy}.$$

The compatibility conditions have the form

$$(10.5) \quad \begin{cases} a_y = b_x, \\ b_y = c_x, \\ c_y = d_x. \end{cases}$$

Besides, we have the relation

$$(10.6) \quad d = b^2 - ac$$

from (10.3).

So, the equation (10.3) is equivalent to the following homogeneous system of hydrodynamic type

$$
(10.7) \qquad \begin{pmatrix} a \\ b \\ c \end{pmatrix}_y = \begin{pmatrix} 0 & 1 & 0 \\ 0 & 0 & 1 \\ -c & 2b & -a \end{pmatrix} \begin{pmatrix} a \\ b \\ c \end{pmatrix}_x.
$$

The system (10.7) is a nondiagonalizable weakly nonlinear homogeneous system of hydrodynamic type. In fact, the integrability of the system (10.7) follows from Dubrovin's results [55], but for the representation (10.7) it can be proved directly by the usual Hamiltonian and differential geometry methods for hydrodynamic type systems.

THEOREM 10.1 [58]. *Equation (10.3) is equivalent to the integrable nondiagonalizable weakly nonlinear homogeneous system of hydrodynamic type (10.7).*

Similarly, for $N = 3$ and for the special solutions (10.2) of second type to the WDVV system, the associativity condition for algebra A_t gives the following Dubrovin equation for the function $f(x, y)$ [54]:

$$
(10.8) \qquad f_{xxx} f_{yyy} - f_{xxy} f_{xyy} = 1.
$$

After introducing new variables a, b, c, d (10.4) from (10.8), we obtain the relation

$$
(10.9) \qquad d = \frac{1 + bc}{a}.
$$

The compatibility conditions (10.5) and the relation (10.9) generate the following homogeneous system of hydrodynamic type

$$
(10.10) \qquad \begin{pmatrix} a \\ b \\ c \end{pmatrix}_y = \begin{pmatrix} 0 & 1 & 0 \\ 0 & 0 & 1 \\ -(1+bc)/a^2 & c/a & b/a \end{pmatrix} \begin{pmatrix} a \\ b \\ c \end{pmatrix}_x.
$$

THEOREM 10.2 [58]. *Equation (10.8) is equivalent to the integrable nondiagonalizable weakly nonlinear homogeneous system of hydrodynamic type (10.10).*

Acknowledgement. This work was partially supported by the Russian Foundation of Fundamental Research (Grant No. 94-01-01478) and the International Science Foundation (Grant No. RKR000).

References

1. B. A. Dubrovin and S. P. Novikov, *Hydrodynamics of soliton lattices*, Part 4, Soviet Sci. Rev. Sect. C: Math. Phys. **9** (1993), 1–136.
2. _____, *Hamiltonian formalism of one-dimensional systems of hydrodynamic type and the Bogolyubov–Whitham averaging method*, Dokl. Akad. Nauk SSSR **270** (1983), no. 4, 781–785; English transl., Soviet Math. Dokl. **27** (1983), 665–669.
3. _____, *On Poisson brackets of hydrodynamic type*, Dokl. Akad. Nauk SSSR **279** (1984), no. 2, 294–297; English transl., Soviet Math. Dokl. **30** (1984), 651–654.

4. _____, *Hydrodynamics of weakly deformed soliton lattices. Differential geometry and Hamiltonian theory*, Uspekhi Mat. Nauk **44** (1989), no. 6, 29–98; English transl., Russian Math. Surveys **44** (1989), no. 6, 35–124.

5. S. P. Tsarev, *On Poisson brackets and one-dimensional Hamiltonian systems of hydrodynamic type*, Dokl. Akad. Nauk SSSR **282** (1985), no. 3, 534–537; English transl., Soviet Math. Dokl. **31** (1985), 488–491.

6. _____, *Geometry of Hamiltonian systems of hydrodynamic type. The generalized hodograph method*, Izv. Akad. Nauk SSSR Ser. Mat. **54** (1990), no. 5, 1048–1068; English transl., Math. USSR-Izv. **37** (1991), no. 2, 397–419.

7. O. I. Mokhov, *Poisson brackets of Dubrovin–Novikov type (DN-brackets)*, Funktsional. Anal. i Prilozhen. **22** (1988), no. 4, 92–93; English transl., Functional Anal. Appl. **22** (1988), no. 4, 336–338.

8. _____, *Symplectic forms on loop space and Riemannian geometry*, Funktsional. Anal. i Prilozhen. **24** (1990), no. 3, 86–87; English transl. in Functional Anal. Appl. **24** (1990).

9. _____, *Homogeneous second order symplectic structures on loop spaces and symplectic connections*, Funktsional. Anal. i Prilozhen. **25** (1991), no. 2, 65–67; English transl. in Functional Anal. Appl. **25** (1991).

10. _____, *Canonical Hamiltonian representation of the Krichever–Novikov equation*, Mat. Zametki **50** (1991), no. 3, 87–96; English transl. Math. Notes **50** (1991), 939–945.

11. _____, *Two-dimensional nonlinear sigma models and symplectic geometry on loop spaces of (pseudo-)Riemannian manifolds,*, Nonlinear Evolution Equations and Dynamical Systems. Proceedings of the 8th Internat. Workshop (NEEDS'92), 6–17 July, 1992, Dubna, Russia (V. G. Makhan'kov, ed.), World Sci. Publishing, Singapore, 1993, pp. 444–456.

12. I. Ya. Dorfman and O. I. Mokhov, *Local symplectic operators and structures related to them*, J. Math. Phys. **32** (1991), no. 12, 3288–3296.

13. O. I. Mokhov, *Geometry of commuting differential operators of rank 3 and Hamiltonian flows*, Ph. D. Thesis (1984), Moscow State University, Moscow.

14. _____, *Symplectic structures on loop spaces of smooth manifolds and Lagrangian systems of the field theory*, Abstracts of the Internat. Geom. Colloquium, May 10–14, 1993, Moscow, pp. 38–39.

15. O. I. Mokhov and E. V. Ferapontov, *On the nonlocal Hamiltonian hydrodynamic type operators connected with constant curvature metrics*, Uspekhi Mat. Nauk **45** (1990), no. 3, 191–192; English transl. in Russian Math. Surveys **45** (1990).

16. O. I. Mokhov, *Hamiltonian systems of hydrodynamic type and constant curvature metrics*, Phys. Lett. A **166** (1992), no. 3–4, 215–216.

17. E. V. Ferapontov, *Differential geometry of nonlocal Hamiltonian operators of hydrodynamic type*, Funktsional. Anal. i Prilozhen. **25** (1991), no. 3, 37–49; English transl. in Functional Anal. Appl. **25** (1991).

18. S. P. Novikov, *Andrejewski Lectures*, Berlin, November–December 1993, Sfb 288 Preprint No. 117.

19. M. V. Pavlov, *Multi-Hamiltonian structures of the Whitham equations*, Dokl. Akad. Nauk SSSR **338** (1994), no. 2, 165–167; English transl. in Soviet Math. Dokl. **50** (1994).

20. V. V. Sokolov, *On Hamiltonian property of the Krichever–Novikov equation*, Dokl. Akad. Nauk SSSR **272** (1983), no. 1, 48–50; English transl. in Soviet Math. Dokl. **28** (1983).

21. I. M. Krichever and S. P. Novikov, *Holomorphic bundles and nonlinear equations. Finite-zone solutions of rank* 2, Dokl. Akad. Nauk SSSR **247** (1979), no. 1, 33–37; English transl. in Soviet Math. Dokl. **20** (1979).

22. I. M. Krichever and S. P. Novikov, *Holomorphic bundles over algebraic curves and nonlinear equations*, Uspekhi Mat. Nauk **35** (1980), no. 6, 47–68; English transl. in Russian Math. Surveys **35** (1980).

23. S. P. Novikov, *Two-dimensional Schrödinger operators in periodic fields*, Itogi Nauki i Tekhniki. Sovremennye Problemy Matematiki, vol. 23, VINITI, Moscow, 1983, pp. 3–32; English transl. in J. Soviet Math. **28** (1985), no. 1.

24. O. I. Mokhov, *Commuting ordinary differential operators of rank 3 corresponding to an elliptic curve*, Uspekhi Mat. Nauk **37** (1982), no. 4, 169–170. (Russian)

25. _____, *Commuting differential operators of rank 3 and nonlinear equations*, Izv. Akad. Nauk SSSR Ser. Mat. **53** (1989), no. 6, 1291–1315; English transl. in Math. USSR-Izv. **35** (1990).

26. S. I. Svinolupov and V. V. Sokolov, *Evolution equations with nontrivial conservation laws*, Funktsional. Anal. i Prilozhen. **16** (1982), no. 4, 86–87; English transl., Functional Anal. Appl. **16** (1982).

27. S. I. Svinolupov, V. V. Sokolov, and R. I. Yamilov, *Bäcklund transformations for integrable evolution equations*, Dokl. Akad. Nauk SSSR **271** (1983), no. 4, 802–805.
28. I. Ya. Dorfman, *Krichever–Novikov equation and local symplectic structures*, Dokl. Akad. Nauk SSSR **302** (1988), no. 4, 792–795; English transl. in Soviet Math. Dokl. **38** (1989).
29. _____, *Dirac structures of integrable evolution equations*, Phys. Lett. A **125** (1987), no. 5, 240–246.
30. _____, *Dirac structures and integrability of nonlinear evolution equations*, Wiley, London, 1993.
31. G. Wilson, *On the quasi-Hamiltonian formalism of the KdV equation*, Phys. Lett. A **132** (1988), no. 8–9, 445–450.
32. A. P. Fordy, Nonlinear Evolution Equations and Dynamical Systems, Proc. of the 8th Internat. Workshop (NEEDS'92), 6–17 July, 1992, Dubna, Russia (V. G. Makhan' kov, ed.), World Sci. Publishing, Singapore, 1993.
33. G. Reeb, *Quelques properietés globales des trajectories de la dynamique dues a l'existence de l'invariant integrale de M. Elie Cartan*, C. R. Acad. Sci. Paris **229** (1949), no. 20, 969–971.
34. _____, *Varietés de Riemann dont toutes les geodesiques sont fermeés*, Bull. Cl. Sci. Acad. Royale Belg. 5 Sér. **36** (1950), no. 4, 324–329.
35. A. Weinstein, *On the volume of manifolds all of whose geodesics are closed*, J. Differential Geom. **9** (1974), no. 4, 513–517.
36. A. Besse, *Manifolds all of whose geodesics are closed*, Springer-Verlag, Berlin, Heidelberg, and New York, 1978.
37. F. Magri, *A simple model of the integrable Hamiltonian equation*, J. Math. Phys. **19** (1978), no. 5, 1156–1162.
38. O. I. Mokhov, *Symplectic geometry on loop spaces of smooth manifolds and nonlinear systems*, Internat. Workshop "Theory of Nonlinear Waves," September 1991, Kaliningrad University, Kaliningrad, Russia.
39. _____, *Symplectic forms on loop spaces of Riemannian manifolds*, Internat. Conference "Differential equations and related problems" in honor of 90-th anniversary of I. G. Petrovsky (1901–1973), May 1991, Moscow State University, Moscow, Russia.
40. _____, *Two-dimensional σ-models in the field theory: symplectic approach*, Abstracts of the 9th Workshop "Modern Group Analysis. Methods and Applications", June 1992, Nizhniĭ Novgorod, Russia.
41. A. G. Meshkov, *Hamiltonian and recursion operators for two-dimensional scalar fields*, Phys. Lett. A **170** (1992), no. 6, 405–408.
42. C. Rogers, *Reciprocal transformations and their applications*, Nonlinear Equations, Proc. 5th Workshop on Nonlinear Evolution Equations and Dynamical Systems (NEEDS'87), France, 1987, pp. 109–123.
43. O. I. Mokhov and Y. Nutku, *Bianchi transformation between the real hyperbolic Monge–Ampère equation and the Born–Infeld equation*, Lett. Math. Phys. **32** (1994), no. 2, 121–123.
44. G. V. Potemin, *On Poisson brackets of differential-geometric type*, Dokl. Akad. Nauk SSSR **286** (1986), no. 1, 39–42; English transl., Soviet Math. Dokl. **33** (1986), 30–33.
45. _____, *Ph. D. Thesis*, Moscow State University, Moscow.
46. P. W. Doyle, *Differential geometric Poisson bivectors in one space variable*, J. Math. Phys. **34** (1993), no. 4, 1314–1338.
47. S. P. Tsarev, Mat. Zametki **46** (1989), no. 1, 105–111; English transl. in Math. Notes **46** (1989).
48. O. I. Mokhov, *On Hamiltonian structure of evolution with respect to the space variable x for the Korteweg-de Vries equation*, Uspekhi Mat. Nauk **45** (1990), no. 1, 181–182; English transl. in Russian Math. Surveys **45** (1990).
49. O. I. Mokhov, *Joint Hamiltonian representation of the Korteweg–de Vries equation and the three-wave equation* (1989).
50. Y. Nutku and Ö. Sarıoğlu, *An integrable family of Monge–Ampère equations and their multi-Hamiltonian structure*, Phys. Lett. A **173** (1993), no. 3, 270–274.
51. B. L. Rozhdestvenskiĭ and N. N. Yanenko, *Systems of quasilinear equations and their applications to gas dynamics*, "Nauka", Moscow, 1978. (Russian)
52. O. I. Mokhov and E. V. Ferapontov, *Hamiltonian pairs associated with skew-symmetric Killing tensors on spaces of constant curvature*, Funktsional. Anal. i Prilozhen. **28** (1994), no. 2, 60–63; English transl. in Functional Anal. Appl. **28** (1994).

53. O. I. Mokhov, *Killing–Poisson bivectors on Riemannian manifolds and integrable systems*, Abstracts of Intern. Congress of Mathematicians, 3–11 August 1994, Zürich, Switzerland, 50.
54. B. A. Dubrovin, *Geometry of* 2D *topological field theories*, Preprint SISSA–89/94/FM (1994).
55. _____, *Integrable systems in topological field theory*, Nucl. Phys. B **379** (1992), 627–689.
56. E. Witten, *On the structure of the topological phase of two-dimensional gravity*, Nucl. Physics B **340** (1990), 281–332.
57. R. Dijkgraaf, E. Verlinde, and H. Verlinde, Nucl. Phys. B **352** (1991); *Notes on topological string theory and* 2D *quantum gravity*, Preprint PUPT–1217, IASSNS-HEP–90/80, November, 1990.
58. O. I. Mokhov, *Differential equations of associativity in* 2D *topological field theories and geometry of nondiagonalizable systems of hydrodynamic type*, Abstracts of Internat. Conference on Integrable Systems "Nonlinearity and Integrability: from Mathematics to Physics", February 21–24, 1995, Montpellier, France.

Translated by THE AUTHOR

DEPARTMENT OF GEOMETRY AND TOPOLOGY STEKLOV MATHEMATICAL INSTITUTE UL. VAVILOVA, 42 MOSCOW, GSP-1, 117966, RUSSIA

E-mail address: mokhov@class.mian.su; mokhov@top.mian.su

Real Nonsingular Finite Zone Solutions of Soliton Equations

S. M. NATANZON

§0. Introduction

The method of finite zone (algebro–geometric) integration of differential equations was discovered by S. P. Novikov, B. A. Dubrovin, A. R. Its, I. M. Krichever, V. B. Matveev, and others in the mid-seventies (see the references in [1]). This method consists of two steps: 1) look for the solutions as theta functions for some algebraic curves; 2) among these solutions, select those important for applications (i.e., real and nonsingular). The details of this scheme vary from one equation to another. The second stage becomes much easier if the finite zone complex solutions associated with a hyperelliptic curve are constructed as in the case of KdV [2] or sine-Gordon equations [3], or in the case of solutions associated with a meromorphic function [4].

Complex algebraic solutions of the KP equation

$$\pm \frac{3}{4} u_{yy} = \frac{\partial}{\partial x} \left[u_t - \frac{1}{4} (6uu_x + u_{xxx}) \right]$$

were constructed by Krichever [5]. The selection of the real solutions was completed in [9], while the selection of the nonsingular solutions was completed in [1, 3].

Potential finite zone two-dimensional Shrödinger operators $L = \partial^2/\partial z \partial \bar{z} + u$ were constructed by Veselov and Novikov in [10, 11]. The same papers treat some sufficient conditions for an operator to be real and nonsingular. A criterion for an operator to be real was found in [12], while the selection of the nonsingular operators (under some additional conditions) was suggested in [13].

In this paper we present certain results on the KP equation and the Shrödinger operators in a self-contained form with more detailed and sometimes new proofs. We also eliminate some inaccuracies.

The problems we are treating are in fact problems of the theory of complex and real algebraic curves. In §1 we prove that the logarithm of the theta function of a complex algebraic curve satisfies a system of differential equation with constant coefficients (this system is the same for all the curves of all genera). Such a system was first introduced in [9] using the Hirota equations. In the present paper we use of an algebro-geometric method, which generalizes the method due to Krichever [5].

1991 *Mathematics Subject Classification.* Primary 58F07; Secondary 14H42.

The combinatorial presentations for the coefficients obtained in this way differ from those obtained in [9] and this yields to nontrivial combinatorial identities. In §2 a similar system of differential equations is constructed for the logarithm of Prym theta function. Here we also make use of the algebro-geometric method. (Some other possible but incomplete approach to the construction of such equations are described in [15, 14].) In §3 we use these systems to prove a criterion for a complex curve (respectively, for a complex curve with involution) to be real. We prove that the curves are real if and only if the values of the logarithms of their Riemann theta functions (respectively of the logarithms of their Prym theta functions) are real on a two-dimensional subvariety of Jacobi variety (respectively on a two-dimensional subvariety of Prym variety). The selection of nonsingular solutions is based on the properties of meromorphic and holomorphic differentials on real curves. These properties are investigated in §4. Here we mainly follow the paper [13], but we give new proofs of the theorems and eliminate some inaccuracies. Some of the results presented here are also contained in [6-8]. However, in §4 we use a new method based the on investigation of spinor bundles. This method is described in [16] in more detail.

The results of §4 are used in §§5 and 6 for the selection of real and imaginary tori in Jacobians and Primians that do not intersect the theta divisor. These tori precisely determine nonsingular solutions of the KP equation and nonsingular Shrödinger operators in generic position.

The author is grateful to the International Science Foundation (Grant MD8000) for partial support of this work.

§1. Differential equations for Riemann θ-functions

1. Let P be a compact Riemann surface of genus $g > 0$. Choose a basis

$$\{a_i, b_i \, (i = 1, \ldots, g)\} \subset H_1(P, \mathbb{Z})$$

with the standard intersection matrix $(a_i, a_j) = (b_i, b_j) = 0$, $(a_i, b_j) = \delta_{ij}$. Choose a basis ω_i, $i = 1, \ldots, g$, in the space of holomorphic differentials on P such that $\oint_{a_m} \omega_j = 2\pi i \delta_{mj}$. The Riemann θ-function

$$\theta(\Delta) = \sum_{n_1, \ldots, n_g = -\infty}^{+\infty} \exp\left\{\frac{1}{2} \sum_{i,j=1}^{g} B_{ij} n_i n_j + \sum_{j=1}^{g} n_j \Delta_j\right\}$$

corresponds to the matrix $B_{ij} = \oint_{b_i} \omega_j$; here $\Delta = (\Delta_1, \ldots, \Delta_g) \in \mathbb{C}^g$.

Let $q \in P$. Consider a local chart $k^{-1}: V \to \mathbb{C}$ with $k^{-1}(q) = 0$. Let Ω_j be a differential of the second kind on P, normalized so that $\oint_{a_i} \Omega_j = 0$, which is holomorphic outside of q and has the following presentation

$$\Omega_j = -j \frac{dk^{-1}}{(k^{-1})^{j+1}} + 2 \sum_{i=1}^{\infty} \alpha_{ij} k^{-i+1} dk^{-1}$$

in V. Set

$$U_j = \left(\oint_{b_1} \Omega_j, \ldots, \oint_{b_g} \Omega_j\right) \in \mathbb{C}^g.$$

For a finite sequence $z = (z_1, z_2, \ldots)$, set

$$v(z) = v(z \mid k^{-1} \mid \Delta \mid c) = -\ln \theta \left(\sum_{i=1}^{\infty} z_i u_i + \Delta \right) + \sum \alpha_{ij} z_i z_j + c.$$

The standard basis $e_i = (\delta_{i1}, \ldots, \delta_{ig})$ and the vectors $B_i = (B_{i1}, \ldots, B_{ig})$ generate a lattice $\Gamma \subset \mathbb{C}^g$. The torus $J = J(P)$ is called the *Jacobian variety* or the *Jacobian* of the surface P. Let $\Phi: \mathbb{C}^g \to J$ be the natural bundle and let $\tilde{\Phi}: J \to \mathbb{C}^g$ be one of its sections.

Consider closed paths $\{\tilde{a}_i, \tilde{b}_i, i = 1, \ldots, g\} \in \pi_1(P, q)$ representing the classes $\{a_i, b_i\}$ and for each $p \in P$ choose a segment $l_{pq} \subset P$ connecting p and q that does not intersect any of the paths $\tilde{a}_i, \tilde{b}_i$. Then the mapping

$$p \mapsto \left(\int_{l_{pq}} \omega_1, \ldots, \int_{l_{pq}} \omega_g \right) \in \mathbb{C}^g$$

determines the inclusion $A: P \to J$ and the mapping $A_J: P^g \to J$, where

$$A_J(p_1, \ldots, p_q) = -\sum_{i=1}^{g} A(p_i) - \mathcal{K}_p$$

and $\mathcal{K}_p$ is the Riemann constant vector [1]. It can be shown that $A_J(P^g) = J$ and A_J is a one-to-one correspondence on a subset $Q \subset P^g$ with $\dim_{\mathbb{C}}(J - A_J(Q)) = g - 1$. The set $J_0 = A_J(Q)$ is called the *set of nonspecial divisors*.

We set

$$\Psi(z, p) = \exp \left(\sum_{j=1}^{\infty} z_j \int_{l_{pq}} \Omega_j \right) \frac{\theta\left(\tilde{\Phi}(A(p)) + \sum_{i=1}^{\infty} z_i U_i + \Delta \right) \theta(\Delta)}{\theta(\tilde{\Phi}(A(p)) + \Delta) \theta\left(\sum_{i=1}^{\infty} z_i U_i + \Delta \right)},$$

where Δ is a nonspecial divisor.

It can be shown [1] that the value of Ψ does not depend on the choice of l_{pq} and $\tilde{\Phi}$. In the local chart V

$$\Psi(z, p) = \exp \left(\sum_{j=1}^{\infty} z_j k^j \right) \left(1 + \sum_{j=1}^{\infty} \xi_j(z) k^{-j} \right),$$

$$\ln \Psi(z, p) = \sum_{j=1}^{\infty} z_j k^j + \sum_{j=1}^{\infty} \eta_j k^{-j},$$

where

(1) $$\xi_j = \sum_{n=1}^{\infty} \frac{1}{n!} \sum_{i_1 + \cdots + i_n = j} \eta_{i_1} \cdots \eta_{i_n}.$$

2. We shall further make use of the numbers $P_s \begin{pmatrix} i_1 & \cdots & i_n \\ j_1 & \cdots & j_n \end{pmatrix}$ that are uniquely determined by the properties:

1. $P_s \begin{pmatrix} i \\ j \end{pmatrix} = \begin{pmatrix} s \\ j \end{pmatrix} = \dfrac{s!}{j!(s-j)!}$;

2. $P_s \begin{pmatrix} i_1 & \cdots & i_n \\ 0 & \cdots & 0 \end{pmatrix} = 0$;

3. $P_s \begin{pmatrix} i_1 & \cdots & i_n \\ j_1 & \cdots & j_n \end{pmatrix} = \dfrac{1}{n!} \dfrac{(j_1 + \cdots + j_n)!}{j_1! \cdots j_n!} \begin{pmatrix} s \\ j_1 + \cdots + j_n \end{pmatrix}$
$$- \sum_{q=1}^{n-1} P_s \begin{pmatrix} i_1 & \cdots & i_q \\ j_1 & \cdots & j_q \end{pmatrix} \dfrac{1}{(n-q)!} \dfrac{(j_{q+1} + \cdots + j_n)!}{j_{q+1}! \cdots j_n!}$$
$$\times \begin{pmatrix} s - (i_1 + \cdots + i_q + j_1 + \cdots + j_q) \\ j_{q+1} + \cdots + j_n \end{pmatrix}.$$

(Here and below $0! = 1$ and $\begin{pmatrix} s \\ t \end{pmatrix} = 0$ for $t > s$.)

Denote by $\begin{bmatrix} i_1 & \cdots & i_n \\ j_1 & \cdots & j_n \end{bmatrix}$ the set of matrices that can be obtained from the matrix $\begin{pmatrix} i_1 & \cdots & i_n \\ j_1 & \cdots & j_n \end{pmatrix}$ by a permutation of the columns. Let $\left\| \begin{matrix} i_1 & \cdots & i_n \\ j_1 & \cdots & j_n \end{matrix} \right\|$ be the number of such (different) matrices. Set

$$P_s \begin{bmatrix} i_1 & \cdots & i_n \\ j_1 & \cdots & j_n \end{bmatrix} = \sum P_s \begin{pmatrix} a_1 & \cdots & a_n \\ b_1 & \cdots & b_n \end{pmatrix},$$

where the sum is taken over all

$$\begin{pmatrix} a_1 & \cdots & a_n \\ b_1 & \cdots & b_n \end{pmatrix} \in \begin{bmatrix} i_1 & \cdots & i_n \\ j_1 & \cdots & j_n \end{bmatrix}.$$

Induction over $m + k$ easily gives

LEMMA 1.1. *Let* $h = i_1 + \cdots + i_m + j_1 + \cdots + j_m$ *and* $i_n \geq 1$, $j_n \geq 1$ *for* $n \leq m$. *Then*

$$P_s \begin{bmatrix} i_1 & \cdots & i_n & j_1 & \cdots & j_n \\ i_{m+1} & \cdots & i_{m+k} & 0 & \cdots & 0 \end{bmatrix}$$
$$= \begin{cases} 0, & \text{if } h \leq s, \\ \dfrac{1}{k!} \left\| \begin{matrix} i_{m+1} & \cdots & i_{m+k} \\ 0 & \cdots & 0 \end{matrix} \right\| P_s \begin{bmatrix} i_1 & \cdots & i_m \\ j_1 & \cdots & j_m \end{bmatrix}, & \text{if } h > s. \end{cases}$$

3. Set $\partial_i = \partial/\partial z_i$, $\partial = \partial_1$ and consider the functions

$$B_s^t = -\sum_{i=1}^{t-1} \binom{s}{i} \partial^i \xi_{t-i} - \sum_{j=2}^{t-1} B_s^j \left(\sum_{i=0}^{t-j-1} \binom{s-j}{i} \partial^i \xi_{t-i-j} \right).$$

Using (1) and Lemma 1.1, one can prove

LEMMA 1.2. *Let $s \geq 2$, $2 \leq t \leq s+1$. Then*

$$B_s^t = -\sum_{n=1}^{\infty} \sum P_s \begin{pmatrix} i_1 & \cdots & i_n \\ j_1 & \cdots & j_n \end{pmatrix} \partial^{j_1} \eta_{i_1} \cdots \partial^{j_n} \eta_{i_n},$$

where the inner sum is taken over all matrices $\begin{pmatrix} i_1 & \cdots & i_n \\ j_1 & \cdots & j_n \end{pmatrix}$ such that

$$i_m \geq 1, \quad j_m \geq 1, \quad \text{and} \quad i_1 + \cdots + i_n + j_1 + \cdots + j_n = t.$$

LEMMA 1.3. *We have*

$$\partial_s \eta_r = \sum_{n=1}^{\infty} \sum P_s \begin{pmatrix} i_1 & \cdots & i_n \\ j_1 & \cdots & j_n \end{pmatrix} \partial^{j_1} \eta_{i_1} \cdots \partial^{j_n} \eta_{i_n},$$

where the internal sum is taken over all matrices $\begin{pmatrix} i_1 & \cdots & i_n \\ j_1 & \cdots & j_n \end{pmatrix}$ such that

$$i_m \geq 1, \quad j_m \geq 1, \quad \text{and} \quad i_1 + \cdots + i_n + j_1 + \cdots + j_n = r + s.$$

PROOF. Let $L_n = \partial^n + \sum_{k=2}^{n} B_n^k \partial^{n-k}$. Then

$$\phi(z, p) = L_n \Psi(z, p) - \partial_n \Psi(z, p)$$

$$= \exp\left(\sum_{j=1}^{\infty} z_j k^j\right) \left(\sum_{i=1}^{\infty} \left(-\partial_n \xi_i + \sum_{j=1}^{\infty} \binom{n}{j} \partial^j \xi_{i+n-j}\right.\right.$$

$$\left.\left. + \sum_{k=2}^{n} B_n^k \sum_{j=0}^{n-k} \binom{n-k}{j} \partial^j \xi_{i+n-j-k}\right) k^{-i}\right).$$

Thus we have $\phi(z, p) = 0$ [1] and it follows that

$$\partial_n \xi_i = \sum_{j=1}^{\infty} \binom{n}{j} \partial^j \xi_{i+n-j} + \sum_{k=2}^{n} B_n^k \sum_{j=0}^{n-k} \binom{n-k}{j} \partial^j \xi_{i+n-j-k}.$$

Relation (1) and Lemmas 1.1, 1.2 then give the statement of Lemma 1.3.

4. Set $F(z) = \ln \theta(\sum_{j=1}^{\infty} z_j U_j + \Delta)$. Using Riemann's theorem [2], we get

$$\ln \Psi = \sum_{j=1}^{\infty} z_j k^j + \sum_{i,j=1}^{\infty} \frac{\alpha_{ij}}{i} z_j k^{-i} + F(z - \varepsilon(k)) - F(z) - F(-\varepsilon(k)) + F(0),$$

where $\varepsilon(k) = (k^{-1}, k^{-2}/2, k^{-3}/3, \ldots)$.
Expanding this relation in powers of k^{-1}, we see that

$$(2) \quad \eta_r(z) = 2 \sum_{j=1}^{\infty} \frac{\alpha_{rj}}{r} z_j + \sum_{i_1 + \cdots + i_n = 1} \frac{(-1)^n}{n! \, i_1 \cdots i_n} (\partial_{i_1} \cdots \partial_{i_n} F(z) - \partial_{i_1} \cdots \partial_{i_n} F(0)).$$

THEOREM 1.1. *Let $v(z) = v(z \mid k^{-1} \mid \Delta \mid c)$, where Δ is a nonspecial divisor. Then there exist rational coefficients $R_r \begin{pmatrix} s_1 & \cdots & s_m \\ t_1 & \cdots & t_m \end{pmatrix}$ and $R_{ij} \begin{pmatrix} s_1 & \cdots & s_m \\ t_1 & \cdots & t_m \end{pmatrix}$ independent of P and such that*

(3) $$\eta_r = \frac{1}{r} \partial_r v + \sum R_r \begin{pmatrix} s_1 & \cdots & s_m \\ t_1 & \cdots & t_m \end{pmatrix} \partial_{s_1} \partial^{t_1} v \cdots \partial_{s_m} \partial^{t_m} v + \mathrm{const},$$

(4) $$\partial_i \partial_j v = \sum R_{ij} \begin{pmatrix} s_1 & \cdots & s_m \\ t_1 & \cdots & t_m \end{pmatrix} \partial_{s_1} \partial^{t_1} v \cdots \partial_{s_m} \partial^{t_m} v,$$

where s_i, $t_i \geqslant 1$ and $s_1 + \cdots + s_m + t_1 + \cdots + t_m$ equals r in the first case and equals $i + j$ in the second case, while

$$R_{ij}\begin{pmatrix} 1 \\ i+j-1 \end{pmatrix} = \frac{ij}{i+j-1}.$$

We shall prove the theorem by simultaneous induction over k and $i + j$. For $r = 1$, $i + j = 2$ the statement follows from (2). Let the statement be valid for $r < k$, $i + j \leqslant k$. Then relation (3) for $r = k$ follows from (2) and the induction hypothesis. Using the induction hypothesis, we find that

$$\partial_j \eta_i = \frac{1}{i} \partial_i \partial_j v + \sum \widetilde{R}_{ij} \begin{pmatrix} s_1 & \cdots & s_m \\ t_1 & \cdots & t_m \end{pmatrix} \partial_{s_1} \partial^{t_1} v \cdots \partial_{s_m} \partial^{t_m} v.$$

According to Lemma 1.3 and the induction hypothesis, relation (4) then follows for $i + j = k$.

REMARK. The first of the equations (4) is the KP equation "integrated over x_1"

$$\partial_2^2 v = \tfrac{4}{3} \partial_3 \partial_1 v - \tfrac{1}{3} \partial_1^4 v + 2(\partial_1^2 v)^2.$$

Relation (4) was first deduced in [9] from the Hirota hierarchy [17]. The coefficients are calculated in [9] according to absolutely different rules. The equalities between the coefficients provide nontrivial combinatorial identities.

§2. Differential equations for Prym θ-functions

1. Let $\alpha \colon P \to P$ be a holomorphic involution having precisely two fixed points q and $\hat{q}$. With the help of A_J, the involution α induces the involution $\alpha_J \colon J \to J$. The subset $\mathrm{Pr} = \{\Delta \in J \mid \Delta + \alpha_J \Delta = 0\}$ is called the *Prym variety* or the *Primian* of the pair (P, α).

Choose a basis $\{a_i, b_i, i = 1, \ldots, g\}$ such that

$$\alpha a_i = -a_{i+\tilde{g}}, \qquad \alpha b_i = -b_{i+\tilde{g}},$$

where $g = 2\tilde{g}$. Set

$$\widetilde{B}_{ij} = B_{ij} + B_{i+\tilde{g},j} \qquad (i, j = 1, \ldots, \tilde{g}).$$

The vectors $\tilde{B}_i = (\tilde{B}_{i_1}, \ldots, \tilde{B}_{i_{\tilde{g}}})$ and the vectors $e_i = (\delta_{i_1}, \ldots, \delta_{i_{\tilde{g}}})$ generate the lattice $\tilde{\Gamma} \subset \mathbb{C}^{\tilde{g}}$ such that $\mathbb{C}^{\tilde{g}}/\tilde{\Gamma} \cong \mathrm{Pr}$. In addition, if $\Delta \in \mathrm{Pr}$, then $\Delta \in (\tilde{\Delta}, \tilde{\Delta})$, where $\tilde{\Delta} \in \mathbb{C}^{\tilde{g}}$ and $\theta(\Delta) = (\theta_{\mathrm{Pr}}(\tilde{\Delta}))^2$, where

$$\theta_{\mathrm{Pr}}(\tilde{\Delta}) = \sum_{n_1,\ldots,n_{\tilde{g}}=-\infty}^{\infty} \exp\left\{\frac{1}{2}\sum_{i,j=1}^{\tilde{g}} \tilde{B}_{ij}n_i n_j + \sum_{i=1}^{g} n_i \tilde{\Delta}_i\right\}$$

is the Prym function [10, 11]. We suppose below that $\Delta \in \mathrm{Pr}$ is a fixed nonspecial divisor.

2. Choose local parameters $k^{-1}: V \to \mathbb{C}$ and $\hat{k}^{-1}: \hat{V} \to \mathbb{C}$ in neighborhoods V, $\hat{V}$ of the points q and $\hat{q}$ such that $k^{-1}(q) = 0 = \hat{k}^{-1}(\hat{q})$ and $k^{-1}(\alpha p) = -k^{-1}(p)$, $\hat{k}^{-1}(\alpha p) = -\hat{k}^{-1}(p)$. Consider the meromorphic differentials Ω_j with a unique pole at the point q of the form

$$\Omega_j = -j\frac{dk^{-1}}{(k^{-1})^{j+1}} + 2\sum_{i=1}^{\infty} \alpha_{ij} k^{-i+1} dk^{-1}$$

normalized so that

$$\oint_{a_i} \Omega_j = 0, \quad i = 1, \ldots, g.$$

Let

$$\Omega_j = \sum_{i=1}^{\infty} \hat{\beta}_{ij} \hat{k}^{-i+1} d\hat{k}^{-1}$$

be their representation in the chart $\hat{k}^{-1}$.

Consider also a similar system of differentials $\hat{\Omega}_j$ having poles only at the point $\hat{q}$ and normalized so that

$$\oint_{a_i} \hat{\Omega}_j = 0, \quad i = 1, \ldots, g.$$

Let

$$\hat{\Omega}_j = -j\frac{d\hat{k}^{-1}}{(\hat{k}^{-1})^{j+1}} + 2\sum_{i=1}^{\infty} \hat{\alpha}_{ij} \hat{k}^{-i+1} d\hat{k}^{-1}, \qquad \hat{\Omega}_j = \sum_{i=1}^{\infty} \beta_{ij} k^{-i+1} dk^{-1}$$

be their representations in the charts $\hat{k}^{-1}$ and k^{-1}.

The involution α induces the change of variables $k^{-1} \mapsto -k^{-1}$, $\hat{k}^{-1} \mapsto -\hat{k}^{-1}$, which maps $\Omega_j \mapsto (-1)^j \Omega_j$, $\hat{\Omega}_j \mapsto (-1)^j \hat{\Omega}_j$. Thus $\alpha_{ij} = \hat{\alpha}_{ij} = \beta_{ij} = \hat{\beta}_{ij} = 0$ for $i+j$ odd.

3. Consider the vectors

$$U_j = \left(\oint_{b_1} \Omega_j, \ldots, \oint_{b_g} \Omega_j\right) \quad \text{and} \quad \hat{U}_j = \left(\oint_{b_1} \hat{\Omega}_j, \ldots, \oint_{b_g} \hat{\Omega}_j\right).$$

For j odd $U_j = (\widetilde{U}_j, \widetilde{\widetilde{U}}_j)$ and $\widehat{U}_j = (\widetilde{\widehat{U}}_j, \widetilde{\widehat{U}}_j)$, where $\widetilde{U}_j, \widetilde{\widehat{U}}_j \in \mathbb{C}^{\tilde{g}}$. The expression

$$\Psi(z, \hat{z}, p) = \Psi(z, \hat{z}, p \mid \Delta) = \exp\left(\sum_{j=1}^{\infty} \left(z_j \int_q^p \Omega_j + \hat{z}_j \left(\int_q^p \widehat{\Omega}_j - A(\hat{q})\right)\right)\right)$$

$$\times \frac{\theta\left(A(p) + \sum_{j=1}^{\infty}(z_j U_j + \hat{z}_j \widehat{U}_j) + \Delta\right)\theta(\Delta)}{\theta(A(p) + \Delta)\theta\left(\sum_{j=1}^{\infty}(z_j U_j + \hat{z}_j \widehat{U}_j)\right)}$$

with fixed $z = (z_1, z_2, \dots)$, $\hat{z} = (\hat{z}_1, \hat{z}_2, \dots)$, $\Delta \in \text{Pr}$ determines a function on P. In the chart V we have

$$\Psi(z, \hat{z}, p) = \exp\left(\sum_{j=1}^{\infty} z_j k^j\right)\left(1 + \sum_{j=1}^{\infty} \xi_j(z, \hat{z}) k^{-j}\right).$$

Denote by E the set of sequences $z = (z_1, z_2, \dots)$ such that $z_i = 0$ for i even.

LEMMA 2.1. *If $(z, \hat{z}) \in E \times \widehat{E}$, then*

$$\Psi(z, \hat{z}, p) = \exp\left(\sum_{j=1}^{\infty} \hat{z}_j k^j\right)\left(1 + \sum_{j=1}^{\infty} \hat{\xi}_j(z, \hat{z}) \hat{k}^{-j}\right)$$

in the chart $\widehat{V}$.

PROOF. For every $\Delta \in \text{Pr}$ there exists a divisor $\mathcal{D} \in P^{\tilde{g}}$ such that $\Delta = A_J(\mathcal{D})$ and $\mathcal{D} + \alpha \mathcal{D}$ is the set of zeros for a differential ω of the third kind with simple poles at the points q and $\hat{q}$. Then

$$\Psi(z, \hat{z}, p \mid \Delta)\Psi(z, \hat{z}, \alpha p \mid -\Delta)\omega$$

is also a differential of the third kind, holomorphic outside the points q and $\hat{q}$. The residues at these points are equal, which is equivalent to the statement of the Lemma.

4. Let us set

$$v(z, \hat{z} \mid \Delta) = v\left(z \mid k^{-1} \mid \sum \hat{z}_i \widehat{U}_i + \Delta \mid \sum_{i,j=1}^{\infty} \hat{\alpha}_{ij} \hat{z}_i z_j + \sum_{i,j=1}^{\infty} \beta_{ij} z_i \hat{z}_j\right)$$

$$= -\ln \theta\left(\sum_{i=1}^{\infty}(z_i U_i + \hat{z}_i \widehat{U}_i) + \Delta\right)$$

$$+ \sum_{i,j=1}^{\infty} \alpha_{ij} z_i z_j + \sum_{i,j=1}^{\infty} \hat{\alpha}_{ij} \hat{z}_i \hat{z}_j + \sum_{i,j=1}^{\infty} \beta_{ij} z_i \hat{z}_j.$$

Let $w(z, \hat{z}) = w(z, \hat{z} \mid k^{-1}, \hat{k}^{-1} \mid \Delta)$ be the restriction of the function $v(z, \hat{z}\mid\Delta)$ to the set $(E \times E \mid \text{Pr})$. Then

$$w(z, \hat{z} \mid \Delta) = -2\ln\theta_{\text{Pr}}\left(\sum_{j=1}^{\infty}(z_j\widetilde{U}_j + \hat{z}_j\widehat{\widetilde{U}}_j) + \widetilde{\Delta}\right) + \sum_{i,j=1}^{\infty}(\alpha_{ij}z_iz_j + \hat{\alpha}_{ij}\hat{z}_i\hat{z}_j + \beta_{ij}z_i\hat{z}_j).$$

As above, set

$$\xi_j = \sum_{n=1}^{\infty}\frac{1}{n!}\sum_{i_1+\cdots+i_n=j}\eta_{i_1}\cdots\eta_{i_n},$$

$$B_s^t = -\sum_{i=1}^{t-1}\binom{s}{i}\partial^i\xi_{t-i} - \sum_{j=2}^{t-1}B_s^j\left(\sum_{i=0}^{t-j-1}\binom{s-j}{i}\partial_1^i\xi_{t-i-j}\right).$$

THEOREM 2.1. *There exist rational constants*

$$R_{ij}^*\begin{pmatrix}s_1 & \cdots & s_m \\ t_1 & \cdots & t_m\end{pmatrix}, \quad R_r^*\begin{pmatrix}s_1 & \cdots & s_m \\ t_1 & \cdots & t_m\end{pmatrix}, \quad B_{ij}^*\begin{pmatrix}s_1 & \cdots & s_m \\ t_1 & \cdots & t_m\end{pmatrix},$$

independent of P, such that for $(z, \hat{z}) \in E \times \widehat{E}$

$$\partial_i\partial_j w = \sum R_{ij}^*\begin{pmatrix}s_1 & \cdots & s_m \\ t_1 & \cdots & t_m\end{pmatrix}\partial_{s_1}\partial^{t_1}w\cdots\partial_{s_m}\partial^{t_m}w,$$

$$\eta_r = \frac{1}{r}\partial_r v + \sum R_r^*\begin{pmatrix}s_1 & \cdots & s_m \\ t_1 & \cdots & t_m\end{pmatrix}\partial_{s_1}\partial^{t_1}w\cdots\partial_{s_m}\partial^{t_m}w + \text{const},$$

$$B_j^i = \sum B_{ij}^*\begin{pmatrix}s_1 & \cdots & s_m \\ t_1 & \cdots & t_m\end{pmatrix}\partial_{s_1}\partial^{t_1}w\cdots\partial_{s_m}\partial^{t_m}w,$$

where the numbers $i, j, s_1, \ldots, s_m$ *are odd, the numbers* $t_1, \ldots, t_m$ *are positive, and the sum* $s_1 + \cdots + s_m + t_1 + \cdots + t_m$ *equals* $i + j$ *in the first and the third equalities, and equals r in the second equality.*

PROOF. According to Lemma 2.2, the Baker–Akhiezer function

$$\phi = \partial_{2n+1}\Psi - \partial^{2n+1}\Psi - \sum_{k=2}^{2n}B_{2n+1}^k\partial^{2n+1-k}\Psi$$

vanishes on $(E \times \widehat{E}, \hat{q} \mid \Delta)$ and this implies that ϕ vanishes on $(E \times \widehat{E}, P \mid \Delta)$. Thus $B_{2n+1}^{2n+1}(E \times \widehat{E} \mid \Delta) = 0$. Then it follows from Lemma 1.2 and Theorem 1.1 that on the set $(E \times \widehat{E} \mid \Delta)$ we have

$$\partial_{2n}\partial v = \sum Q_n\begin{pmatrix}s_1 & \cdots & s_m \\ t_1 & \cdots & t_m\end{pmatrix}\partial_{s_1}\partial^{t_1}v\cdots\partial_{s_m}\partial^{t_m}v,$$

where $Q_n\begin{pmatrix}s_1 & \cdots & s_m \\ t_1 & \cdots & t_m\end{pmatrix}$ are rational coefficients independent of P and the numbers s_i and t_i satisfy $s_1 + \cdots + s_m + t_1 + \cdots + t_m = 2n + 1$. Comparing this equality with Theorem 1.1, we obtain Theorem 2.1.

Now let us set

$$\hat{\partial}_i = \frac{\partial}{\partial \hat{z}_i}, \qquad \hat{\partial} = \hat{\partial}_1,$$

$$\widehat{B}_s^t = \sum_{i=1}^{t-1} \binom{s}{i} \hat{\partial} \hat{\xi}_{t-i} - \sum_{j=2}^{t-1} \widehat{B}_s^t \left(\sum_{i=0}^{t-j-1} \binom{s-j}{i} \hat{\partial}^i \xi_{t-i-j} \right),$$

$$\hat{\xi}_j = \sum_{n=1}^{\infty} \frac{1}{n!} \sum_{i_1 + \cdots + i_n = j} \hat{\eta}_{i_1} \cdots \hat{\eta}_{i_n}.$$

Repeating the proof of Theorem 2.1, we obtain

THEOREM 2.2. *For* $(z, \hat{z}) \in (E \times \widehat{E})$ *we have*

$$\hat{\partial}_i \hat{\partial}_j w = \sum R_{ij}^* \begin{pmatrix} s_1 & \cdots & s_m \\ t_1 & \cdots & t_m \end{pmatrix} \hat{\partial}_{s_1} \hat{\partial}^{t_1} w \cdots \hat{\partial}_{s_m} \hat{\partial}^{t_m} w,$$

$$\hat{\eta}_r = \frac{1}{r} \hat{\partial}_r v + \sum R_r^* \begin{pmatrix} s_1 & \cdots & s_m \\ t_1 & \cdots & t_m \end{pmatrix} \hat{\partial}_{s_1} \hat{\partial}^{t_1} w \cdots \hat{\partial}_{s_m} \hat{\partial}^{t_m} w$$

$$+ \sum_{j=1}^{\infty} (\hat{\beta}_{ij} \hat{z}_i \hat{z}_j - \beta_{ij} z_i \hat{z}_j) + \text{const},$$

$$\widehat{B}_j^i = \sum B_{ij}^* \begin{pmatrix} s_1 & \cdots & s_m \\ t_1 & \cdots & t_m \end{pmatrix} \hat{\partial}_{s_1} \hat{\partial}^{t_1} w \cdots \hat{\partial}_{s_m} \hat{\partial}^{t_m} w,$$

where the numbers i, j, $s_1, \ldots, s_m$ *are odd, the numbers* $t_1, \ldots, t_m$ *are positive, and the sum* $s_1 + \cdots + s_m + t_1 + \cdots + t_m$ *equals* $i + j$ *in the first and the third equalities, and equals* r *in the second equality.*

LEMMA 2.2. *For* $(z, \hat{z}) \in (E \times \widehat{E})$ *we have*

$$\hat{\partial}_i \eta_r = \sum_{k=0}^{i-1} \widehat{B}_i^k \sum_{n=1}^{\infty} \frac{1}{n!} \sum_{\substack{j_1 + \cdots + j_n = i-k \\ i_1 + \cdots + i_n = r, \, j_s \geq 1}} \hat{\partial}^{j_1} \eta_{i_1} \cdots \hat{\partial}^{j_n} \eta_{i_n},$$

$$\widehat{B}_i^0 = 1, \qquad \widehat{B}_i^1 = 0.$$

PROOF. Consider Baker–Akhiezer function

$$\phi = \hat{\partial}_i \Psi - \hat{\partial}^i \Psi - \sum_{k=2}^{i-1} \widehat{B}_i^k \hat{\partial}^i \Psi.$$

This function vanishes at the point q and consequently it vanishes on P. Thus

$$\hat{\partial}_i \xi_j - \sum_{k=0}^{i-1} \widehat{B}_i^k \hat{\partial}^{i-k} \xi_j = 0.$$

Let us show by induction on j that this relation together with the defining relation

$$\xi_j = \sum_{n=1}^{\infty} \frac{1}{n!} \sum_{i_1+\cdots+i_n=j} \eta_{i_1}\cdots\eta_{i_n}$$

is equivalent to the statement of the lemma.

For $j = 1$ the statement is obvious. Suppose the statement is true for $j < r$. Then

$$0 = \eta_{t_1}\cdots\eta_{t_s}(\hat{\partial}_i\eta_j) - \eta_{t_1}\cdots\eta_{t_s}\left(\sum_{k=0}^{i-1}\widehat{B}_i^k \sum_{m=1}^{\infty}\frac{1}{m!}\sum \partial^{j_1}\eta_{i_1}\ldots\partial^{j_m}\eta_{i_m}\right),$$

and it follows that

$$0 = \hat{\partial}_i\xi_r - \sum_{k=0}^{i-1}\widehat{B}_i^k\hat{\partial}^{i-k}\xi_r$$

$$= \hat{\partial}_i\left(\sum_{n=1}^{\infty}\frac{1}{n!}\sum_{i_1+\cdots+i_n=r}\eta_{i_1}\cdots\eta_{i_n}\right) - \sum_{k=0}^{i-1}\widehat{B}_i^k\hat{\partial}^{i-k}\left(\sum_{n=1}^{\infty}\frac{1}{n!}\sum_{i_1+\cdots+i_n=r}\eta_{i_1}\cdots\eta_{i_n}\right)$$

$$= \hat{\partial}_i\eta_r - \sum_{k=0}^{i-1}\widehat{B}_i^k\sum_{n=1}^{\infty}\frac{1}{n!}\sum_{\substack{j_1+\cdots+j_n=i-k \\ i_1+\cdots+i_n=r,\ j_s\geqslant 1}}\hat{\partial}^{j_1}\eta_{i_1}\cdots\hat{\partial}^{j_n}\eta_{i_n}.$$

LEMMA 2.3. *For* $(z,\hat{z}) \in E \times \widehat{E}$,

$$\hat{\partial}\eta_{j+1} + \hat{\partial}\partial\eta_j + \sum_{i_1+i_2=j}\hat{\partial}\eta_{i_1}\partial\eta_{i_2} = 0.$$

PROOF. Consider the Baker–Akhiezer function

$$\phi = \hat{\partial}\partial\Psi - \partial\xi_1\Psi$$

$$= \exp\left(\sum_{j=1}^{\infty}z^jk^j\right)\left(\sum_{j=1}^{\infty}\hat{\partial}\xi_{j+1}k^{-j} + \sum_{j=1}^{\infty}\hat{\partial}\partial\xi_j - \partial\xi_1\sum_{j=1}^{\infty}\xi_jk^{-j}\right) = 0.$$

Thus we have $\hat{\partial}\xi_{j+1} + \hat{\partial}\partial\xi_j - \xi_j\hat{\partial}\xi_1 = 0$. Using induction over j, let us prove that the last equality is equivalent to the statement of Lemma 2.3. For $j = 1$ this is obvious:

$$0 = \hat{\partial}\xi_2 + \hat{\partial}\partial\xi_1 - \xi_1\hat{\partial}\xi_1 = \hat{\partial}(\eta_2 + \tfrac{1}{2}\eta_1^2) + \hat{\partial}\partial\eta_1 - \eta_1\hat{\partial}\eta_1$$
$$= \hat{\partial}\eta_2 + \eta_1\hat{\partial}\eta_1 + \hat{\partial}\partial\eta_1 - \eta_1\hat{\partial}\eta_1 = \hat{\partial}\eta_2 + \hat{\partial}\partial\eta_1.$$

Suppose the statement is proved for $j < k$. Then

$$\eta_{i_1}\cdots\eta_{i_m}(\hat{\partial}\eta_{j+1})\eta_{i_{m+1}}\cdots\eta_{i_n} + \eta_{i_1}\cdots\eta_{i_m}(\hat{\partial}\partial\eta_j)\eta_{i_{m+1}}\cdots\eta_{i_n}$$
$$- \eta_{i_1}\cdots\eta_{i_m}\left(\sum_{j_1+j_2=j}\hat{\partial}\eta_{j_1}\partial\eta_{j_2}\right)\eta_{i_{m+1}}\cdots\eta_{i_n} = 0.$$

Having this in mind, we conclude

$$0 = \hat{\partial}\xi_{k+1} + \hat{\partial}\partial\xi_k - \xi_k\partial\xi_1$$

$$= \hat{\partial}\left(\sum_{n=1}^{\infty}\frac{1}{n!}\sum_{i_1+\cdots+i_n=k+1}\eta_{i_1}\cdots\eta_{i_n}\right) + \hat{\partial}\partial\left(\sum_{n=1}^{\infty}\frac{1}{n!}\sum_{i_1+\cdots+i_n=k}\eta_{i_1}\cdots\eta_{i_n}\right)$$

$$- \hat{\partial}\eta_1\left(\sum_{n=1}^{\infty}\frac{1}{n!}\sum_{i_1+\cdots+i_n=k}\eta_{i_1}\cdots\eta_{i_n}\right)$$

$$= \hat{\partial}\eta_{k+1} + \hat{\partial}\partial\eta_k + \sum_{i_1+i_2=k}\hat{\partial}\eta_{i_1}\partial\eta_{i_2}.$$

LEMMA 2.4. *There exist rational constants*

$$N_j\begin{pmatrix} i_1 & \ldots & i_n & | & s_1 & \ldots & s_m \\ \hat{i}_1 & \ldots & \hat{i}_n & | & t_1 & \ldots & t_m \end{pmatrix}$$

independent of P such that for $(z, \hat{z}) \in E \times \widehat{E}$,

$$\hat{\partial}\partial_j v = \sum N_j\begin{pmatrix} i_1 & \ldots & i_n & | & s_1 & \ldots & s_m \\ \hat{i}_1 & \ldots & \hat{i}_n & | & t_1 & \ldots & t_m \end{pmatrix}\partial^{i_1}\hat{\partial}^{\hat{i}_1}w\cdots\partial^{i_n}\hat{\partial}^{\hat{i}_n}w\partial_{s_1}\partial^{t_1}w\cdots\partial_{s_m}\partial^{t_m}w,$$

where the numbers s_k are odd, the numbers t_k are positive, $i_k + \hat{i}_k \geq 2$, and

$$i_1+\cdots+i_n+s_1+\cdots+s_m+t_1+\cdots+t_m = j, \qquad \hat{i}_1+\cdots+\hat{i}_n = 1.$$

PROOF. We proceed by induction on j. For $j = 1$ the statement is obvious. Suppose the statement is true for $j < r$. Then, according to Lemma 2.3 and Theorem 2.1,

$$\hat{\partial}\eta_r = \hat{\partial}\left(\frac{1}{r}\partial_r v + \sum_{(s_i<r)} R_r^*\begin{pmatrix} s_1 & \ldots & s_m \\ t_1 & \ldots & t_m \end{pmatrix}\partial_{s_1}\partial^{t_1}w\cdots\partial_{s_m}\partial^{t_m}w\right)$$

$$= -\hat{\partial}\partial\eta_{r-1} - \sum_{i_1+i_2=r-1}\hat{\partial}\eta_{i_1}\partial\eta_{i_2}$$

$$= \hat{\partial}\partial\left(\frac{1}{r-1}\partial_{r-1}v + \sum R_{r-1}^*\begin{pmatrix} s_1 & \ldots & s_m \\ t_1 & \ldots & t_m \end{pmatrix}\partial_{s_1}\partial^{t_1}w\cdots\partial_{s_m}\partial^{t_m}w\right)$$

$$- \sum_{i_1+i_2=r-1}\hat{\partial}\left(\frac{1}{i_1}\partial_{i_1}v + \sum_{(s_i<r)} R_{i_1}^*\begin{pmatrix} s_1 & \ldots & s_m \\ t_1 & \ldots & t_m \end{pmatrix}\partial_{s_1}\partial^{t_1}w\cdots\partial_{s_m}\partial^{t_m}w\right)$$

$$\times \partial\left(\frac{1}{i_2}\partial_{i_2}v + \sum_{(s_i<r)} R_{i_2}^*\begin{pmatrix} s_1 & \ldots & s_m \\ t_1 & \ldots & t_m \end{pmatrix}\partial_{s_1}\partial^{t_1}w\cdots\partial_{s_m}\partial^{t_m}w\right),$$

and the statement of Lemma 2.4 now follows from the induction hypothesis.

THEOREM 2.3. *There exist rational constants*

$$N_{j\hat{j}}\begin{pmatrix} s_1 & \ldots & s_m & | & i_1 & \ldots & i_n & | & \hat{s}_1 & \ldots & \hat{s}_m \\ t_1 & \ldots & t_m & | & \hat{i}_1 & \ldots & \hat{i}_n & | & \hat{t}_1 & \ldots & \hat{t}_m \end{pmatrix},$$

independent of P such that

$$\hat{\partial}_{\hat{j}}\partial_j w = \sum N_{j\hat{j}}\begin{pmatrix} s_1 & \ldots & s_m & | & i_1 & \ldots & i_n & | & \hat{s}_1 & \ldots & \hat{s}_m \\ t_1 & \ldots & t_m & | & \hat{i}_1 & \ldots & \hat{i}_n & | & \hat{t}_1 & \ldots & \hat{t}_m \end{pmatrix}$$
$$\times \partial_{s_1}\partial^{t_1}w\ldots\partial_{s_m}\partial^{t_m}w\partial^{i_1}\hat{\partial}^{\hat{i}_1}w\cdots\partial^{i_n}\hat{\partial}^{\hat{i}_n}w\hat{\partial}_{\hat{s}_1}\hat{\partial}^{\hat{t}_1}w\cdots\hat{\partial}_{\hat{s}_{\hat{m}}}\hat{\partial}^{\hat{t}_{\hat{m}}}w,$$

where the numbers $\hat{j}$, j, s_k, $\hat{s}_k$ *are odd, the numbers* t_k, $\hat{t}_k$ *are positive, and*

$$i_k + \hat{i}_k \geqslant 2, \qquad s_1 + \cdots + s_m + t_1 + \cdots + t_m + i_1 + \cdots + i_n = j,$$
$$\hat{s}_1 + \cdots + \hat{s}_{\hat{m}} + \hat{t}_1 + \cdots + \hat{t}_{\hat{m}} + \hat{i}_1 + \cdots + \hat{i}_n = \hat{j}.$$

PROOF. Let us prove the statement by induction on j. According to Theorem 2.1, $\eta_1 = \partial w + \text{const}$. By Lemma 2.2, it follows that

$$\hat{\partial}_{\hat{j}}\partial w = \sum_{k=0}^{j-1} \widehat{B}_{\hat{j}}^k \hat{\partial}^{\hat{j}-k}\eta_1 = \sum_{k=0}^{j-1} \widehat{B}_{\hat{j}}^k \partial\hat{\partial}^{\hat{j}-k}w.$$

Comparing this with Theorem 2.2, we obtain the statement of Theorem 2.3 for $j = 1$. Suppose the statement is true for $j < r$. Then, according to Theorem 2.1 and Lemma 2.2, we have

$$\hat{\partial}_{\hat{j}}\eta_r = \hat{\partial}_{\hat{j}}\left(\frac{1}{r}\partial_r w + \sum_{(s_i < r)} R_r^*\begin{pmatrix} s_1 & \ldots & s_m \\ t_1 & \ldots & t_m \end{pmatrix}\partial_{s_1}\partial^{t_1}w\cdots\partial_{s_m}\partial^{t_m}w\right)$$

$$= \sum_{k=0}^{j-1} \widehat{B}_{\hat{j}}^k \sum_{n=1}^{\infty}\frac{1}{n!}\sum \hat{\partial}^{j_1}\eta_{i_1}\cdots\hat{\partial}^{j_n}\eta_{i_n}$$

$$= \sum_{k=0}^{j-1} \widehat{B}_{\hat{j}}^k \sum_{n=1}^{\infty}\frac{1}{n!}\sum \hat{\partial}^{j_1}\left(\frac{\partial_{i_1}w}{i_1} + \sum R_{i_1}^*\begin{pmatrix} s_1 & \ldots & s_m \\ t_1 & \ldots & t_m \end{pmatrix}\partial_{s_1}\partial^{t_1}w\cdots\partial_{s_m}\partial^{t_m}w\right)$$

$$\times \cdots \times \hat{\partial}^{j_n}\left(\frac{\partial_{i_n}w}{i_n} + \sum R_{i_n}^*\begin{pmatrix} s_1 & \ldots & s_m \\ t_1 & \ldots & t_m \end{pmatrix}\partial_{s_1}\partial^{t_1}w\cdots\partial_{s_m}\partial^{t_m}w\right).$$

The statement of the theorem for $j = r$ now follows from Lemma 2.4, Theorem 2.2, and the induction hypothesis.

6. Consider the function

$$\widetilde{w}(z, \hat{z} \mid \Delta) = w(z, \hat{z} \mid \Delta) - \sum_{i,j=1}^{\infty}(\beta_{ij}z_i\hat{z}_j - \hat{\beta}_{ij}z_i\hat{z}_j).$$

Substituting $\partial \leftrightarrow \hat{\partial}$, $q \leftrightarrow \hat{q}$, $k \leftrightarrow \hat{k}$, $z \leftrightarrow \hat{z}$ and repeating the previous considerations, we conclude that the function w satisfies the equations of Theorem 2.3. This means that $\beta_{ij} = \hat{\beta}_{ji}$. Theorem 2.2 implies

COROLLARY 2.1. *For* $(z, \hat{z}) \in E \times \widehat{E}$,

$$\hat{\eta}_r = \frac{1}{r}\hat{\partial}_r v + \sum R_r^*\begin{pmatrix} s_1 & \ldots & s_m \\ t_1 & \ldots & t_m \end{pmatrix}\hat{\partial}_{s_1}\hat{\partial}^{t_1}w\cdots\hat{\partial}_{s_m}\hat{\partial}^{t_m}w + \text{const},$$

where $s_1 + \cdots + s_m + t_1 + \cdots + t_m = r$, *the numbers* $s_1, \ldots, s_m$ *are odd and* $t_i \geqslant 1$.

§3. A criterion for certain functions to be real

1. First we find a criterion for the function $v(z) = v(z \mid k^{-1} \mid \Delta \mid c)$ generated by the surface P (see §1) to be real.

Consider the change of local parameter $k = k_0 + ak_0^{-m}$. This change maps the functions

$$\ln(\Psi) = \sum_{j=1}^{\infty} z_j k^j + \sum_{j=1}^{\infty} \eta_j k^{-j}$$

and $v(z)$ to the functions

$$\ln(\Psi^0) = \sum_{j=1}^{\infty} z_j^0 k_0^j + \sum_{j=1}^{\infty} \eta_j^0 k_0^{-j}$$

and $v_0(z^0)$ respectively, where the variables $z^0 = (z_1^0, z_2^0, \ldots)$ are determined by the conditions

$$\sum_{j=1}^{\infty} z_j k^{-j} = \sum_{j=1}^{\infty} z_j^0 k_0^{-j}.$$

Set $\partial_i^0 = \partial/\partial z_i^0$ and $\partial^0 = \partial_1^0$.

LEMMA 3.1. $\partial_i^0 \partial^0 v^0 = \partial_i \partial v$ for $i < m$ and $\partial_m^0 \partial^0 v^0 = \partial_m \partial v + ma$.

PROOF. By definition, we have $z_n^0 = z_n + \sum_{i>m+1} b_i z_i$ and $\eta_1^0 = \eta_1 + z_m am + \sum_{i>m} c_i z_i$. Therefore, for $i < m$ we have

$$\partial_i \eta_1 = \partial_i \eta_1^0 = \sum_{n=1}^{\infty} \partial_n^0 \eta_1^0 \partial_i z_n^0 = \partial_i^0 \eta_1^0.$$

Similarly,

$$\partial_m \eta_1 = \partial_m \eta_1^0 - am = \partial_m^0 \eta_1^0 - am.$$

In addition, according to Theorem 1.1,

$$\eta_1 = \partial v + \text{const} \quad \text{and} \quad \eta_1^0 = \partial^0 v^0 + \text{const}.$$

THEOREM 3.1. *The functions $\partial^2 v(z)$ and $\partial_2 \partial v(z)$ are real for $z_1, z_2 \in \mathbb{R}$, $z_3 = z_4 = \cdots = 0$ if and only if there exists an antiholomorphic involution $\tau: P \to P$ such that $\tau q = q$, $k^{-1}\tau = \overline{k}^{-1} + o(\overline{k}^{-4})$ and $\tau_J \Delta = \overline{\Delta}$, where $\tau_J = A_J \tau^* A_J^{-1}$ and $\tau^*(p_1, \ldots, p_g) = (\tau p_1, \ldots, \tau p_g)$.*

PROOF. First we prove the necessity of the conditions. Cover the surface P by an atlas of holomorphic charts $\varepsilon_\alpha: U_\alpha \to \mathbb{C}$, one of the charts being k^{-1}. The functions $\overline{\varepsilon}_\alpha: U_\alpha \to \mathbb{C}$ determine a new complex structure generating the Riemann surface $\overline{P}$ and the antiholomorphic mapping $\sigma: P \to \overline{P}$. Set $\tilde{v}(z) = v(z \mid k^{-1} \mid \overline{\Delta})$. Then $\tilde{v}(z) = \overline{v(\overline{z})}$ and therefore $\partial^2 \tilde{v}(z) = \partial^2 v(z)$ and $\partial_2 \partial \tilde{v}(z) = \partial_2 \partial v(z)$ on the set $\mathcal{D} = \mathbb{R} \times \mathbb{R} \times 0 \times 0 \times \cdots$.

Now we prove that there exists a local parameter change

$$\tilde{k} = \overline{k} + \sum_{j=3}^{\infty} c_j(\overline{k})^{-j}$$

such that $\partial_i \partial \tilde{v} = \partial_i \partial v$ on $\mathcal{D}$ for all i. To achieve this goal, we show that if the change of variables

$$\tilde{k} = \overline{k} + \sum_{j=3}^{N-1} c_j(\overline{k})^{-j}$$

gives $\partial_i \partial \tilde{v} = \partial_i \partial v$ for $i < N$, then there exists a constant c_N such that the change of variables

$$\tilde{k} = \overline{k} + \sum_{j=3}^{N} c_j(\overline{k})^{-j}$$

leads to the same equality for all $i \leqslant N$.

According to Theorem 1.1,

$$\partial_2^{N-1} v = Q \partial_N \partial^{N-2} v + \sum Q_n \begin{pmatrix} i_1 & \cdots & i_n \\ j_1 & \cdots & j_n \end{pmatrix} \partial_{i_1} \partial^{j_1} v \cdots \partial_{i_n} \partial^{j_n} v,$$

where $Q_n \in \mathbb{R}$, $Q \in \mathbb{R} - 0$ and the sum is taken over i_k, $j_k \geqslant 1$ such that

$$i_1 + \cdots + i_n + j_1 + \cdots + j_n = 2N - 2.$$

Thus $\partial_N \partial^{N-2} \tilde{v} = \partial_N \partial^{N-2} v$. Since $\partial_N \partial \tilde{v}$ and $\partial_N \partial v$ are quasiperiodic this implies that $\partial_N \partial \tilde{v} = \partial_N \partial v + \text{const}$. Applying Lemma 3.1, we obtain the required constant c_N.

The Baker–Akhiezer function $\tilde{\Psi}(z, p)$ on $\overline{P}$ is assigned to the mapping $\tilde{\varepsilon} = \tilde{k}^{-1} \colon \tilde{v} \to \mathbb{C}$ and to the vector $\overline{\Delta}$. According to Theorem 1.1, for $p \in \overline{V}$ we have $z \in \Delta$ and

$$\partial \ln \overline{\Psi}(z, p) = \tilde{k} + \sum_{j=1}^{\infty} \partial \tilde{\eta}_j \tilde{k}^{-j} = \tilde{k} + \sum_{j=1}^{\infty} \partial \eta_j \tilde{k}^{-j} = \partial \ln \Psi(z, \varepsilon^{-1} \tilde{\varepsilon}(p)),$$

where $\Psi(z, p)$ is the Baker–Akhiezer function on P corresponding to the local chart $\varepsilon = k^{-1} \colon V \to \mathbb{C}$ and the vector Δ. The mapping $\varepsilon^{-1} \tilde{\varepsilon} \colon \tilde{V} \to V$ then extends to a biholomorphic mapping $\phi \colon \overline{P} \to P$ such that

$$\partial \ln \tilde{\Psi}(z, p) = \partial \ln \Psi(z, \phi(p)).$$

It then follows that the mapping $\tau = \phi \sigma$ is an antiholomorphic involution of the surface P with $\tau q = q$, $k^{-1} \tau = \tilde{k}^{-1} = \overline{k}^{-1} + O(\overline{k}^{-4})$ and $\overline{\Psi(z, \tau p)} = \Psi(z, p)$. In particular, the set of poles of the function $\Psi(z, p)$ is invariant under τ, which implies $\tau_J \Delta = \overline{\Delta}$.

Now we prove that the conditions are sufficient. Let $\tau \colon P \to P$ be an antiholomorphic involution, $\tau q = q$, $k^{-1} \tau = \overline{k}^{-1} + o(\overline{k}^{-4})$, $\tau_J(\Delta) = \overline{\Delta}$ and $\Psi(z, p)$ be the Baker–Akhiezer function associated to the chart k^{-1} and to the vector Δ. According to Lemma 3.1, it is sufficient to prove the required statement for the local chart $k^{-1} \tau = \overline{k}^{-1}$. The equality $\overline{\Psi(\overline{z}, \tau p)} = \Psi(z, p)$ implies in this case $\overline{\eta_1(\overline{z})} = \eta_1(z)$ and, according to Theorem 1.1, it then follows that $\overline{\partial^2 v(\overline{z})} = \partial^2 v(z)$ and $\overline{\partial_2 \partial v(\overline{z})} = \partial_2 \partial v(z)$.

THEOREM 3.2. *The functions $\partial^2 v(z)$ and $\partial_2 \partial v(z)$ are real for z_1, $iz_2 \in \mathbb{R}$, $z_3 = z_4 = \cdots = 0$ if and only if there exists an antiholomorphic involution $\tau\colon P \to P$ such that $\tau q = q$, $k^{-1}\tau = -\overline{k}^{-1} + o(\overline{k}^{-4})$ and $\tau_J \Delta = -\Delta$.*

PROOF. First we prove that the conditions are necessary. Following the proof of Theorem 3.1, consider the surface $\overline{P}$. The function

$$\tilde{v}(z_1, z_2, z_3, \ldots) = \overline{v(\overline{z_1}, -\overline{z_2}, \overline{z_3}, -\overline{z_4}, \ldots)}$$

is associated to the vector $-\overline{\Delta}$ and to the local parameter $-\overline{k}^{-1}$. By repeating the proof of necessity in Theorem 3.1, we obtain the necessity of the conditions $\tau q = q$, $k^{-1}\tau = -\overline{k}^{-1} + o(\overline{k}^{-4})$, and $\tau_y \Delta = -\Delta$.

Now let us prove that these conditions are sufficient for the functions $\partial^2 v(z)$ and $\partial_2 \partial v(z)$ to be real on $\widetilde{\mathcal{D}} = \mathbb{R} \times i\mathbb{R} \times 0 \times \ldots$. Consider the following Baker–Akhiezer functions on P:

(1) $\Psi(z, p)$, associated to the pair (k^{-1}, Δ),
(2) $\widetilde{\Psi}(z, p)$, associated to the pair $(k^{-1}, -\overline{\Delta})$.

According to Lemma 3.1, it is sufficient to prove the required conditions under the assumption $k^{-1}\tau = -\overline{k}^{-1}$. In this case

$$\widetilde{\Psi}(z, p) = \overline{\Psi(z, \tau p)} = \exp(-z_1 k + z_2 k^2)\left(1 + \sum_{j=1}^{\infty} \overline{\xi}_j (-k^{-1})^j\right)$$

and

$$\Psi(z, p)\widetilde{\Psi}(z, p) = (1 + (\xi_1 - \overline{\xi}_1)k^{-1} + o(k^{-1})).$$

The condition $\tau_J \Delta + \Delta = 0$ means that there exists a meromorphic differential of the second kind having zeros at the poles of the function v and having its only pole of order 2 at the point q. Thus

$$\Psi(z, p)\widetilde{\Psi}(z, p)\omega = (k^2 + (\xi_1 - \overline{\xi}_1)k + c + O(k^{-1}))$$

is a meromorphic differential with a unique pole at q, which implies that $\xi_1 = \overline{\xi}_1$. According to Theorem 1.1, it then follows that $\partial^2 v(z)$ and $\partial_2 \partial v(z)$ are real on $\widetilde{\mathcal{D}}$.

2. Now let us consider the criterion for the function

$$w(z, \hat{z}) = w(z, \hat{z} \mid k^{-1}, \hat{k}^{-1} \mid \Delta)$$

generated by the Riemann surface P of genus g and by the holomorphic involution $\alpha\colon P \to P$ (see §2) to be real.

LEMMA 3.2. *There exists a rational function $F(y_1, \ldots, y_s)$ with rational coefficients depending on g such that*

$$\partial^2 w = F(\partial^{i_1}\hat{\partial}^{j_1} w, \ldots, \partial^{i_s}\hat{\partial}^{j_s} w),$$

where i_k, $j_k \geqslant 1$.

PROOF. The linear relation $\widehat{U}_1 = \sum_{j=1}^{g} \gamma_j U_j$ implies that

$$\partial \hat{\partial}^2 w = \sum_{j=1}^{g} \gamma_j \partial \hat{\partial} \partial_j w.$$

Transforming this equation and applying $\hat{\partial}$, we can eliminate the constants γ_j and obtain the equation

$$F_1(\partial\partial_{i_1}\hat{\partial}^{j_1}w, \ldots, \partial\partial_{i_m}\hat{\partial}^{j_m}w) = 0,$$

where $j_t \geq 1$. Transforming this equality by applying $\hat{\partial}$ and using Theorem 2.3, we obtain

$$F_2(\partial^{i_1}\hat{\partial}^{j_1}w, \ldots, \partial^{i_m}\hat{\partial}^{j_m}w) = 0,$$

where F_2 is a polynomial with rational coefficients and

$$i_k \geq 1, \quad i_1 = 2, \quad j_1 = 0, \quad \frac{\partial F_2}{\partial x_1} \neq 0.$$

Transforming the last expressions and applying $\hat{\partial}$, we obtain the statement of the Lemma.

THEOREM 3.3. *For $w(z, \hat{z}) = w(z, \hat{z} \mid k^{-1}, \hat{k}^{-1} \mid \Delta)$ the function $\partial\hat{\partial}w(z, \bar{z})$ is real at $z_2 = z_3 = \cdots = 0$ if and only if there exists an antiholomorphic involution $\tau: P \to P$ such that $\tau q = \hat{q}$, $\tau\alpha = \alpha\tau$, $k^{-1}\tau = (\overline{\hat{k}})^{-1} + o((\overline{\hat{k}})^{-4})$, and $\tau_J\Delta = \overline{\Delta}$.*

PROOF. First we prove the necessity of the conditions. Following the proof of Theorem 3.1, consider the surface $\overline{P}$ and the function

$$\widetilde{w}(z, \hat{z}) = w(z, \hat{z} \mid (\overline{\hat{k}})^{-1}, \overline{k}^{-1} \mid \overline{\Delta}).$$

Then $\widetilde{w}(z, \hat{z}) = \overline{w}(\overline{\hat{z}}, \overline{z})$ and therefore $\partial\hat{\partial}\widetilde{w}(z, \hat{z}) = \partial\hat{\partial}w(z, \hat{z})$ on the set

$$\mathcal{D}^2 = \{(z, \hat{z}) \mid \hat{z}_1 = \overline{z}_1, z_2 = \hat{z}_2 = \cdots = 0\}.$$

According to Lemma 3.2, it follows then that $\partial^2\widetilde{w} = \partial^2 w$ on $\mathcal{D}^2$.

Let $\Psi(z, \hat{z}, p)$ be the Baker–Akhiezer function on P associated with the local chart $\varepsilon: k^{-1}: V \to \mathbb{C}$. Repeating the arguments used in the proof of Theorem 3.1 (using Theorems 2.1 and 2.2 instead of Theorem 1.1), we find a local parameter

$$\tilde{k}^{-1} = (\overline{\hat{k}})^{-1} + \sum_{j=3}^{\infty} c_j(\overline{\hat{k}})^{-j}$$

such that the Baker–Akhiezer function $\tilde{\Psi}(z, \hat{z}, p)$ for $p \in V$ (where $\tilde{\varepsilon}: \tilde{k}^{-1}: \widetilde{V} \to \mathbb{C}$ is a local chart) associated with this parameter satisfies the following condition on $\mathcal{D}^2$:

$$\partial \ln \tilde{\Psi}(z, \hat{z}, p) = \partial \ln \Psi(z, \hat{z}, \varepsilon^{-1}\tilde{\varepsilon}(p)).$$

Since the functions $\partial \ln \tilde{\Psi}$ and $\partial \ln \Psi$ are meromorphic on $\overline{P}$ and P respectively, it follows that the mapping $\varepsilon^{-1}\tilde{\varepsilon}: \widetilde{V} \to V$ extends to a biholomorphic mapping $\phi: \overline{P} \to P$. This means that an antiholomorphic mapping $\tau = \phi\sigma$ such that $\tau_J\Delta = \overline{\Delta}$, $\tau q = \hat{q}$, and $k^{-1}\tau = (\overline{\hat{k}})^{-1} + o((\overline{\hat{k}})^{-3}) = (\tilde{k})^{-1}$ is determined; this implies $\tau\alpha = \alpha\tau$ and $\tau^2 = 1$.

Now let us prove that if there exists an antiholomorphic involution $\tau: P \to P$ satisfying the stated conditions, then $\partial\hat{\partial}w(z, \hat{z})$ is real on $\mathcal{D}^2$. According to Lemma 3.1, it is sufficient to prove this statement under the assumption $k^{-1}\tau = (\bar{\hat{k}})^{-1}$. Let $\phi(z, \hat{z}, p)$ be the Baker–Akhiezer function associated to the pair $(k^{-1}, \bar{\hat{k}}^{-1}\tau)$. Then $\Psi(z, \hat{z}, p) = \overline{\Psi(\bar{\hat{z}}, \bar{z}, \tau p)}$ implying that $\xi_1(z, \hat{z}) = \overline{\hat{\xi}_1(\bar{\hat{z}}, \bar{z})}$. According to Theorem 2.1 and Corollary 2.1, it then follows that $\partial w(z, \hat{z}) = \overline{\hat{\partial} w(\bar{\hat{z}}, \bar{z})} + \mathrm{const}$ and therefore

$$w(z, \hat{z}) = \overline{w(\bar{\hat{z}}, \bar{z})} + \sum_{i=1}^{\infty}(a_i z_i + b_i \bar{z}_i).$$

Thus $\partial\hat{\partial}w(z, \hat{z}) = \overline{\partial\hat{\partial}w(\bar{\hat{z}}, \bar{z})}$ and, therefore $\partial\hat{\partial}w(z, \bar{z}) = \overline{\partial\hat{\partial}w(z, \bar{z})}$.

§4. Spinors and differentials on real algebraic curves

In the previous section we have shown that real solutions of the KP equation and real Schrödinger potentials are related to real algebraic curves, i.e., to pairs (P, τ), where P is a compact Riemann surface of genus g and $\tau: P \to P$ is an antiholomorphic involution.

We suppose that the involution τ has fixed points (for solutions of KP this follows from Theorem 2.1 and 2.2). The fixed points form the set of real points of the curve (P, τ). This set decomposes into simple closed curves $a_0, \ldots, a_k$, the so-called *ovals*. A curve is called *orientable* if the ovals split P into two parts P_1 and P_2. In this case $k \leqslant g$ and $k \equiv g \pmod 2$. Otherwise the curve is called *nonorientable* and $k \leqslant g - 1$. For an orientable curve the orientation on P_1 induces an orientation on ovals, which is called the *standard orientation*. The standard orientation on $\{a_0, \ldots, a_k\}$ is determined up to the simultaneous change of orientation on all ovals.

The selection of nonsingular solutions among the real ones requires investigation of the properties of real spinors and differentials on (P, τ).

1. Recall that a line bundle over P is called a *spinor*, or a *theta-characteristic*, if its tensor square coincides with the cotangent bundle on P [18]. We suppose below that the fundamental group $\pi_1(P)$ is nonabelian.

Recall that the group $\mathrm{PSL}(2, \mathbb{R})$ acts on the upper half plane $H = \{z \in \mathbb{C} \mid \mathrm{Im}\, z > 0\}$ by the automorphisms

$$\begin{pmatrix} a & b \\ c & d \end{pmatrix}(z) = \frac{az+b}{cz+d}.$$

Let the group $\mathrm{SL}(2, \mathbb{R})$ act on $H \times \mathbb{C}$ according to the rule

$$\begin{pmatrix} a & b \\ c & d \end{pmatrix}(z, \theta) = \left(\frac{az+b}{cz+d}, (cz+d)\theta \right).$$

A subgroup $\Gamma \subset \mathrm{SL}(2, \mathbb{R})$ is called SL-*Fuchsian* if the natural projection $\phi: \mathrm{SL}(2, \mathbb{R}) \to \mathrm{PSL}(2, \mathbb{R})$ maps Γ to a Fuchsian group acting on H without fixed points.

Assign to an SL-Fuchsian group Γ the spinor bundle $E_\Gamma \colon B_\Gamma \to P_\Gamma$, where $P_\Gamma = H/\phi(\Gamma)$, $B = (H \times \mathbb{C})/\Gamma$, and E_Γ is the mapping generated by the natural projection $H \times \mathbb{C} \to H$.

Let $E \colon B \to P$ be a spinor bundle and $P = H/\widetilde{\Gamma}$ its uniformization. Then $B = (H \times \mathbb{C})/f(\widetilde{\Gamma})$, where

$$f(\tilde{\gamma})(z, \theta) = (\tilde{\gamma}z, \xi_{\tilde{\gamma}}(z)\theta), \quad \tilde{\gamma}(z) = \frac{\tilde{a}z + \tilde{b}}{\tilde{c}z + \tilde{d}}, \quad \xi_{\tilde{\gamma}}^2(z) = (\tilde{a}\tilde{b} - \tilde{c}\tilde{d})(\tilde{c}z + \tilde{d})^2.$$

Therefore there exists an element

$$h(\tilde{\gamma}) = \begin{pmatrix} a & b \\ c & d \end{pmatrix} \in \mathrm{SL}(2, \mathbb{R})$$

such that

$$f(\tilde{\gamma})(z, \theta) = \left(\frac{az + b}{cz + d}, (cz + d)\theta \right) \quad \text{and} \quad E \colon B \to P \cong (E_\Gamma \colon B_\Gamma \to P_\Gamma),$$

where $\Gamma = h(\widetilde{\Gamma})$.

For $\gamma = \begin{pmatrix} a & b \\ c & d \end{pmatrix} \in \Gamma$ we set $\omega_\Gamma(\gamma) = 0$ if $a + d < 0$ and $\omega_\Gamma(\gamma) = 1$ if $a + d > 0$. Since $|a + d| > 2$, we obtain the mapping $\omega_\Gamma \colon \Gamma \to \mathbb{Z}_2 = \mathbb{Z}/2\mathbb{Z}$ and, therefore, the mapping $\omega_\pi \colon \pi_1(P) \to \mathbb{Z}_2$. One can prove that the mapping ω_π determines the mapping $\omega = \omega_E \colon H_1(P, \mathbb{Z}_2) \to \mathbb{Z}_2$, which is an Arf function, i.e., $\omega(a+b) = \omega(a) + \omega(b) + (a, b)$, where $(a, b) \in \mathbb{Z}_2$ is the parity of the intersection index of a and b [19]. It can be also proved [19] that the mapping $E \to \omega_E$ establishes a one-to-one correspondence between the spinor bundles $E \colon B \to P$ and the Arf functions $\omega \colon H_1(P, \mathbb{Z}_2) \to \mathbb{Z}_2$.

2. A *real bundle* over (P, τ) is the pair (E, β), where $E \colon B \to P$ is a bundle and $\beta \colon B \to B$ is an antiholomorphic involution, which maps $E^{-1}(p)$ to $E^{-1}(\tau p)$ linearly.

Let $\mathrm{SL}_\pm(2, \mathbb{R}) \subset \mathrm{GL}(2, \mathbb{R})$ be the group of matrices with determinant ± 1. A subgroup $\Gamma \subset \mathrm{SL}_\pm(2, \mathbb{R})$ is called $\mathrm{SL}_\pm$-*Fuchsian* if $\Gamma_+ = \Gamma \cap \mathrm{SL}(2, \mathbb{R})$ is a SL-Fuchsian group and $\Gamma_+ \neq \Gamma$. For

$$\begin{pmatrix} a & b \\ c & d \end{pmatrix} \in \Gamma - \Gamma_+, \quad \text{set} \quad \begin{pmatrix} a & b \\ c & d \end{pmatrix}(z, \theta) = \left(\frac{a\bar{z} + b}{c\bar{z} + d}, (c\bar{z} + d)\bar{\theta} \right).$$

Then $\Gamma - \Gamma_+$ induces involutions $\beta_\Gamma \colon B_{\Gamma_+} \to B_{\Gamma_+}$ and $\tau_\Gamma \colon P_{\Gamma_+} \to P_{\Gamma_+}$ and generates the real spinor bundle $(E_{\Gamma_+}, \beta_\Gamma)$ over $(P_{\Gamma_+}, \tau_\Gamma)$. It can be proved that (up to isomorphism) all real spinor bundles are of this form [20]. We suppose below that $(E, \beta) = (E_{\Gamma_+}, \beta_\Gamma)$ and $(P, \tau) = (P_{\Gamma_+}, \tau_\Gamma)$.

Let a be an oval of the involution $\tau = \tau_\Gamma$. Changing the group Γ to a conjugate subgroup in $\mathrm{SL}(2, \mathbb{R})$, we can suppose that $I = \{z \in \mathbb{C} \mid \mathrm{Im}\, z > 0\}$ maps to a under the natural projection $\phi \colon H \to P = H/\Gamma_+$ and $\beta_\Gamma(z, \theta) = (-\bar{z}, \bar{\theta})$. The orientation of I by increasing values of $|z|$ determines the orientation of a, which we call the *spinor orientation*, following [20].

LEMMA 4.1. *Let a_1, a_2 be two ovals of (P, τ) with the spinor orientation and suppose $c \subset P$ is a simple closed orientable path on P such that $\tau c = -c$, $a_i \cap c \neq \varnothing$ ($i = 1, 2$). Then c has coinciding intersection indices with a_1 and a_2 if and only if $\omega_E(c) = 1$.*

PROOF. Replacing Γ by a conjugate subgroup, we can suppose that the path c corresponds to the matrix

$$C = \sigma \begin{pmatrix} \lambda & 0 \\ 0 & \lambda^{-1} \end{pmatrix} \in \Gamma_+, \quad \text{where } \lambda > 0, \ \sigma = \pm 1.$$

In this case the oval a_i corresponds to a matrix of the form

$$\sigma_i \begin{pmatrix} \alpha_i(\lambda_i + 1) & \alpha_i^2(\lambda_i - 1) \\ (\lambda_i - 1) & \alpha_i(\lambda_i + 1) \end{pmatrix} \in \Gamma_+, \quad \text{where } \lambda_i > 1, \ \alpha_1, \alpha_2 \text{ and } \sigma_1, \sigma_2 = \pm 1.$$

Since the sets a_i are ovals, it follows that

$$C = C_1 C_2, \quad \text{where } C_i = \sigma_i \begin{pmatrix} -\alpha_i & 0 \\ 0 & \alpha_i^{-1} \end{pmatrix}.$$

Thus $\sigma = \sigma_1 \sigma_2$. But, on the other hand, the intersection indices of c with a_1 and a_2 coincide if and only if $\sigma_1 = \sigma_2$.

A holomorphic section ξ of a bundle (E, β) is called a *real spinor* if $\xi(\tau p) = \beta \xi(p)$.

LEMMA 4.2. *Let a be an oval on (P, τ). Then $\omega_E(a) = 1$ if and only if the sum of orders of zeros of the section ξ on a is even.*

PROOF. Replacing Γ by a conjugate subgroup, we can suppose that the matrix $A = \sigma \begin{pmatrix} \lambda & 0 \\ 0 & \lambda^{-1} \end{pmatrix}$, where $\lambda > 0$, corresponds to the oval a, and to the spinor ξ there corresponds a function $\eta: H \to \mathbb{C}$ such that $\eta(-\bar{z}) = \overline{\eta(z)}$. In this case $\eta(I) \subset \mathbb{R}$ and $\eta(az) = \sigma\sqrt{\lambda}\eta(z)$, where $\sqrt{\lambda} > 0$. Thus, the number of zeros of ξ on a equals the number of zeros of η on $[i, \lambda i]$, which is even if and only if $\sigma = 1$, i.e., if and only if $\omega_E(a) = 1$.

3. A *differential* over P is a section of the cotangent bundle $E: B \to P$. The antiholomorphic involution $\tau: P \to P$ induces an antiholomorphic involution $\beta: E \to E$ such that $\tau\beta = \beta\tau$. A differential ξ is called *real* if $\beta\xi(p) = \xi\tau(p)$.

A local trivialization $f: B_0 \to U_0 \times \mathbb{C}$, $U_0 \subset P$, is called *real* if $f\beta f^{-1}(z, \theta) = (\bar{z}, \bar{\theta})$. In this case $Ef^{-1}(U_0 \cap \mathbb{R})$ is a part of the oval $a \subset P$. The orientation of $\mathbb{R}$ by increasing numbers generates the orientation on a. We call this orientation the *orientation generated by f*.

Let a be an oriented oval and suppose f is a real trivialization, generating the same orientation. Then $f(b) = (z, \xi_f(z)\theta)$, where $\xi_f(\bar{z}) = \overline{\xi_f(z)}$ and $\xi_f(U_0 \cap a) \subset \mathbb{R}$. We call the differential ξ *positive* (respectively, *negative*) at a point $p \in U_0$ if $\xi_f(p) > 0$ (respectively, $\xi_f(p) < 0$). This definition does not depend on the choice of f. We call the differential *positive* (respectively, *negative*) on an oriented oval if it is positive (respectively, negative) at all points of the oval.

In what follows we assume that all the ovals $a_0, a_1, \ldots, a_k$ of the curve (P, τ) are oriented; this orientation is assumed to be induced from the orientation of P_1 in the case the ovals split the surface into parts P_1 and P_2.

THEOREM 4.1. *Let P be a nonorientable real curve. Then for all $0 < n \leq k$ and $0 < m \leq k$ there exists a real holomorphic spinor ξ such that the real differential $\eta = \xi^2$ has an odd number of zeros on the ovals $a_1, \ldots, a_n$, an even number of zeros on the ovals $a_{n+1}, \ldots, a_k$, the differential η being nonnegative on $a_0, \ldots, a_m$ and nonpositive on $a_{m+1}, \ldots, a_k$.*

PROOF. According to [21], the uniformization group $\widetilde{\Gamma} \subset \mathrm{PSL}_{\pm}(2, \mathbb{R})$ of the surface $P/\langle \tau \rangle = H/\widetilde{\Gamma}$ has a system of generators A_i, B_i ($i = 1, \ldots, g$), S such that

(1) $A_1, \ldots, A_k$ correspond to $a_1, \ldots, a_k$ and $A_0 = (\prod_{i=1}^{g} A_i)^{-1}$ corresponds to a_0;
(2) the group $\Gamma_+ = \Gamma \cap \mathrm{PSL}(2, \mathbb{R})$ is generated by A_i, B_i ($i = 1, \ldots, g$) with the defining relations $A_1, \ldots, A_g B_g A_g^{-1} B_g^{-1} \cdots A_1 B_1 A_1^{-1} B_1^{-1} = 1$;
(3) $Sz = -\bar{z}$, $SA_0 S = A_0$, $SA_i S = B_i A_i B_i^{-1}$, $SB_i S = B_i^{-1}$ for $i \leq k$ and $A_i = (SB_i)^2$ for $i > k$.

To the element A_i let us assign the matrix

$$M(A_i) = \begin{pmatrix} a_i^A & b_i^A \\ c_i^A & d_i^A \end{pmatrix} \in \mathrm{SL}(2, \mathbb{R})$$

such that

$$A_i(z) = \frac{a_i^A z + b_i^A}{c_i^A z + d_i^A}, \quad a_i^A + d_i^A < 0 \text{ for } 0 < i \leq n \text{ and } a_i^A + d_i^A > 0 \text{ for } i > n.$$

To the element B_i assign the matrix

$$M(B_i) = \begin{pmatrix} a_i^B & b_i^B \\ c_i^B & d_i^B \end{pmatrix} \in \mathrm{SL}(2, \mathbb{R})$$

such that $i(z) = \frac{a_i^B z + b_i^B}{c_i^B z + d_i^B}$, $a_i^B + d_i^B < 0$ for $i \leq m$ and $a_i^B + d_i^B > 0$ for $m < i \leq g - 1$. Denote by $M(B_g)$ the matrix

$$M(B_g) = \begin{pmatrix} a_g^B & b_g^B \\ c_g^B & d_g^B \end{pmatrix}$$

such that $B_g(z) = \frac{a_g^B z + b_g^B}{c_g^B z + d_g^B}$ and, besides, among the pairs $((a_i^A + d_i^A)(a_i^B + d_i^B))$ there is an odd number of pairs for which $(a_i^A + d_i^A) > 0$ and $(a_i^B + d_i^B) > 0$. (This situation is possible because $(a_g^A + d_g^A) > 0$.) Set $M(S) = \begin{pmatrix} -1 & 0 \\ 0 & 1 \end{pmatrix}$. The matrices $\{M(A_i), M(B_i), M(S)\}$ generate the group Γ. The defining relations coincide with those for the generators $\{A_i, B_i, S\}$ in the group $\widetilde{\Gamma}$.

If the orientation of a_0 coincides with the spinor orientation with respect to $(E_{\Gamma_+}, \beta_\Gamma)$, we set $(E, \beta) = (E_{\Gamma_+}, \beta_\Gamma)$. Otherwise we set $(E, \beta) = (E_{\Gamma_+}, -\beta_\Gamma)$.

Let $a_i, b_i \in H_1(P, \mathbb{Z}_2)$ correspond to A_i, B_i and let ω be the Arf function generated by the bundle (E, β). Then

$$\sum_{i=1}^{g} \omega(a_i)\omega(b_i) \equiv 1 \pmod{2}$$

and, according to [18], the bundle (E, β) has a holomorphic nonzero section $\tilde{\xi}$. One of the sections $(\tilde{\xi} + \beta\tilde{\xi})$ or $(i\tilde{\xi} + \beta(i\tilde{\xi}))$ is a real nonzero section ξ of the bundle (E, β). For $0 < i \leqslant k$ we have $a_i^A + d_i^A < 0$, therefore $\omega(a_i) = 0$ and Lemma 4.2 implies that ξ has an odd number of zeros on a_i. It is easy to see that $\eta = \xi^2$ is a real holomorphic differential which is either nonpositive or nonnegative on each oval. Besides, we have $a_i^B + d_i^B < 0$ for $i \leqslant m$ and therefore $\omega(b_i) = 0$. We know also that the orientation of the oval a_0 coincides with the spinor orientation. Lemma 4.1 then implies that the orientations of the ovals a_i, $i \leqslant m$ coincide with the spinor orientations as well. It follows then that η is nonnegative on a_i for $0 \leqslant i \leqslant m$. One proves similarly that η is nonpositive on a_i for $m < i \leqslant k$.

THEOREM 4.2. *Let (P, τ) be an orientable real curve and $q \in a_0$. Then for all $0 < n \leqslant k$ and $0 \leqslant m \leqslant k$ there exists a real holomorphic spinor ξ such that the real differential $\eta = \xi^2$ has an odd number of zeros on ovals $a_1, \ldots, a_n$, an even number of zeros on the ovals $a_{n+1}, \ldots, a_k$, it is nonnegative on ovals a_i for $i < m$, and it is nonpositive on ovals a_i for $i \geqslant m$. There also exists a meromorphic differential, holomorphic outside at q, having a pole of order two at q, and nonnegative on all ovals.*

PROOF. According to [21], the uniformization group $\tilde{\Gamma} \subset \mathrm{PSL}_{\pm}(2, \mathbb{R})$ of the surface $P/\langle \tau \rangle = H/\tilde{\Gamma}$ has a system of generators A_i, B_i ($i = 1, \ldots, k$), C_j, D_j ($j = 1, \ldots, m$) and S such that

(1) $A_1, \ldots, A_k$ correspond to $a_1, \ldots, a_k$ and

$$A_0^{-1} = \prod_{i=1}^{k} A_i \prod_{j=1}^{m} [C_j, D_j] \quad \text{corresponds to } a_0^{-1};$$

(2) the group $\Gamma_+ = \Gamma \cap \mathrm{PSL}(2, \mathbb{R})$ is generated by the elements A_i, B_i ($i = 1, \ldots, k$), C_j, D_j ($j = 1, \ldots, m$) and $\tilde{C}_j = SC_j^{-1}S$, $\tilde{D}_j = SD_j^{-1}S$ with the defining condition

$$\prod_{i=1}^{k} A_i \prod_{j=1}^{m} [C_j, D_j] \prod_{j=m}^{1} [\tilde{C}_j, \tilde{D}_j] \prod_{i=k}^{1} B_i A_i^{-1} B_i^{-1} = 1;$$

(3) $Sz = -\overline{z}$, $SA_0S = A_0$, $SA_iS = B_iA_iB_i^{-1}$, $SB_iS = B_i^{-1}$.

The remaining part of the proof of the existence of ξ almost literally follows the proof of Theorem 4.1. The existence of the meromorphic differential is proved similarly, but in this case it is based on the uniformization of the surface $P - q$.

4. Now assume that (P, τ) is an M-curve, i.e., P has $g + 1$ ovals $a_0, a_1, \ldots, a_g$. The set of ovals splits the surface P into two spheres with holes P_1 and P_2. Supply the ovals with the orientation induced by the orientation, called *standard*, of the surface P_1.

LEMMA 4.3. *For each $0 \leqslant n < g$ there exists a real holomorphic spinor ξ such that the real differential $\eta = \xi^2$ is positive on a_0, nonnegative on $a_1, \ldots, a_n$, and nonpositive on $a_{n+1}, \ldots, a_g$.*

PROOF. According to [21], there exists a system of generators A_i, B_i ($i = 1, \ldots, g$), S of the uniformizing group $\widetilde{\Gamma}$, where $P/\langle \tau \rangle = H/\widetilde{\Gamma}$, such that
(1) the element A_i correspond to the ovals a_i ($i = 1, \ldots, g$), the element $A_0^{-1} = \prod_{i=1}^{g} A_i$ corresponds to a_0^{-1}, and the orientation of a_i is generated by the orientation of A_i for all i;
(2) $Sz = -\bar{z}$, $SA_iS = B_iA_iB_i^{-1}$, $SB_iS = B_i^{-1}$ for $0 < i \leqslant g$, $SA_0S = A_0$.

Assign to A_i the matrix
$$M(A_i) = \begin{pmatrix} a_i^A & b_i^A \\ c_i^A & d_i^A \end{pmatrix} \in \mathrm{SL}(2, \mathbb{R})$$

such that $A_i(z) = \frac{a_i^A z + b_i^A}{c_i^A z + d_i^A}$ and $\delta(a_i^A + d_i^A) > 0$, where $\delta = -1$ for $1 \leqslant i < g$ and $\delta = 1$ for $i = 0, g$. Assign to B_i the matrix

$$M(B_i) = \begin{pmatrix} a_i^B & b_i^B \\ c_i^B & d_i^B \end{pmatrix} \in \mathrm{SL}(2, \mathbb{R})$$

such that $B_i(z) = \frac{a_i^B z + b_i^B}{c_i^B z + d_i^B}$ and $\delta(a_i^B + d_i^B) > 0$, where $\delta = -1$ for $i \leqslant n$ and $\delta = 1$ for $i > n$. Set $M(S) = \begin{pmatrix} -1 & 0 \\ 0 & 1 \end{pmatrix}$. Then the matrices $M(A_i)$, $M(B_i)$, $M(S)$ generate the $\mathrm{SL}_\pm$-Fuchsian group Γ with defining relations coinciding with those for the generators A_i, B_i, S for the group $\widetilde{\Gamma}$. The end of the proof of Lemma 4.3 coincides with the end of the proof of Theorem 4.1.

This lemma immediately implies

LEMMA 4.4. *For all $0 \leqslant n < g$ there exists a real holomorphic differential η that is positive on $a_0, \ldots, a_n$ and negative on $a_{n+1}, \ldots, a_g$.*

Now consider the moduli space M_g consisting of the curves (P, τ) of genus g with orientation and an order on the set of its ovals $a_0, \ldots, a_k$. For orientable curves, we consider only the standard orientations of ovals. Consider the bundle $E_g: B_g \to M_g$ with fiber $E_g^{-1}(P, \tau)$ consisting of all real holomorphic differentials on (P, τ). Consider a basis on $E_g^{-1}(P, \tau)$ consisting of the differentials $\omega_i = \omega_i(P, \tau)$, $i = 1, \ldots, g$, such that $\oint_{a_j} \omega_i = \delta_{ij}$. The mapping $\omega_i(P, \tau) \mapsto \omega(P', \tau')$ determines a connection Φ on E_g.

THEOREM 4.3. *Equip the ovals of a M-curve (P, τ) with the standard orientation. Then for any real holomorphic differential on (P, τ) there exists an oval on which this differential is positive and an oval on which it is negative.*

PROOF. Let $M_g' \subset M_g$ be the subset consisting of all (P, τ) such that there exists a differential η which is nonnegative on each oval. Using Lemma 4.4, we can assume that η has only simple zeros. Applying the connection Φ to η, we conclude that M_g' is an open set. The same arguments prove that $M_g - M_g'$ is also an open set. But according to [21], M_g is a connected set. It then follows that either $M_g = M_g'$ or $M_g' = \varnothing$.

Let us prove that the set $M_g - M'_g$ contains hyperelliptic curves (P, τ), where $P = \{(x, y) \in \overline{\mathbb{C}}^2 \mid y^2 = f(x)\}$ and $f(x) = (x - a_1) \cdots (x - a_{2g+2})$, $a_i \in \mathbb{R}$, $a_i \neq a_j$, and $\tau(x, y) = (\overline{x}, \overline{y})$. Indeed, the real holomorphic differentials for these curves have the form $h(x)\, dx/\sqrt{f(x)}$, where $h(x)$ is a polynomial of degree $\leq g - 1$ with real coefficients. It then easily follows that there exists an oval on which η is negative.

5. Now consider arbitrary oriented curves (P, τ) of type (g, k), i.e., the curves of genus g with k ovals. Equip the ovals with the standard orientation.

THEOREM 4.4. *For all (g, k) with $k \equiv g + 1 \pmod{2}$, $k \leq g + 1$, there exists a separating curve of type (g, k) with ovals $a_1, \ldots, a_k$ such that there does not exist a real holomorphic differential η positive on the ovals $a_1, \ldots, a_{[(k+1)/2]}$ and positive or having zeros on other ovals.*

PROOF. Consider the Riemann surface P of the curve

$$y^4 - 2y^2[(x - \beta_1) \cdots (x - \beta_m) - (x - \alpha_1) \cdots (x - \alpha_n)]$$
$$+ [(x - \beta_1) \cdots (x - \beta_m) + (x - \alpha_1) \cdots (x - \alpha_n)]^2 = 0,$$

where $\alpha_1 < \cdots < \alpha_k < \beta_1 < \cdots < \beta_m \in \mathbb{R}$, $k > 0$, $m > 2$, n, $m \equiv 0 \pmod{2}$ with the involution $\tau \colon P \to P$ induced by the complex conjugation $(x, y) \mapsto (\overline{x}, \overline{y})$. The projection $(x, y) \mapsto x$ maps the ovals to the segments $[\alpha_{2i-1}, \alpha_{2i}]$. The preimage of each segment, except possibly one, consists of two segments. Denote the oval corresponding to the exceptional segment by $a_{[n/4+1]}$ (in this case $n \equiv 2 \pmod{4}$). Choose the numeration for the other ovals so that the ovals a_i for $i \leq n/4$ are projected to different segments.

The surface P consists of the pairs (x, y) for which

$$y = \pm\sqrt{A} \pm \sqrt{B}, \qquad A = -(x - \alpha_1) \cdots (x - \alpha_n), \quad B = (x - \beta_1) \cdots (x - \beta_m).$$

Consider the involution

$$\tau_\beta(x, \pm\sqrt{A} \pm \sqrt{B}) = (x, \pm\sqrt{A} \mp \sqrt{B}).$$

The pair (P', τ'), where $P' = P/\langle\tau_\beta\rangle$, $\tau' = \tau/\langle\tau_\beta\rangle$, is an M-curve corresponding to the function $y^2 = A$. Suppose there exists a real differential η on P with the properties stated in the theorem. Then the differential $\omega = \eta + \tau_\beta\eta$ induces on P' the real differential ω', which is nonnegative on all ovals, but this contradicts Theorem 4.3.

§5. Real nonsingular solutions of the KP equation

According to §3, real solutions of the KP1 and KP2 equations are constructed from a real curve (P, τ) with ovals $a_0, a_1, \ldots, a_k$ and a point $q \in P$. We assume the ovals to be oriented, and in the case when they split P into two parts P_1 and P_2, their orientation is induced by the orientation of P_1.

1. We start with the KP2 equation. According to Theorem 3.1, for a fixed local chart $V \ni q$ the real solution of the equation

$$u = \partial^2 v = \partial^2 \ln \theta(z_1 U_1 + z_2 U_2 + z_3 U_3 + \Delta)$$

is determined by the point $\Delta \in J$ such that $\tau_J \Delta = \Delta$. The set $\operatorname{Re} J = \{\Delta \in J \mid \tau_J \Delta = \Delta\}$ is called the *real part* of the Jacobian. We call a solution $\tilde{u}$ *similar* to the solution u if it is determined by the same data (P, τ), q and any point $\tilde{\Delta} \in \operatorname{Re} J$ belonging to the same connected component of the space $\operatorname{Re} J$ as Δ. We call a solution u *strongly nonsingular* if all the solutions similar to u are nonsingular.

We claim that a nonsingular solution in the general position is strongly nonsingular. Indeed, the set of values of such a solution $u = \partial^2 \ln \theta(z_1 U_1 + z_2 U_2 + z_3 U_3 + \Delta)$ consists of the points $\partial^2 \theta(r)$, where $r \in \tilde{R}$ and

$$\tilde{R} = \{r \in J \mid r = \Delta + z_1 U_1 + z_2 U_2 + z_3 U_3\}.$$

The functions

$$\partial^2 \ln \theta(z_1 U_1 + z_2 U_2 + z_3 U_3 + (\Delta + z_1^0 U_1 + z_2^0 U_2 + z_3^0 U_3))$$

also are real solutions of KP2. It then follows that $\tilde{R} \in \operatorname{Re} J$. If the solution u is in general position, then the closure of $\tilde{R}$ coincides with the connected component J^0 of the set $\operatorname{Re} J$. This solution is nonsingular if and only if $\tilde{R} \cap (\theta) = \varnothing$, where $(\theta) = \{z \in J \mid \theta(z) = 0\}$ is the theta-divisor. For a solution in general position, this condition is equivalent to $J^0 \cap (\theta) = \varnothing$. The last condition is the same for all similar solutions.

Note that periodic nonsingular real solutions are also strongly nonsingular for a different reason [9].

Now let us describe the connected components of the space J_R [6–8, 22, 23]. The mapping A_J^{-1} takes $\operatorname{Re} J$ to the subset $\operatorname{Re} P^g = \{D \in P^g \mid \tau D = D\}$. One can easily see that the connected components of the set $\operatorname{Re} P^g$ are of the form

$$J_R(s_1, \ldots, s_k) = \{D \in \operatorname{Re} P^g \mid \#(D \cap a_i) = s_i \ (i = 1, \ldots, k)\},$$

where $s_i \in \mathbb{Z}_2$ and $\#(D \cap a_i)$ is the parity of the number of the points in the set $D \cap a_i$ (considered with multiplicities). The number of such components equals 2^k. The mapping A_J takes these components to different connected components of the space J_R.

THEOREM 5.1. *The real solution of the KP2 equation*

$$u = \partial^2 v(z_1 U_1 + z_2 U_2 + z_3 U_3 + \Delta),$$

associated with (P, τ), *is strongly nonsingular if and only if* (P, τ) *is an M-curve and* $\Delta = A_J(J_R(1, \ldots, 1))$.

PROOF. Let $J^0 = J_R(s_1, \ldots, s_k)$ be the connected component containing $A_J^{-1}\Delta$. This component is strongly nonsingular if and only if $J^0 \cap (\theta) = \varnothing$. According to the Riemann theorem on zeros of theta-functions, this condition holds if and only if $q \notin D$ for all $D \in J^0$. The last condition is evidently equivalent to the conditions $k = g$, $J^0 = J_R(1, \ldots, 1)$.

2. Now consider the KP1 equation. According to the results of §3, for a given real curve (P, τ), a point $q \in a_0$, and a local chart $V \ni q$, the solution $u = \partial^2 v$ is determined by a point $\Delta \in J$ such that $\tau_J \Delta + \Delta = 0$. The set

$$\operatorname{Im} J = \{\Delta \in J \mid \Delta + \tau \Delta = 0\}$$

is called the imaginary part of the Jacobian. We call the solution

$$u(z_1, z_2, z_3) = \partial^2 v = \partial^2 \ln \theta(z_1 U_1 + z_2 U_2 + z_3 U_3 + \Delta)$$

strongly nonsingular if all the solutions

$$\tilde{u}(z_1, z_2, z_3) = \partial^2 \ln \theta(z_1 U_1 + z_2 U_2 + z_3 U_3 + \tilde{\Delta})$$

are nonsingular for $\tilde{\Delta}$ in the same connected component of the set $\operatorname{Im} J$ as Δ. A generic solution of the KP1 equation is strongly nonsingular for the same reason as in the case of the KP2 equation.

Now let us describe the connected components of the set $\operatorname{Im} J$ [6–8, 22, 23]. The mapping A_J^{-1} takes $\operatorname{Im} J$ to the set $\operatorname{Im} P^g$. Let the set $J_I(s_1, \ldots, s_k)$ consist of divisors $D \in \operatorname{Im} P^g$ such that there exists a meromorphic differential ω_D with zeros $D + \tau D$ having at q a pole of order not more than 2 and holomorphic outside of q, nonnegative on a_0, and nonnegative on a_i if and only if $s_i = 1$. The differential ω_D is nonpositive on other ovals in this case.

LEMMA 5.1. *The set $\operatorname{Im} J$ splits into 2^k connected components of the form*

$$A_J(J_I(s_1, \ldots, s_k)).$$

PROOF. First let us prove that the sets $A_J(J_I(s_1, \ldots, s_k))$ do are disjoint for different I. The set $Q \subset P^g$ on which the mapping A_J is one-to-one is dense in J and $\operatorname{codim} Q = 1$. It is therefore sufficient to prove the statement for the set $A_J(J_I(s_1, \ldots, s_k)) \cap Q$ for which it is obvious. Set $\beta_J = -\tau_J$. Then $\operatorname{Im} J = \{\Delta \in J \mid \beta_J \Delta = \Delta\}$. The involution $A_J^{-1} \beta_J A_J = \beta \colon P^g \to P^g$ is determined by the condition $\beta D + D \sim \mathcal{K}$, where $\mathcal{K}$ is the canonical class. Hence, $\operatorname{Im} J = \bigcup J_I(s_1, \ldots, s_k)$. Besides, $J = \mathbb{C}^g / \Gamma$, where τ_J is generated by the involution $(z_1, \ldots, z_g) \mapsto (\bar{z}_1, \ldots, \bar{z}_g)$. It then easily follows that the number of connected components of the set $\operatorname{Im} J$ coincides with that of $\operatorname{Re} J = \{\Delta \in J \mid \beta \Delta = \Delta\}$, i.e., with 2^k.

The number of the sets $A_J(J_I(s_1, \ldots, s_k))$ also equals 2^k and all these sets are nonempty. Therefore the set $A_J(J_I(s_1, \ldots, s_k))$ is connected.

THEOREM 5.2. *The solution of the KP1 equation*

$$u = \partial^2 v(z_1 U_1 + z_2 U_2 + z_3 U_3 + \Delta),$$

associated with (P, τ), is strongly nonsingular if and only if the surface $P/\langle \tau \rangle$ is orientable and $\Delta \in A_J(J_I(1, \ldots, 1))$.

PROOF. Let $J^0 = J_I(s_1, \ldots, s_k)$ be the connected component that contains $A_J^{-1}\Delta$. This component is strongly nonsingular if and only if $A_J(J^0) \cap (\theta) = \varnothing$.

According to the Riemann theorem on zeros of the theta-function, this condition holds if and only if $q \notin D$ for all $D \in J^0$. According to Theorems 4.1 and 4.2, the last condition breaks down if $P/\langle \tau \rangle$ is nonorientable or $J^0 \neq J_I(1, \ldots, 1)$.

On the other hand, if $D = A^{-1}\Delta \in J_I(1, \ldots, 1)$, then

$$\sum_{i=0}^{k} \oint_{a_i} \omega_D > 0,$$

which is impossible for a holomorphic ω_D (i.e., $q \in D$) if the ovals split P (i.e., $P/\langle \tau \rangle$ is orientable). Thus, for an orientable $P/\langle \tau \rangle$ and $\Delta \in A_J(J_I(1, \ldots, 1))$, the solution u is strongly nonsingular.

§6. Real nonsingular two-dimensional Schrödinger operators

1. According to §3, real Schrödinger operators are associated to real algebraic curves (P, τ) with holomorphic involution $\alpha: P \to P$ having exactly two fixed points $q, \hat{q}$ and such that $\alpha\tau = \tau\alpha$, $\tau q = \hat{q}$. The ovals of the involution τ are separated into two classes: $c_1, \ldots, c_{2t}$, where $\alpha c_i = c_{2t+1-i}$, and $d_1, \ldots, d_r$, where $\alpha d_i = d_i$. The ovals of the real curve $(P, \hat{\tau})$, where $\hat{\tau} = \tau\alpha$, are also separated into two classes: $\hat{c}_1, \ldots, \hat{c}_{2\hat{t}}$, where $\alpha\hat{c}_i = \hat{c}_{2\hat{t}+1-i}$ and $\hat{d}_1, \ldots, \hat{d}_{\hat{r}}$, where $\alpha\hat{d}_i = \hat{d}_i$. The ovals $c_i, d_i, \hat{c}_i, \hat{d}_i$ are pairwise disjoint. The natural projection $\phi: P \to \widetilde{P} = P/\langle \alpha \rangle$ maps them to the set of ovals of the real curve $(\widetilde{P}, \widetilde{\tau})$, where $\widetilde{\tau} = \tau/\langle \alpha \rangle$.

According to [23], the genus $\tilde{g}$ of the curve $\widetilde{P}$, the numbers $t, r, \hat{t}, \hat{r}$ and the orientability of the surface $\widetilde{P}/\langle \widetilde{\tau} \rangle$ form a complete set of topological invariants of the triple (P, τ, α). These invariants take arbitrary values satisfying the following conditions: $r + \hat{r} \equiv 1 \pmod{2}$, $\tilde{k} \leqslant \tilde{g} + 1$, where $\tilde{k} = t + r + \hat{t} + \hat{r}$, $\tilde{k} \equiv \tilde{g} + 1 \pmod{2}$, if $\widetilde{P}/\langle \widetilde{\tau} \rangle$ is orientable, and $\tilde{k} \leqslant \tilde{g}$ otherwise. The tuple $(t, r, \hat{t}, \hat{r}, \varepsilon)$, where $\varepsilon = 1$ if $\widetilde{P}/\langle \widetilde{\tau} \rangle$ is orientable and $\varepsilon = 0$ otherwise, is called the *triple type* of (P, τ, α).

The surface $\widetilde{P}/\langle \widetilde{\tau} \rangle$ is orientable if and only if the collection of ovals $c_i, d_i, \hat{c}_i, \hat{d}_i$ splits P into two components P_1 and P_2. In this case let us equip the ovals with the orientation coming from the orientation of P_1. If $\widetilde{P}/\langle \widetilde{\tau} \rangle$ is nonorientable, equip the ovals with an arbitrary orientation.

2. Following [13, 23], let us describe the connected components of the set

$$\operatorname{Re} P_r(P, \tau, \alpha) = \{\Delta \in J(P) \mid \tau\Delta = \Delta, \ \Delta + \alpha\Delta = 0\}.$$

This set consists of all divisors $D \in P^{2\tilde{g}}$ such that $\tau D = D$ and there exists a meromorphic differential with the divisor $D + \hat{\tau}D - q - \hat{q}$. Let ω be one of these differentials and $p_i \in d_i - (D + \hat{D})$. Assign to this pair an element $\delta(\omega, p_i) \in \mathbb{Z}_2$ according to the following rule. If ω is negative in p_i, then $\delta(\omega, p_i)$ is the parity of the number of points in D_i belonging to the arc of $d_i - (p_i \cup \alpha p_i)$ going from p_i to αp_i. If ω is positive in p_i, then $\delta(\omega, p_i)$ is the opposite parity. The numbers $\delta(\omega, p_i)$ are the same for all $p_i \in d_i - (D + \hat{\tau}D)$, since the set $(D \cap d_i) \cap (\alpha D \cap d_i)$ consists of second order zeros of the differential ω. Set $s_{d_i}(\omega) = \delta(\omega, p_i)$.

By construction, ω is either nonnegative or nonpositive on the ovals of the involution $\hat{\tau}$. For $a \in \{\hat{c}_1, \ldots, \hat{c}_{\hat{t}}, \hat{d}_1, \ldots, \hat{d}_{\hat{r}}\}$ set $s_a(\omega) = 0$ if ω is nonpositive on a, and $s_a(\omega) = 1$ if ω is nonnegative on a. Denote by $s_{c_i} \in \mathbb{Z}_2$ the parity of the number of points of the set $D \cap c_i$. Changing ω to $-\omega$, we change $s_{d_i}(\omega)$, $s_{\hat{c}_i}(\omega)$, $s_{\hat{d}_i}(\omega)$ to opposite values. For $r > 0$ choose ω so that $s_{\hat{d}_i}(\omega) = 1$. If $r = 0$, then $\hat{r} > 0$. In this case choose ω so that $s_{\hat{d}_1}(\omega) = 1$. In both cases set

$$s_a = s(\omega) \quad \text{for } a \in \{d_1, \ldots, d_r, \hat{c}_1, \ldots, \hat{c}_{\hat{t}}, \hat{d}_1, \ldots, \hat{d}_{\hat{r}}\}.$$

Finally, set

$$J(s_1, \ldots, s_{\tilde{k}}) = \{D \in A_J^{-1}(\operatorname{Re}\operatorname{Pr}(P, \tau, \alpha)) \mid s_i = s_{c_i} \text{ for } i \leq t,$$
$$s_i = s_{d_{i-t}} \text{ for } t < i \leq t+r,$$
$$s_i = s_{\hat{d}_{i-t-r}} \text{ for } t+r < i \leq t+r+\hat{r},$$
$$s_i = s_{\hat{c}_{i-t-r-\hat{r}}} \text{ for } i > t+r+\hat{r}\}.$$

3. Now let us describe the procedure for constructing $D \in J(s_1, \ldots, s_{\tilde{k}})$. Denote by $\tilde{J}(s_1, \ldots, s_{\tilde{k}})$ the set of real meromorphic differentials $\tilde{\omega}$ on $(\tilde{P}, \tilde{\tau})$ satisfying the following conditions:

(1) $\tilde{\omega}$ is holomorphic outside of q and $\hat{q}$; it has a pole of order not greater than one at these points;
(2) $\tilde{\omega}$ has a zero on the path $\phi(c_i)$ if and only if $s_i = 1$;
(3) let $t < i \leq t+r$; then for $s_i = 1$ the differential $\tilde{\omega}$ either has a zero on $\phi(d_i)$ or is nonnegative on $\phi(d_i)$; for $s_i = 0$, the differential $\tilde{\omega}$ either has a zero on $\phi(d_{i-t})$ or is nonpositive on $\phi(d_{i-t})$;
(4) for $t+r < i \leq t+r+\hat{r}$, $\tilde{\omega}$ is nonnegative on $\phi(\hat{d}_{i-t-r})$ for $s_i = 1$ and is nonpositive for $s_i = 0$;
(5) for $i > t+r+\hat{r}$, $\tilde{\omega}$ is nonnegative on $\phi(\hat{c}_{i-t-r-\hat{r}})$ for $s_i = 1$ and is nonpositive for $s_i = 0$.

LEMMA 6.1. *Let ω be a differential on P with divisor $D - q - \hat{q}$, where $D \in J(s_1, \ldots, s_{\tilde{k}})$. Then the projection $\phi: P \to \tilde{P}$ maps this differential to $\tilde{\omega} \in \tilde{J}(s_1, \ldots, s_{\tilde{k}})$. For any differential $\tilde{\omega} \in \tilde{J}(s_1, \ldots, s_{\tilde{k}})$ there exists a divisor $D \in J(s_1, \ldots, s_{\tilde{k}})$ and a differential ω with divisor $D - q - \hat{q}$, which is mapped to $\tilde{\omega}$ under the projection ϕ.*

PROOF. The first statement is a consequence of the definition. Let us prove the second statement. Consider the differential ω on P corresponding to $\tilde{\omega}$. Its divisor $D^+ - q - \hat{q}$ satisfies the condition $\tau D^+ = D^+$, $\alpha D^+ = D^+$. One can easily see that it is possible to choose a half D of the divisor D^+ so that $D \in J(s_1, \ldots, s_{\tilde{k}})$.

LEMMA 6.2. *The set $\operatorname{Re}\operatorname{Pr}(P, \tau, \alpha)$ splits into $2^{\tilde{k}-1}$ connected components of the type $A_J(J(s_1, \ldots, s_{\tilde{k}}))$, where $s_{t+1} = 1$.*

PROOF. The topological type of the triple (P, τ, α) determines the action of τ, $\hat{\tau}$ and α on $H_1(P)$ and, therefore, on Γ, where $J = \mathbb{C}^{2\tilde{g}}/\Gamma$. This allows to prove that $\operatorname{Re}\operatorname{Pr}(P, \tau, \alpha)$ splits into $2^{\tilde{k}-1}$ tori [13]. Now prove that different sets of the type $A_J(J(s_1, \ldots, s_{\tilde{k}}))$, where $s_{t+1} = 1$, are disjoint. The set $Q \in P^g$, where the mapping A_J is one-to-one, is dense in J and codim $Q = 1$. It is then evidently sufficient to prove the statement for the sets $A_J(J(s_1, \ldots, s_{\tilde{k}}) \cap Q)$, for which it is obvious. The number of different sets $A_J(J(s_1, \ldots, s_{\tilde{k}}))$ equals $2^{\tilde{k}-1}$, i.e., it is equal to the number of connected components of the set $\operatorname{Re}\operatorname{Pr}(P, \tau, \alpha)$. To complete the proof, it is sufficient to show that all these sets are nonempty. According to Lemma 6.1, this follows from the nonemptiness of the set $\tilde{J}(s_1, \ldots, s_{\tilde{k}})$, but the last statement is a corollary of Theorems 4.1 and 4.2.

4. According to Theorem 3.3, for a fixed local chart $V \ni q$, the real potential

$$u = \partial \hat{\partial} \ln \theta_{\operatorname{Pr}}(z_1 U_1 + \overline{z}_1 \overline{U}_1 + \Delta)$$

is determined by a point $\Delta \in \operatorname{Re}\operatorname{Pr} = \operatorname{Pr} \cap \operatorname{Re} J$. We call the potential u *strongly nonsingular* if all the potentials

$$\tilde{u} = \partial \hat{\partial} \ln \theta_{\operatorname{Pr}}(z_1 U_1 \overline{z}_1 + \overline{U}_1 + \tilde{\Delta})$$

for $\tilde{\Delta}$ belonging to the same connected component of the space $\operatorname{Re}\operatorname{Pr}$ as Δ are nonsingular. For reasons similar to those for the case of KP, a nonsingular potential in general position is strongly nonsingular. An operator with a strongly nonsingular potential is called a *strongly nonsingular operator*.

LEMMA 6.3. *Let $A_J(J(s_1, \ldots, s_{\tilde{k}}))$ be the connected component of the set $\operatorname{Re}\operatorname{Pr}$ containing Δ. Then the potential u is strongly nonsingular if and only if the set $J(s_1, \ldots, s_{\tilde{k}})$ does not contain holomorphic differentials.*

PROOF. The potential u is strongly nonsingular if and only if $A_J(J(s_1, \ldots, s_{\tilde{k}})) \cap (\theta) = \varnothing$. According to the Riemann theorem on zeros of theta-functions,

$$A_J(J(s_1, \ldots, s_{\tilde{k}})) \cap (\theta) \neq \varnothing$$

if and only if there exists a $D \in J(s_1, \ldots, s_{\tilde{k}})$ such that $q, \hat{q} \in D$. According to Lemma 6.1, this is possible if and only if $J(s_1, \ldots, s_{\tilde{k}})$ contains a holomorphic differential.

THEOREM 6.1. *Let the real Schrödinger operator*

$$L = \partial \hat{\partial} + \partial \hat{\partial} \ln \theta_{\operatorname{Pr}}(z_1 U_1 + \overline{z}_1 \overline{U}_1 + \Delta)$$

associated to (P, τ, α) be strongly nonsingular. Then the surface $P/\langle \tau, \alpha \rangle$ is orientable and $\Delta \in A_J(J(1, \ldots, 1))$.

PROOF. Let $A_J^{-1}(J(s_1, \ldots, s_{\tilde{k}}))$ be the connected component of the set $\operatorname{Re}\operatorname{Pr}$ containing Δ. According to Lemma 6.1, the operator L is strongly nonsingular if and only if $\tilde{J}(s_1, \ldots, s_{\tilde{k}})$ does not contain holomorphic differentials. But according to Theorems 4.1 and 4.2, such differentials always exist if $P/\langle \tau, \alpha \rangle = \widetilde{P}/\langle \widehat{\tau} \rangle$ is nonorientable or $(s_1, \ldots, s_{\tilde{k}}) \neq (1, \ldots, 1)$.

THEOREM 6.2. *Let the number $\tilde{k}$ of the ovals of the curve $(\widetilde{P}, \hat{\tau})$ be equal to $\tilde{g}+1$. Then the real Schrödinger operator $L = \partial\hat{\partial} + \partial\hat{\partial}\ln\theta_{\mathrm{Pr}}(z_1 U_1 + \bar{z}_1 \overline{U}_1 + \Delta)$ associated with (P, τ, α) is strongly nonsingular if and only if $\Delta \in A_J(J(1, \ldots, 1))$.*

PROOF. Let $A_J(J(s_1, \ldots, s_{\tilde{k}}))$ be the connected component of the set $\operatorname{Re}\operatorname{Pr}$ containing Δ. According to Lemma 6.3, the operator L is strongly nonsingular if and only if the set $J(s_1, \ldots, s_{\tilde{k}})$ does not contain holomorphic differentials. According to Theorem 4.3, this condition is satisfied if $s_1 = \cdots = s_{\tilde{k}} = 1$. In the other cases L is not strongly nonsingular because of Theorem 6.1.

Consider the moduli space $M(t, r, \hat{t}, \hat{r})$ of all the triples (P, τ, α) of the type $(t, r, \hat{t}, \hat{r}, 1)$.

THEOREM 6.3. *For $t + r \leqslant \hat{t} + \hat{r}$ there exists an open subset*

$$M^0(t, r, \hat{t}, \hat{r}) \subset M(t, r, \hat{t}, \hat{r})$$

such that the operator $L = \partial\hat{\partial} + \partial\hat{\partial}\ln\theta_{\mathrm{Pr}}(z_1 U_1 + \bar{z}_1 \overline{U}_1 + \Delta)$ associated with $(P, \tau, \alpha) \in M^0(t, r, \hat{t}, \hat{r})$ is strongly nonsingular for $\Delta \in A_J(J(1, \ldots, 1))$. For this subset we have

$$M^0(0, 0, \hat{t}, \hat{r}) = M(0, 0, \hat{t}, \hat{r}).$$

PROOF. According to Theorem 4.4, there exists a real oriented curve $(\widetilde{P}, \hat{\tau})$ with $\tilde{k} = t + r + \hat{t} + \hat{r}$ ovals $\tilde{a}_1, \ldots, \tilde{a}_{\tilde{k}}$ such that there are no real holomorphic differentials that have zeros on $\tilde{a}_1, \ldots, \tilde{a}_{t+r}$ and are nonnegative on the other ovals. For $t + r = 0$, all the oriented curves with $\tilde{k}$ ovals satisfy this condition, because the sum of the ovals is homologous to zero. Let $(P, \tau, \alpha) \in M(t, r, \hat{t}, \hat{r})$ be such that $(\widetilde{P}, \hat{\tau}) = (P/\alpha, \tau/\alpha)$ and the natural projection $\phi: P \to \widetilde{P}$ maps the ovals of the involution τ to $a_1, \ldots, a_{t+r}$. Then the set $\tilde{J}(1, \ldots, 1)$ does not contain holomorphic differentials. According to Lemma 6.3, this implies that the operator L associated with (P, τ, α) is strongly nonsingular for $\Delta \in A_J(J(1, \ldots, 1))$. This condition is equivalent to the condition $A_J(J(1, \ldots, 1)) \cap (\theta) = \varnothing$, which remains valid under small deformations of $P(\tau, \alpha)$.

References

1. B. A. Dubrovin, *Theta-functions and nonlinear equations*, Uspekhi Mat. Nauk; English transl. in Russian Math. Surveys **36** (1981), no. 2, 47–62.
2. B. A. Dubrovin, V. B. Matveev, and S. P. Novikov, *Non-linear equations of the KdV type, finite zone linear operators and Abelian varieties*, Uspekhi Mat. Nauk **31** (1976), no. 1, 55–136; English transl. Russian Math. Surveys **1** (1976).
3. B. A. Dubrovin and S. M. Natanzon, *Real two-zone solutions of the sine-Gordon equation*, Functional Anal. Appl. **16** (1982), 21–33.
4. B. A. Dubrovin, *Matrix finite zone operators*, Itogi Nauki i Tekhniki. Sovremennye Problemy Matematiki, vol. 23, VINITI, Moscow, 1983, pp. 33–78; Englsih transl. in J. Soviet Math. **28** (1985), no. 1.
5. I. M. Krichever, *Algebro-geometric construction of Zakharov–Shabat equations and their periodic solutions*, Dokl. Akad. Nauk SSSR **227** (1976), no. 2, 291–294; English transl. in Soviet Math. Dokl. **17** (1976).
6. J. D. Fay, *Theta function of Riemann surfaces*, Lecture Notes in Math., vol. 352, Springer-Verlag, Berlin–Heidelberg–New York, 1973.
7. B. H. Gross and J. Harris, *Real algebraic curves*, Ann. Sci. École Norm. Sup. (4) **14** (1981), 157–182.

8. V. Vinnikov, *Self-adjoint determinantal representations of real plane curves*, Math. Ann. **296** (1993), 453–478.
9. B. A. Dubrovin and S. M. Natanzon, *Real theta-functional solutions of KP equation*, Math. USSR-Izv. **52** (1988), no. 26, 267–286.
10. A. P. Veselov and S. P. Novikov, *Finite zone two-dimensional Schrödinger operators. Explicit formulae and evolutionary equations*, Dokl. Akad. Nauk SSSR **279** (1984), no. 4, 20–24; English transl. in Soviet Math. Dokl. **30** (1984).
11. _____, *Finite zone two-dimensional Schrödinger operators. Potential operators*, Dokl. Akad. Nauk SSSR **279** (1984), no. 4, 784–788; English transl. in Soviet Math. Dokl. **30** (1984).
12. S. M. Natanzon, *Differential equations for Prym theta-functions. A criterion for two-dimensional finite zone potential Schrödinger operators to be real*, Funktsional. Anal. i Prilozhen. **26** (1992), no. 1, 17–26; English transl. in Functional Anal. Appl. **26** (1992).
13. _____, *Primians of real curves and their applications to effectivization of Schrödinger operators*, Funktsional. Anal. i Prilozhen. **23** (1989), no. 1, 41–55; English transl. in Functional Anal. Appl. **23** (1989).
14. E. Date, M. Jimbo, M. Kashiwara, and T. Miwa, *Transformation groups for soliton equations. IV. A new hierarchy of soliton equations of KP-type*, vol. 359, Preprint RIMS Kyoto: RIMS6, 1981.
15. I. A. Taimanov, *Prym varieties of ramified coverings and nonlinear equations*, Mat. Sb. **181** (1990), no. 7, 934–950; English transl. in Math. USSR-Sb. **70** (1991).
16. S. M. Natanzon, *Spinor and differentials of real algebraic curves*, Advances in Soviet Mathematics.
17. E. Date, M. Kashiwara, M. Jimbo, and T. Miwa, *Transformation Groups for Soliton Equations.*, Proceedings of RIMS Symposium on Non-Linear Integrable Systems, World Science, Singapore, 1983, pp. 39–119.
18. D. Mumford, *Tata lectures on theta*, Progress in Mathematics, vol. 28, Birkhäuser, Boston–Basel–Stuttgart, 1983; vol. 43, 1984.
19. S. M. Natanzon, *Moduli spaces of Riemann $N = 1$ and $N = 2$ supersurfaces*, J. Geometry and Physics **12** (1993), 35–54.
20. _____, *Moduli spaces of Riemann and Klein supersurfaces*, Development in Mathematics. The Moscow School (1993), Chapman & Hall, 100–130.
21. _____, *Moduli spaces of real curves*, Trudy Moskov. Mat. Obshch. **37** (1987), 219–253; English transl., Trans. Moscow Math. Soc. **1980**, no. 1, 233–272.
22. SR. Silhol, *Real Abelian varieties and the theory of Comessatti*, Math. Z. **181** (1982), no. 3, 233–272.
23. S. M. Natanzon, *Klein surfaces*, Uspekhi Mat. Nauk **45** (1990), no. 6, 47–90; English transl. in Russian Math. Surveys **45** (1990).

Translated by S. LANDO

Representations of Krichever–Novikov Algebras

O. K. SHEINMAN

Introduction

Two different classes of Lie algebras are called Krichever–Novikov algebras; both were introduced by I. Krichever and S. Novikov in [1–3]. These are the class $\mathcal{A}(\Gamma^*)$ of Lie algebras of affine type, and the class $\mathcal{V}(\Gamma^*)$ of Lie algebras of Virasoro type, both related to a compact Riemann surface Γ with two marked points $P_\pm$ (where $\Gamma^* = \Gamma \setminus \{P_\pm\}$). The class $\mathcal{A}(\Gamma^*)$ contains the subclass $\mathcal{H}(\Gamma^*)$ of algebras of Heisenberg type, also considered by A. Jaffe, S. Klimek, and A. Lesniewsky [4]. In the recent papers of M. Schlichenmaier (see [12] for the references) and M. Bremner [13], the multipoint generalization of Virasoro-type algebras and the universal central extensions of current algebras on Riemann surfaces respectively are considered. Krichever–Novikov algebras of Virasoro type, as well as those of Heisenberg type, arise as the basis of the operator quantization formalism for multiloop string diagrams [2–3].

In [5–7] monodromy groups appear in the framework of the orbital approach originally developed by I. Frenkel [8] and G. Segal [9] for affine Kac–Moody algebras. The modules with highest weight (= modules corresponding to the orbits with commutative monodromy groups) are constructed. The abelian differentials on Γ appear as weights and the Baker–Akhiezer functions turn out to be (in model cases) weight vectors. This theory is sufficiently well developed, in particular, it includes a Weil–Kac-type formula for characters [7].

This paper is devoted to the further examination of the highest weight representations, as well as to more general representations of the algebras in the classes $\mathcal{A}(\Gamma^*)$ and $\mathcal{V}(\Gamma^*)$, namely to representations with noncommutative monodromy groups. It should be noted that the latest have no analogs in the representation theory of affine Kac–Moody algebras.

Let us give some preliminary definitions. Let Γ be a compact algebraic curve over $\mathbb{C}$ of genus g with two marked points $P_\pm$, $\mathcal{A}^\Gamma$ be the algebra of meromorphic functions on Γ that are regular outside the points $P_\pm$. Let $\mathfrak{g}$ be a semisimple finite-dimensional Lie algebra over $\mathbb{C}$ and $\mathfrak{h}$ be its Cartan subalgebra. We call the algebra

(I.1) $$\mathcal{T} = \mathfrak{g} \otimes_{\mathbb{C}} \mathcal{A}^\Gamma$$

1991 *Mathematics Subject Classification.* Primary 17B68; Secondary 58F07.
Supported in part by International Science Foundation (Grant MD8000).

with the usual bracket the *current algebra* on Γ [1].

Let $c_0, \ldots, c_{2g}$ be the basis of 1-homology of the Riemann surface Γ^* (it is clear that $\dim H_1(\Gamma^*, \mathbb{C}) = 2g + 1$). Here $c_1, \ldots, c_g$ are basic a-cycles, $c_{g+1}, \ldots, c_{2g}$ are basic b-cycles, and c_0 is the so-called *separating cycle* introduced in [1]. The separating cycle is defined as a cycle homologous to a small circle around P_+ or P_- on $\Gamma \setminus \{P_\pm\}$. Let the cycles $c_\pm$ be these small circles themselves.

Let us consider the following cocycle γ [1] on the algebra T:

$$(I.2) \qquad \gamma(xA, yB) = \frac{(x, y)}{2\pi i} \oint_{c_0} A \, dB.$$

Here $x, y \in \mathfrak{g}$, $A, B \in \mathcal{A}^\Gamma$, (x, y) is the Cartan–Killing form of the Lie algebra $\mathfrak{g}$.

DEFINITION I.1. The central extension $\mathcal{G}$ of the current algebra T by means the one-dimensional space Z (with formal generator c) and the cocycle Γ is called the *affine Lie algebra on the Riemann surface* Γ^*:

$$\mathcal{G} = T \oplus Z.$$

Let us introduce, following [1], a quasigraded Lie algebra structure on G.

DEFINITION I.2 [1]. The Lie algebra G is said to be *quasigraded* if it has a decomposition into the direct sum of linear spaces G_j, where $j \in \mathbb{Z} + g/2$ and there exists a natural number N such that

$$[G_i, G_j] \subseteq \bigoplus_{k=i+j-N}^{i+j+N} G_k.$$

Following to [1], let us define the function $\tilde{A}_j \in \mathcal{A}^\Gamma$, where $|j| > g/2$, as the unique function up to scalar factor with the following asymptotic behavior

$$(I.3) \qquad \tilde{A}_j(z_\pm) = \alpha_j^\pm z_\pm^{\pm j - g/2}(1 + O(z_\pm)), \qquad \alpha_j^+ = 1,$$

at the points $P_\pm$ (here $z_\pm$ are local parameters in neighborhoods of the points $P_\pm$). In the case $|j| \leq g/2$, $j \neq g/2$, we put

$$(I.4) \qquad \tilde{A}_j = \alpha_j^\pm z_\pm^{\pm j - g/2 \pm 1/2 - 1/2}(1 + O(z_\pm)), \qquad \alpha_j^+ = 1,$$

and $\tilde{A}_{g/2} \equiv 1$.

PROPOSITION I.1 [1]. *The functions* $\tilde{A}_j$, $j \equiv g/2 \pmod{\mathbb{Z}}$ *form a basis of the algebra* $\mathcal{A}^\Gamma$ *regarded as a linear space.*

We modify the notation and in the sequel use the basis $\{A_j\}$ instead $\{\tilde{A}_j\}$, where $A_j = \tilde{A}_{j-g/2}$ if $j < 0$ or $j > g$, $A_0 = \tilde{A}_{g/2} \equiv 1$, $A_j = \tilde{A}_{j-g/2-1}$ if $1 \leq j \leq g$. Let $\mathcal{G}_j = \mathfrak{g} A_j$ for $j \neq g$ and $\mathcal{G}_g = \mathfrak{g} \oplus \mathbb{C}c$.

PROPOSITION I.2. *The following decomposition holds*:

$$\mathcal{G} = \bigoplus_{j \in \mathbb{Z}} \mathcal{G}_j.$$

This decomposition defines the structure of a quasigraded Lie algebra on $\mathcal{G}$.

The proof follows from Proposition I.1 and the multiplication formulas for the functions A_j [1, Section 3].

For the sequel we need some supplementary notation. Let $\mathcal{A}_{\pm}^{\Gamma} \subset \mathcal{A}^{\Gamma}$ be the subalgebras of functions from $\mathcal{A}^{\Gamma}$ that are regular at the points $P_{\pm}$ respectively. It is clear that the functions A_j with $j > g$ form a basis in $\mathcal{A}_{+}^{\Gamma}$ and the functions A_j with $j < 0$ form a basis in $\mathcal{A}_{-}^{\Gamma}$. We also denote by $\mathcal{A}_0^{\Gamma}$ the subspace of $\mathcal{A}^{\Gamma}$ spanned by $\{A_j : 0 \leqslant j \leqslant g\}$ and put $\mathcal{G}_{\pm} = \mathfrak{g} \otimes \mathcal{A}_{\pm}^{\Gamma}$ and $\mathcal{G}_0 = \mathfrak{g} \otimes \mathcal{A}_0^{\Gamma}$.

PROPOSITION I.3. $1°. \; \mathcal{A}^{\Gamma} = \mathcal{A}_{+}^{\Gamma} \oplus \mathcal{A}_0^{\Gamma} \oplus \mathcal{A}_{-}^{\Gamma}.$
$2°. \; T = \mathcal{G}_{+} \oplus \mathcal{G}_0 \oplus \mathcal{G}_{-}.$

The proof follows from the definitions. Decomposition I.3.1° was introduced in [1].

Heisenberg-type algebras $\mathcal{H} \in \mathcal{H}(\Gamma^*)$ appear as the particular case of $\mathcal{G}$ when $\mathfrak{g}$ is a (pseudo-)Euclidean linear space with the trivial bracket operation.

Finally, the *Virasoro-type algebra* $\mathcal{L}^{\Gamma} \in \mathcal{V}(\Gamma^*)$ is defined as follows [1]. Let $\mathcal{L}^{\Gamma} = \mathcal{L}(\Gamma, P_{\pm})$ be the algebra of meromorphic vector fields on Γ which are holomorphic outside the points $P_{\pm}$. In the case $g \neq 1$, the basis of the space $\mathcal{L}^{\Gamma}$ is formed by the vector fields e_j with the following asymptotic behavior at the points $P_{\pm}$ [1]:

(I.5) $\qquad e_j = \varepsilon_j^{\pm} z_{\pm}^{\pm j - g_0 + 1}(1 + O(z_{\pm}))\, \partial/\partial z_{\pm}, \qquad g_0 = 3g/2, \; \varepsilon_j^{+} = 1.$

In the case $g = 1$, one has $e_j = A_j \, \partial/\partial z$, where z is the global coordinate and A_j are defined above.

Like affine-type algebras, the algebra $\mathcal{L}^{\Gamma}$ has the decomposition

(I.6) $\qquad\qquad\qquad\qquad \mathcal{L}^{\Gamma} = \mathcal{L}_{-}^{\Gamma} \oplus \mathcal{L}_0^{\Gamma} \oplus \mathcal{L}_{+}^{\Gamma},$

where the sets $\mathcal{L}_{\pm}^{\Gamma}$ are generated (as linear spaces) by e_j with $j > g_0$ ($j < -g_0$) respectively and $\mathcal{L}_0^{\Gamma}$ is generated (as a linear space) by e_j with $|j| \leqslant g_0$. The subspaces $\mathcal{L}_{\pm}^{\Gamma}$ are Lie subalgebras of $\mathcal{L}^{\Gamma}$, while $\mathcal{L}_0^{\Gamma}$ is not.

In what follows an important role is played by the notion of monodromy group. Let $\pi_1(\Gamma^*)$ denote the fundamental group of Γ^*. For affine-type (Heisenberg-type) algebras by *monodromy group* we mean the image of any homomorphism $\pi_1(\Gamma^*) \to \exp \mathfrak{g}$ (defined up to conjugation). For the algebra $\mathcal{L}^{\Gamma}$, the group $\exp \mathfrak{g}$ must be replaced by $SL_2(\mathbb{C})$. Equivalently, we can consider the equivalence classes of certain representations of the fundamental group $\pi_1(\Gamma^*)$. In what follows they are referred to as the *standard representations*.

There is a remarkable correspondence between the orbits of the coadjoint action on the one hand and monodromy groups on the other hand.

PROPOSITION I.4. *The orbits of coadjoint action are in one-to-one correspondence with conjugation classes of monodromy groups.*

This statement was proved by I. Frenkel [8] and G. Segal [9] for affine Kac–Moody algebras, and by the author [5, 6] for Krichever–Novikov algebras of affine type and abelian monodromy groups. In its full generality, this proposition reduces to the Riemann-Hilbert problem on a Riemann surface. By Kirillov's orbit method and Proposition I.4, a generic irreducible representation of a Krichever–Novikov algebra must depend on some monodromy group. For abelian monodromy groups, the corresponding construction was suggested in [7], as we mentioned above. Note that the problem of nonabelian monodromy groups does not arise for Kac–Moody and classical Virasoro algebras. In what follows we study the representations of Krichever–Novikov algebras corresponding to arbitrary (in particular nonabelian) monodromy groups.

The author is grateful to I. Krichever for his interest in this work.

§1. Representations of affine-type algebras with arbitrary monodromy groups

Let us begin with the case of an affine Lie algebra $\mathcal{G}$ and take some monodromy group M with generators $\mu_0, \ldots, \mu_{2g}$. Suppose that $\mu_k = \gamma_k h_k \gamma_k^{-1}$, where $h_k \in H = \exp\mathfrak{h}$, $\gamma_k \in G = \exp\mathfrak{g}$ for each $k = 0, \ldots, 2g$ and $\gamma_0 = 1$. Also put $\mathfrak{h}_k = \gamma_k \mathfrak{h} \gamma_k^{-1}$, $\mathfrak{n}_k = \gamma_k \mathfrak{n}_+ \gamma_k^{-1}$, and $\mathfrak{b}_k = \gamma_k \mathfrak{b}_0 \gamma_k^{-1}$, where $\mathfrak{n}_+ \in \mathfrak{g}$ is the upper nilpotent subalgebra in $\mathfrak{g}$ and $\mathfrak{b}_0 = \mathfrak{h} \oplus \mathfrak{n}_+$ is the corresponding Borelian subalgebra. First of all we want to assign a representation T of the group B to each set $\{\mu_k \mid k = 0, \ldots, 2g\}$, where B is equal to the direct product of $B_k = \exp\mathfrak{b}_k$ ($k = 0, \ldots, 2g$), i.e., $B = B_0 \times \cdots \times B_{2g}$. To each h_k ($k = 0, \ldots, 2g$) we assign a character $\chi_k \colon H_k \to S^1$ of the group $H_k = \gamma_k H \gamma_k^{-1}$ as follows: we consider $\ln h_k$ and then, using duality, assign to it the Lie algebra character $\dot{\chi}_k \in \mathfrak{h}_k^*$ and finally put $\chi_k = \exp\dot{\chi}_k$.

REMARK 1.1. Both $\ln h_k$ and $\dot{\chi}_k$ are defined up to a shift by an element of the corresponding lattices, which led in [7] (also see the Appendix below) to the notion of *Bloch weight* of representation (which is the analog in representation theory of the notion of Bloch spectrum in soliton theory [10, 11]).

Now we define the representation T_k of the group B_k in a one-dimensional linear space L_k as follows: if v_k is the generator of L_k, then for each $k = 0, \ldots, 2g$ we set

$$(1.1) \qquad (\exp\mathfrak{n}_k)v_k = v_k \quad \text{and} \quad \beta_k v_k = \chi_k(\beta_k)v_k \qquad (\beta_k \in H_k).$$

By definition, each element $\beta \in B$ is of the form $\beta = \beta_0 \cdots \beta_{2g}$ ($\beta_k \in B_k$, $k = 0, \ldots, 2g$). We put

$$(1.2) \qquad L = L_0 \otimes \cdots \otimes L_{2g},$$
$$(1.3) \qquad T(\beta) = T_0(\beta_0) \otimes \cdots \otimes T_{2g}(\beta_{2g}).$$

The corresponding representation of the Lie algebra $\mathfrak{b} = \mathfrak{b}_0 \oplus \cdots \oplus \mathfrak{b}_{2g}$ is given by

$$(1.4) \qquad \dot{T}(b_k) = 1 \otimes \cdots \otimes \dot{T}_k(b_k) \otimes \cdots \otimes 1,$$

where $b_k \in \mathfrak{b}_k$ ($k = 0, \ldots, 2g$) and $\dot{T}_k$ is the differential of the representation T_k.

We shall consider the Lie algebra $\mathcal{B} = \mathfrak{b} \oplus Z \oplus \mathcal{G}_+$ as the analog of the Borelian subalgebra in $\mathcal{G}$. First we extend the action of $\mathfrak{b}$ in L to the action of $\mathcal{B}$, setting

$$(1.5) \qquad \mathcal{G}_+ v_k = 0, \quad c v_k \doteq \zeta v_k \qquad (k = 0, \ldots, 2g).$$

Both the universal enveloping algebra $U(\mathcal{G})$ and the linear space L are modules over $\mathcal{B}$, so we can define the space

$$(1.6) \qquad V_M^{\mathcal{G}} = U(\mathcal{G}) \otimes_{U(\mathcal{B})} L.$$

The action of Borelian subalgebra $\mathfrak{b} = \mathfrak{h} \oplus \mathfrak{n}_+ \in \mathfrak{g}$ on $V_M^{\mathcal{G}}$ is identified with the action of $\mathfrak{b}_0$.

DEFINITION 1.1. The linear space $V_M = V_M^{\mathcal{G}}$ with the natural $\mathcal{G}$-module structure (defined by action of $\mathcal{G}$ on the first tensor cofactor) is called the *Verma module* of the Lie algebra $\mathcal{G}$ with monodromy group M.

The module $V_M^{\mathcal{G}}$ admits different kinds of (poly-)quasigradings. One of them is defined in [7] (also see the Appendix below). Here we shall define another kind of quasigrading, which is appropriate for the purposes of the Virasoro construction (see §3). For $v \in V_M^{\mathcal{G}}$, the polygrading will be denoted by $\deg v = (\deg_0 v, \ldots, \deg_g v)$. In the zero component, this polygrading will be a quasigrading in the sense of Definition I.2, and in the other components it will be the usual grading. Let us fix

$$v = v_0 \otimes \cdots \otimes v_{2g},$$

and then modify the "natural basis" in $V_M^{\mathcal{G}}$ by changing the basis elements $(h_i A_k) v$ and $(h_i A_k)(h_j A_l) v$ ($i, j = 1, \ldots, \dim \mathfrak{h}; k, l = 1, \ldots, g; h_i \in \mathfrak{h}_k, h_j \in \mathfrak{h}_l$) to the elements $w_{ik} = (h_i A_k) v - \dot{\chi}_k(h_i) v$ and $w_{ik,jl} = (h_i A_k)(h_j A_l) v - \dot{\chi}_k(h_i) \dot{\chi}_l(h_j) v$ respectively. Put $\deg w_{ik} = \deg v - (0, \ldots, 1, \ldots, 0)$ (1 at the kth position), $\deg w_{ik,jl} = \deg v - (0, \ldots, 0, 1, 0, \ldots, 0, 1, 0, \ldots, 0)$ (1 at the kth and lth positions or 2 in the kth position when $k = l$). Then for each $h \in \mathfrak{h}_k, h' \in \mathfrak{h}_l$, one has the relations

$$(1.7) \qquad (h A_k) v = \dot{\chi}_k(h) v + \ldots,$$

$$(1.8) \qquad (h A_k)(h' A_l) v = \dot{\chi}_k(h) \dot{\chi}_l(h') v + \ldots,$$

where ... denotes elements of lower degree (where the order is defined component by component). Now we introduce monomials of the form

$$(1.9) \qquad (x_1 A_{j_1}) \cdots (x_s A_{j_s}) w,$$

where $x_p \in \mathfrak{g}$ ($p = 1, \ldots, s$), $j_1 \leqslant \cdots \leqslant j_s < g$, $w = v$ or $w = w_{ik}$ or $w = w_{ik,jl}$ ($i, j = 1, \ldots, \dim \mathfrak{h}, k, l = 1, \ldots, g$). We shall regard them as quasihomogeneous elements in $\mathcal{G}$ of degree $\deg w + \deg A_{j_1} + \cdots + \deg A_{j_s}$, where $\deg A_j = (j, 0, \ldots, 0)$ if $j < 0$ and $\deg A_j = (0, \ldots, -1, \ldots, 0)$ (-1 at jth position) if $1 \leqslant j \leqslant g$. So we have defined the quasihomogeneous subspaces of V_M and the quasigrading together with them.

REMARK 1.2. If $\dot{\chi}_0 \neq 0$, then (1.7) follows from (1.8) and (1.1).

REMARK 1.3. Relations (1.7), (1.8) can be assumed in addition to (1.1), (1.4), (1.5) if we think of the elements $A_j \in \mathcal{A}_0^\Gamma$ as elements of zero degree. Then one does not need to change the "natural basis" in $V_M^{\mathcal{G}}$.

PROPOSITION 1.1. *If M is an abelian monodromy group in general position, then $V_M^{\mathcal{G}}$ is endowed with the structure of an N-module for each abelian monodromy group N which commutes with M.*

The proof is analogous to the proof of Proposition 1.2 below.

Now let us consider the case of nonabelian M. In this case there is no natural N-action in $V_M^{\mathcal{G}}$ for an arbitrary monodromy group N, but there is an extension $\widetilde{V}_M^{\mathcal{G}} \supseteq V_M^{\mathcal{G}}$ endowed with such an action. In the simplest case of dominant (i.e., integer, nonnegative) weights $\dot{\chi}_0, \dots, \dot{\chi}_{2g}$, this extension coincides with $V_M^{\mathcal{G}}$ (otherwise the representations L and $\widetilde{L}$ are infinite-dimensional and the Lie groups and the corresponding Lie algebras act in different linear spaces in general). The extension $\widetilde{V}_M$ is constructed as follows.

Let $\widetilde{T}_k$ be an irreducible representation of the group G in the finite-dimensional linear space $\widetilde{L}_k$ generated by the highest vector v_k satisfying conditions (1.1). We put

$$(1.10) \qquad \widetilde{L} = \widetilde{L}_0 \otimes \cdots \otimes \widetilde{L}_{2g}, \qquad \widetilde{V}_M = U(\mathcal{G}) \otimes_{U(\widetilde{\mathfrak{g}})} \widetilde{L},$$

where $\widetilde{\mathfrak{g}} = \mathfrak{g} \oplus \cdots \oplus \mathfrak{g} \oplus Z \oplus \mathcal{G}_+$ ($\mathfrak{g}$ is repeated $2g+1$ times). The action of the subalgebra $\mathfrak{g} \in \mathcal{G}$ is identified with the action of the first $\mathfrak{g}$ from the definition of $\widetilde{\mathfrak{g}}$.

PROPOSITION 1.2. *1°. If $\dot{\chi}_0, \dots, \dot{\chi}_{2g}$ is the set of dominant weights, then $V_M \cong \widetilde{V}_M$.*

2°. $\widetilde{V}_M$ is equipped with the natural structure of a module over an arbitrary monodromy group, in particular over the group M.

PROOF. Part 1° follows from a simple comparison of (1.6) and (1.10).

Let N be an arbitrary monodromy group with generators $v_0, \dots, v_{2g}$. For the basis element $l = l_0 \otimes \cdots \otimes l_{2g}$ ($l_k \in \widetilde{L}_k, k = 0, \dots, 2g$), we put

$$v_k l = l_0 \otimes \cdots \otimes (v_k l_k) \otimes \cdots \otimes l_{2g}.$$

In this way we have defined an N-module structure on $\widetilde{L}$. As to $U(\mathcal{G})$, it is endowed with the natural action of N. For $u = (x_1 A_{j_1}) \cdots (x_n A_{j_n}) \in U(\mathcal{G})$, we put

$$vuv^{-1} = (x_1^v A_{j_1}) \cdots (x_n^v A_{j_n}),$$

where $x_s^v = (\operatorname{Ad} v) x_s$ ($s = 1, \dots, n$). Now we can define the action of N on the elements of the form $u \otimes l$ ($u \in U(\mathcal{G}), l \in \widetilde{L}$) which span all of $\widetilde{V}_M$:

$$v(u \otimes l) = (vuv^{-1}) \otimes vl.$$

The proposition is proved.

REMARK 1.4. If there are nondominant weights among $\dot{\chi}_0, \dots, \dot{\chi}_{2g}$, we must take some infinite-dimensional representations as the corresponding $\widetilde{T}_k$. The representation $\widetilde{T}_k$ corresponding to a nondominant $\dot{\chi}_k$ is induced by the character χ_k of the Borelian subgroup B_k. We do not deal with this case in this paper. In what follows we consider either abelian monodromy groups or nonabelian ones but with a dominant sets of weights.

Now let us turn to the definition of the generic representation of $\mathcal{G}$ with highest vector. This definition is motivated by Propositions 1.2, 1.3.

DEFINITION 1.2. A $\mathcal{G}$-module V is said to be a *module with monodromy group M and highest vector v_M* if it is quasigraded and the following conditions hold:

$1°$. $v_M \in V$ is a singular vector (i.e., (1.5) holds for v_M) and $V = U(\mathcal{G})v_M$.

$2°$. V is N-module for each monodromy group N. For each $v \in N$, $u \in U(\mathcal{G})$, $v \in V$,
$$v(uv) = (vuv^{-1})vv.$$

$3°$. Denote by $v_0, \ldots, v_{2g}$ some generators of monodromy group N and let $v_k = (v_k)^{(-)}h_k(v_k)^{(+)}$, where $(v_k)^{(\pm)} \in \gamma_k(\exp \mathfrak{n}_\pm)\gamma_k^{-1}$, $h_k \in H_k$, then $v_k v_M = (v_k)^{(-)}\chi_k(h_k)v_M$.

REMARK 1.5. The requirement $3°$ of Definition 2 is motivated by the mode of action of v_k on the highest vector in (1.10).

COROLLARY 1.1. *The action of the group M itself on the element v_M is as follows:* $\mu_k v_M = \chi_k(h_k)v_M$ $(k = 0, \ldots, 2g)$.

COROLLARY 1.2. *For each monodromy group N, the quasihomogeneous subspaces of the module V are N-invariant and finite-dimensional.*

For each set $s = (s_0, \ldots, s_g)$, let $V_s \subset V$ be a quasihomogeneous subspace of degree s, and let $S = \{s\}$. In the sequel it is assumed that $\dim V_s < \infty$.

DEFINITION 1.3. Let a be a linear operator acting in V such that for each $s \in S$ the space V_s is a-invariant. The formal series
$$\operatorname{Tr}_V a = \sum_{s \in S}(\operatorname{tr}_{V_s} a) A^s$$
is said to be the *graded trace* of the operator a.

Let $\mathcal{O}$ denote either the space of all monodromy groups or the space of abelian monodromy groups, depending on the class of modules under consideration. Consider each element of the space $\mathcal{O}$ as a class of homomorphisms $\sigma: \pi_1(\Gamma^*) \to G$ (defined up to conjugation).

DEFINITION 1.4. The function $\operatorname{ch} V$ on $\pi_1(\Gamma^*) \times \mathcal{O}$ defined by
$$(\operatorname{ch} V)(p, \sigma) = \operatorname{Tr}_V \sigma(p)$$
is said to be the *reduced character* of the module V.

In the case of KN-algebras, the theory of characters is sufficiently well developed only for the highest weight modules. Representation theory for the later is considered in detail in [7]. A brief outline of that theory is contained in the Appendix. In what follows it is enough to consider a *highest weight module* for a KN-algebra of affine type as a module with abelian monodromy group. Suppose V is such a module. Since the monodromy group M is defined up to conjugation, one can consider its generators $\mu_0, \ldots, \mu_{2g}$ as elements of the same Cartan subgroup H. In the notation introduced above, this means that $\gamma_0 = \cdots = \gamma_{2g} = 1$, and $\chi_0, \ldots, \chi_{2g}$ can be regarded as linear functionals on the same commutative Lie algebra $\mathfrak{h}$. Let ω_0 be some meromorphic differential on Γ^* having only two poles of orders ± 1

at the points $P_\pm$ respectively and let $\omega_1, \ldots, \omega_g$ be the basis holomorphic differentials corresponding to some choice of canonical a-cycles. We assign the abelian differential ω with values in $\mathfrak{h}^*$ to the set $\dot\chi_0, \ldots, \dot\chi_{2g}$, where

$$\omega = \dot\chi_0 \omega_0 + \cdots + \dot\chi_g \omega_g. \tag{1.11}$$

DEFINITION 1.5. The differential ω defined above is said to be the *reduced highest weight* of the module V.

REMARK 1.6. The full (*not reduced*) weights include $\dot\chi_{g+1}, \ldots, \dot\chi_{2g}$ and the meromorphic differentials as well [7].

§2. Highest weight modules over algebras of Virasoro type

DEFINITION 2.1. A *weight* of a Lie algebra $\mathcal{L}^\Gamma$ is the set of complex numbers $\lambda = \{\lambda_j : j \equiv g/2 \pmod{\mathbb{Z}}, |j| \leqslant g_0\}$.

To each weight λ we assign the quadratic differential

$$\Omega = \sum_{k=-g_0}^{g_0} \lambda_k \Omega_k, \tag{2.1}$$

where Ω_k is uniquely defined by the requirement to be meromorphic, not to have any poles outside the points $P_\pm$, and by a certain asymptotic behavior [1, 2] at these points. The differentials Ω_k satisfy the following condition:

$$\frac{1}{2\pi i} \oint_{c_0} e_l \Omega_k = \delta_{lk}, \tag{2.2}$$

where δ_{lk} is the Kronecker symbol. The differential Ω will be called the *weight* of Lie algebra $\mathcal{L}^\Gamma$ as well.

The construction of Verma modules for $\mathcal{L}^\Gamma$ is quite analogous to the one for $\mathcal{G}$. Let L be a one-dimensional linear space with a formal generator v and suppose the *trivial action of* $\mathcal{L}^\Gamma_+$, denoted Ω and defined by (2.1), is a weight. Let us introduce the linear space

$$V = U(\mathcal{L}^\Gamma) \otimes_{U(\mathcal{L}^\Gamma_+)} L \tag{2.3}$$

and define a polygrading on it (which is the quasigrading in the component corresponding to e_j, $j < -g_0$ and the usual grading in the components corresponding to $e_j, |j| \leqslant g_0$). In order to do that, let us introduce $w_i = e_i v - \lambda_i v$ ($i = -g_0, \ldots, g_0$) and declare w_i to be the generators of quasihomogeneous subspaces of degree $\deg v - (0, \ldots, 1, \ldots, 0)$ (1 in the ith position). Equivalently, one can require the following relation to hold:

$$e_i v = \lambda_i v + \ldots, \tag{2.4}$$

where $\ldots$ denotes terms of lower degree.

DEFINITION 2.2. The linear space V defined by (2.3) with the natural action of $\mathcal{L}^\Gamma$ and quasigrading defined by (2.4) is said to be the *Verma module of $\mathcal{L}^\Gamma$ of weight* Ω.

The definition of a generic highest weight module for $\mathcal{L}^\Gamma$ is quite analogous to the corresponding definition for $\mathcal{G}$ (see the Appendix) with only one difference, namely if $-g_0 \leqslant j \leqslant g_0$, then the action of e_j causes a shift of the weight by Ω_j, but if $j < -g_0$, then the action of e_j does not change the weight, but only the degree of the element of the module.

§3. Virasoro construction

Let us consider a generalized Heisenberg algebra $\mathcal{H}$ on the Riemann surface Γ. It is obtained by the substitution of a commutative Lie algebra $\mathfrak{h}$ equipped with an inner product $\langle \cdot\,,\,\cdot \rangle$ for $\mathfrak{g}$ in Definition I.1 and in formula (I.1). In [1, 2], an analog of the Virasoro construction on a Riemann surface is suggested; given a representation of $\mathcal{H}$, this construction yields the corresponding representation of $\mathcal{L}^\Gamma$. Our main purpose in this section is to obtain the relation between the highest weights of these representations.

Let α_m, α_n be the representation operators of two quasihomogeneous elements of $\mathcal{H}$ of degrees m and n respectively and $\Sigma^\pm$ be a decomposition of the two-dimensional lattice $\{(m, n)\}$ into two parts such that $\Sigma^\pm$ can differ from $\Sigma_0^\pm = \{(m, n) \mid m \leqslant n \, (m \geqslant n)\}$ only for $0 \leqslant m, n \leqslant g$. According to I. Krichever and S. Novikov [2, 3], each decomposition of this kind defines a normal ordering in $\mathcal{H}$ such that

$$(3.1) \qquad :\alpha_m \alpha_n: = \begin{cases} \alpha_m \alpha_n, & (m, n) \in \Sigma^+, \\ \alpha_n \alpha_m, & (m, n) \in \Sigma^-. \end{cases}$$

This allows one to define the *energy-momentum tensor* T as follows:

$$(3.2) \qquad T = \frac{1}{2} \sum :\alpha_m \alpha_n: \omega_m \omega_n,$$

where ω_n, $0 \leqslant n \leqslant g$, are introduced in §1, and ω_n for $n < 0$ and $n > g$ are the basis meromorphic differentials on Γ^* introduced in [1–3]. The explicit form of the latter is inessential now. The Virasoro operators L_k on the Riemann surface are defined by the decomposition [2]

$$(3.3) \qquad T = \sum L_k \Omega_k.$$

THEOREM 3.1 [2]. *The operators L_k define a representation of $\mathcal{L}^\Gamma$.*

LEMMA 3.1. *The highest weight Ω of the representation of $\mathcal{L}^\Gamma$ just obtained, the highest weight vector v, and the energy-momentum tensor are related as follows:*

$$(3.4) \qquad Tv = \Omega v + \ldots,$$

where ... denotes terms of lower degree.

PROOF. Since v is the highest vector, by (3.3) and for reasons related to the grading, one has

$$Tv = \sum_{k=-g_0}^{g_0} L_k \Omega_k v,$$

where from now on ... denotes the sum of terms of degree less then $\deg v$. Next, by (2.4) one has $L_k v = \lambda_k v + \ldots$, so

$$Tv = \sum_{k=-g_0}^{g_0} \lambda_k \Omega_k v + \ldots.$$

Taking in consideration (2.1), one obtains (3.4).

In the rest of this section, we shall suppose that the properties of the quasigrading of the $\mathcal{H}$-module V under consideration are similar to the corresponding properties of the algebra $\mathcal{A}^\Gamma$. This means that for each pair $i < 0$, $j < 0$ and any element $v_j \in V$ such that $\deg_0 v_j = j$ one has

$$\left\{ k : A_i v_j = \sum_k d_{ij}^k v_k, \; d_{ij}^k \neq 0 \right\} \subset \left\{ k : A_i A_j = \sum_k c_{ij}^k A_k, \; c_{ij}^k \neq 0 \right\}.$$

In addition, we suppose that the highest vector $v \in V$ is situated in the sphere of this agreement, i.e., $\deg_0 v < 0$.

The relation between the weight of the representation of $\mathcal{H}$ and the weight of the corresponding representation of $\mathcal{L}^\Gamma$ is given by following theorem.

THEOREM 3.2. *Let ω be the reduced highest weight of the representation of $\mathcal{H}$ and Ω be the weight of the corresponding representation of $\mathcal{L}^\Gamma$. Then*

$$\Omega = \tfrac{1}{2} \langle \omega, \omega \rangle.$$

PROOF. By Lemma 3.1, it is enough to find the term of highest degree in Tv, where T is given by formula (3.2). Note that $:\alpha_m \alpha_n: v$ has a nonzero projection onto $\mathbb{C}v$ only if $0 \leqslant n, m \leqslant g$. In fact, suppose that $:\alpha_m \alpha_n: = \alpha_m \alpha_n$. If $n > g$, then $\alpha_n v = 0$. If $n < 0$, then by the properties of the class of normal orderings under consideration one has $m \leqslant n$ and so $m < 0$. In this case we should have $\deg \alpha_m \alpha_n v < \deg v$. So we have proved that $0 \leqslant n \leqslant g$. By (1.7) one can conclude that $\alpha_n v = \dot{\chi}_n v + \ldots$. This implies $0 \leqslant m$ because in the converse case $\deg \alpha_m \alpha_n v < \deg v$. Now one can conclude that

$$(3.5) \qquad Tv = \left(\frac{1}{2} \sum_{0 \leqslant m, n \leqslant g} :\alpha_m \alpha_n: \omega_m \omega_n \right) v + \ldots.$$

Together with the relations

$$(3.6) \qquad :\alpha_m \alpha_n: v = \dot{\chi}_m \dot{\chi}_n v + \ldots,$$

which follow from (1.8) for $0 \leqslant m, n \leqslant g$, the relation (3.5) means that

$$(3.7) \qquad Tv = \left(\frac{1}{2} \sum_{0 \leqslant m, n \leqslant g} (\dot{\chi}_m \omega_m)(\dot{\chi}_n \omega_n) \right) v + \ldots.$$

Because of (1.11), this relation is equivalent to the statement to be proved.

Appendix. Highest weight representations for Krichever–Novikov algebras of affine type

Let V be a $\mathcal{G}$-module. Here we shall consider modules with an action of the group $\mathfrak{h}_a^{g+1} \oplus Q^g$ (which corresponds to the space of abelian monodromy groups, see Proposition 1.1). Here Q is the lattice generated by the roots of the Lie algebra $\mathfrak{g}$. Suppose $P = \bigoplus_{k=0}^{2g} \mathfrak{h}_k^*$ is the space of the weights, $P_0 = \bigoplus_{k=0}^{2g} Q_k^\vee$, where $\mathfrak{h}_k \cong \mathfrak{h}_a$ ($k = 0, \ldots, g$), and $\mathfrak{h}_k \cong \mathfrak{h}$ ($k = g+1, \ldots, 2g$), $Q_k^\vee$ is the dual root lattice in $\mathfrak{h}_k$ (i.e., $Q_0^\vee, \ldots, Q_g^\vee$ are the dual affine root lattices). The elements of the linear space P are said to be the *weights*. In the sequel $\mathfrak{h}_k^*$ and $\mathfrak{h}_k$ are identified by the Cartan–Killing form, so P_0 turns out to be identified with the lattice of integer weights. An integer weight with nonnegative components is said to be a *dominant weight*. An element v of a $\mathcal{G}$-module V is said to be a *Bloch weight vector with Bloch weight* λ if for each $k = 0, \ldots, g$ we have

$$hv = \lambda_k(h)v \qquad (h \in \mathfrak{h}_k)$$

and for each $k = g+1, \ldots, 2g$

$$T_q v = \{\exp 2\pi i \lambda_k(q)\} v \qquad (q \in Q_k),$$

where T_q denotes the action of the element $q \in Q_k$. The space of the weight vectors of weight λ is denoted by V_λ.

DEFINITION A.1. A $\mathcal{G}$-module V is said to be a *module with highest weight* χ if it is quasigraded over $\mathcal{G}_-$ and the following conditions are fulfilled:

$1°$. $V = \bigoplus_{\lambda \in P} V_\lambda$, where V_λ is a space of weight λ and $\dim V_\lambda < \infty$;

$2°$. if $0 \leq j \leq g$, α is a finite root and $\lambda_0 = 0, \ldots, \lambda_{j-1} = 0$, then $(x_\alpha A_j) V_\lambda \subset V_{\lambda'}$, where $\lambda_i' = \lambda_i$ ($i \neq j$) and $\lambda_j' = \lambda_j + (\alpha + 1)$;

$3°$. if $j \leq -1$, then $(x_\alpha A_j) V_\lambda \subset \bigoplus V_{\lambda'}$, where $\lambda_i' = \lambda_i$ ($i = 1, \ldots, 2g$), $\lambda_0' \in \{\lambda_0 + (\alpha + k) : k = j - g, \ldots, j\}$;

$4°$. a highest vector v_χ with Bloch weight χ such that $V = U(\mathcal{G})v_\chi$ is given.

For $\lambda \in P$, let us denote $m(\lambda) = \dim V_\lambda$. We say that λ is the *weight of the module* V if $m(\lambda) \neq 0$.

DEFINITION A.2. A $\mathcal{G}$-invariant subspace $V' \subseteq V$ is said to be a *submodule* if $V' = \bigoplus V_\lambda'$, where $V_\lambda' = V' \cap V_\lambda$.

For each module V with highest weight, let us denote by $V_{\max}$ its maximal submodule, i.e., the sum of all nontrivial submodules. Let M_χ be a Verma module with a highest weight χ.

PROPOSITION A.1. *The quotient module $L_\chi = M_\chi/(M_\chi)_{\max}$ is a nontrivial irreducible module of highest weight χ.*

DEFINITION A.3. The *character* of a module V is the formal sum

$$\operatorname{ch} V = \sum_{\dim V_\lambda \neq 0} m(\lambda) e^\lambda.$$

If one excludes the sum over the last g components of the weight, one will obtain a reduced character in the sense of §1.

Let us introduce the following standard notation: R is the root system of the Lie algebra $\mathfrak{g}$, R_a is the corresponding affine root system, d_α is the multiplicity of the root α, $\rho = (\rho_0, \ldots, \rho_{2g})$ where each ρ_k is the sum of the fundamental affine weights. We write $R_{a,j}$ instead of R_a (respectively R_k instead of R) if the corresponding root system lies in $\mathfrak{h}_j$ ($j = 0, \ldots, g$) (respectively in $\mathfrak{h}_k$ ($k = g+1, \ldots, 2g$)). Let us also denote

$$K = \prod_{j=0}^{g} \prod_{\alpha \in R_{a,j}^+} (1 - e^{-\alpha})^{-d_\alpha} \Big/ \prod_{k=1}^{g} \prod_{\beta \in R_k^+} (1 - e^{-\beta})^{-d_\beta}.$$

PROPOSITION A.2 [7, Lemma 3.4]. *For the reduced character of a Verma module M_χ with reduced highest weight χ, one has* $\operatorname{ch} M_\chi = e^\chi K$.

The following theorem shows that for Krichever–Novikov algebras, the usual relation between the character of an irreducible module and the characters of the corresponding Verma modules holds.

Let us denote the Weil group of the Lie algebra $\mathfrak{g}$ by $\overline{W}$. It is well known that the corresponding affine Weil group W_a can be regarded as a subgroup of $GL(\mathfrak{h}_a^*)$. On the other hand, W_a can be naturally identified with the semidirect product $\overline{W} Q^\vee$ and thus be embedded into the group $GA(\mathfrak{h}^*)$ of affine transformations of the space $\mathfrak{h}^*$. So each element $w \in W_a$ is identified with some $\bar{w} T_q$ ($\bar{w} \in \overline{W}$, $q \in Q^\vee$, T_q is the corresponding translation in $\mathfrak{h}^*$). We shall refer to $\bar{w}$ as the *finite part* of the element w (see Remark 3.1).

DEFINITION A.4. The group W consisting of elements of the form $w = (w_0, \ldots, w_{2g})$, where $w_k \in W_a \subset GL(\mathfrak{h}_a^*)$ ($k = 0, \ldots, g$), $w_k \in \overline{W} Q^\vee \subset GA(\mathfrak{h}^*)$ ($k = g+1, \ldots, 2g$), and all w_k have the same finite part, is said to be the *Weil group* of the Lie algebra G.

The action of Weil group on the weights is natural: if $\lambda = (\lambda_0, \ldots, \lambda_{2g})$, then $(w\lambda)_i = w_i \lambda_i$ ($i = 0, \ldots, 2g$).

THEOREM A.1 [7, Theorem 3.1]. *The character of an irreducible module L_χ with dominant highest weight χ and the characters of Verma modules are related as follows:*

$$\operatorname{ch} L_\chi = \sum_{w \in W} (-1)^w \operatorname{ch} M_{w(\chi+\rho)-\rho}.$$

The explicit analog of the Weil–Kac formula for the characters of irreducible modules of Krichever–Novikov algebras of affine type is as follows:

$$\operatorname{ch} L_\chi = K \sum_{w \in W} (-1)^w \exp(w(\chi+\rho) - \rho).$$

References

1. I. Krichever and S. Novikov, *Virasoro-type algebras, Riemann surfaces and soliton theory*, Funktsional. Anal. i Prilozhen. **21** (1987), no. 2, 46–63; English transl. in Functional Anal. Appl. **21** (1987), no. 2.
2. _____, *Virasoro-type algebras, Riemann surfaces and strings in Minkowski space*, Funktsional. Anal. i Prilozhen. **21** (1987), no. 4, 47–61; English transl. in Functional Anal. Appl. **21** (1987), no. 4.
3. _____, *Virasoro-type algebras, stress–energy tensor and operator products on Riemann surfaces*, Funktsional. Anal. i Prilozhen. **23** (1989), no. 1, 24–40; English transl. in Functional Anal. Appl. **23** (1989), no. 1.
4. A. Jaffe, S. Klimek, and A. Lesniewsky, *Representations of the Heisenberg algebra on a Riemann surface*, Harvard University Preprint HUTMP 89/B293 (1989).
5. O. Sheinver, *Elliptic affine Lie algebras*, Funktsional. Anal. i Prilozhen. **24** (1990), no. 3, 51–61; English transl. in Functional Anal. Appl. **24** (1990), no. 3.
6. _____, *Affine Lie algebras on the Riemannian surfaces*, Funktsional. Anal. i Prilozhen. **27** (1993), no. 4; English transl. in Functional Anal. Appl. **27** (1993), no. 4.
7. _____, *Modules with highest weight for affine Lie algebras on Riemann surfaces*, Funktsional. Anal. i Prilozhen. **29** (1995), no. 1, 56–71; English transl. in Functional Anal. Appl. **29** (1995), no. 1.
8. I. B. Frenkel, *Orbital theory of affine Lie algebras*, Inv. Math. **77** (1984), no. 2, 301–352.
9. G. Segal, *Unitary representations of some infinite-dimensional groups*, Comm. Math. Phys. **80** (1981), no. 3, 301–342.
10. S. Novikov, *Two-dimensional Schrödinger operators in periodic fields*, Itogi Nauki i Tehniki. Sovremennye Problemy Matematiki, vol. 23, VINITI, Moscow, 1983, pp. 3–32; English transl. in J. Societ Math. **28** (1985), no. 1.
11. I. Krichever, *Nonlinear equations and elliptic curves*, In: Itogi Nauki i Tehniki. Sovremennye Problemy Matematiki, vol. 23, VINITI, Moscow, 1983, pp. 79–136; English transl. in J. Societ Math. **28** (1985), no. 1.
12. M. Schlichenmaier, *Differential operator algebras on compact Riemann surfaces*, Mannheimer Manuskripte 164, hep-th/9311036 (1993).
13. M. Bremner, *Universal central extensions of elliptic affine Lie algebras*, J. Math. Phys. (to appear).

Translated by THE AUTHOR

Krzhizhanovsky Power Engineering Institute, Leninskiĭ prospekt 19, Moscow, 117071, Russia
E-mail address: sheinman@cpd.landau.ac.ru

Huygens' Principle and Algebraic Schrödinger Operators

A. P. VESELOV

ABSTRACT. The relationship between Hadamard's problem of the description of all second-order hyperbolic equations satisfying Huygens' Principle and "finite-gap" or algebraic Schrödinger operators is discussed.

It is well known that the behavior of the solutions of wave equations in two- and three-dimensional spaces are quite different. A pointwise perturbation at the initial moment will produce a pure spherical wave in $\mathbb{R}^3$, while in $\mathbb{R}^2$ the whole region inside the circle will be disturbed. In the first case we say, following Hadamard, that we have *Huygens' Principle* (HP), in the second, that we deal with the *diffusion of waves*.

The first phenomenon plays a very important role in our life: it is because of HP that the clear transmission of signals is possible and we can hear each other and listen to music. A two-dimensional world would have been drastically different from this point of view.

More precisely, we say that a second-order hyperbolic equation satisfies HP (in Hadamard's sense) if the solution of the corresponding Cauchy problem at some point x depends only on the initial data in an arbitrary small vicinity of the characteristic conoid with vertex x. This means that the interior domain bounded by this conoid is a *lacuna* for the considered equation, or equivalently, the fundamental solution of the corresponding Cauchy problem is located on the characteristic conoid.

The problem of describing all second-order hyperbolic equations satisfying HP is known as the *Hadamard problem*. Hadamard [2] proved that HP is possible only in the case when the number of spatial variables is odd and found a criterion for the validity of HP. Unfortunately, this criterion turned out not to be effective enough to solve the Hadamard problem, which is still open (for a survey, see e.g. [3]).

In the present paper we discuss some results concerning the Hadamard problem, which were discovered recently by Yu. Berest, O. Chalykh, and the author [4–6] using the theory of "finite-gap" or algebraic Schrödinger operators [7–11].

The starting point for us was the following remarkable result of K. Stellmacher and J. Lagnese [12–13]. They found all hyperbolic equations of the form

$$(1) \qquad (\Box + u(x))\varphi = 0,$$

1991 *Mathematics Subject Classification*. Primary 35L15, 35Q40.

where $\Box = \partial_0^2 - \partial_1^2 - \cdots - \partial_n^2$, $\partial_i = \partial/\partial x_i$, in the case when the potential $u(x)$ depends only on one variable, say x_1. All such potentials turn out to be rational and can be obtained from $u = 0$ by a transformation now usually known as the Darboux transformation.

In the simplest case one has equation (1) with potential $u(x) = 2x_1^{-2}$ and $n = 5$, which was found in 1953 by Stellmacher as the first example of a nontrivial equation with HP.

The same potentials were later investigated in detail by Airault, McKean, Moser, and Adler (see [15]) in the theory of the Korteweg–de Vries equation. In fact, they are nothing else but rational decreasing solutions of the stationary higher KdV equations (Novikov equations, see, e.g., [14]). As is shown by S. P. Novikov in [16], the general solution of these equations has the following remarkable property: a one-dimensional Schrödinger operator L with such a potential has only finite number of gaps in its spectrum. At present, after the works of Novikov, Dubrovin, Lax, Matveev, Its, Marchenko, McKean, and van Moerbeke (see [7] and references therein), there exists a beautiful theory of such operators, called "finite-gap".

The characteristic property of all such operators is the existence of a commuting differential operator A of odd order: $[L, A] = 0$. This can be used to define the notion of *algebraic integrability* of the Schrödinger equation or *algebraic Schrödinger operator* in dimension n, demanding the existence of more than n commuting partial differential operators with some assumptions on the highest symbols (for the details, see [8–10]). Recall that the existence of n commuting operators means ordinary integrability of an n-dimensional quantum system.

In contrast to the one-dimensional case [7], the complete theory of algebraic Schrödinger operators in dimensions more than 1 does not exist. In dimension 2, an alternative approach was proposed by Dubrovin, Krichever, and Novikov [17], who considered operators that are algebraic only on one energy level (see also [18]). One of the first problems posed to the author by Novikov in 1976 was to investigate the possibility of changing the energy level in this construction. The answer turned out to be negative in general, although in the special case when the potential is the sum of one-dimensional finite-gap potentials, one obviously has an operator which is algebraic on each energy level.

The natural question whether other examples of such operators exist or not was open until 1987, when the author, motivated by the investigations of commuting polynomial and rational mappings [19], put forward the conjecture that the following generalized Lamé operators and their trigonometric and rational degenerations are algebraic:

$$(2) \qquad L = -\Delta + \sum_{\alpha \in \mathfrak{R}_+} m_\alpha (m_\alpha + 1)(\alpha, \alpha) f((\alpha, x))$$

and moreover, under certain additional assumption, there are no other algebraic operators in dimension more than 1.

Here Δ is the Laplacian in $\mathbb{R}^n$, $x \in \mathbb{R}^n$, $f(z)$ is the Weierstrass elliptic function $\wp(z)$ or its degenerations: $f(z) = \sin^{-2} z$, $f(z) = \sinh^{-2} z$ (trigonometric case) and $f(z) = z^{-2}$ (rational case), $\mathfrak{R}$ denotes a root system in $\mathbb{R}^n$, $\mathfrak{R}_+$ is its positive part, the numbers m_α are integers and the corresponding function m on $\mathfrak{R}$ is invariant under the Weyl group. In the rational case, one can consider an arbitrary Coxeter group and its root system.

The Schrödinger operator (2) without the condition that m_α are integers was first considered by M. A. Olshanetsky and A. M. Perelomov [20] as a generalization of the Calogero–Moser operator

$$L = -\Delta + g \sum_{i \neq j} \frac{1}{(x_i - x_j)^2},$$

which corresponds to the particular case when the root system is of type $\mathcal{A}_n$. The importance of the integer-valuedness condition for m_α was clarified by O. Chalykh and the author [9], who proved the existence of additional commuting integrals for Calogero–Moser systems with an arbitrary number of particles and therefore established a part of the above-mentioned conjecture.

Later Chalykh, Styrkas, and the author [11] extended this result to an arbitrary root system both in the rational and trigonometric cases and proved that the corresponding operators (2) are indeed algebraic. The question whether this is true in the elliptic case is still open.[1] One of the possible approaches is to consider the elliptic case as the deformation of the trigonometric one. This is the object of our current investigations with M. Schmidt.

The following construction [9–11] plays a crucial role in our study of the rational and trigonometric cases. Let A be a finite set of vectors α of n-dimensional Euclidean space with prescribed integer multiplicities m_α. Let the function $\psi(k, x)$, $k, x \in \mathbb{R}^n$, have the following properties:

1) The function ψ has the form

$$(3) \qquad \psi = P(k, x) \exp(k, x),$$

where $P(k, x)$ is some polynomial in k with the leading term

$$A(k) = \prod_{\alpha \in A} (k, \alpha)^{m_\alpha}.$$

2) The following relation

$$(4) \qquad \psi(k + s\alpha, x) \equiv \psi(k - s\alpha, x)$$

holds for all $\alpha \in A$, $s = 1, \ldots, m_\alpha$ and for all k belonging to the corresponding hyperplanes $(k, \alpha) = 0$.

It is easy to prove that there exists at most one function with these properties, so the main problem is the existence of such a function.

---

[1] In the paper [21] J. Feldman, H. Knörrer, and E. Trubowitz proved that in dimension two there are no nonsingular algebraic operators except for sums of one-dimensional ones. This does not contradict the above conjecture, because all generalized Lamé operators (2) are singular on the reals and only for the root system of type $\mathcal{A}_1$ (i.e., for the classical Lamé operator) is it possible to make the operator nonsingular by an imaginary shift.

THEOREM 1. *Let A be the set of positive roots of a simple complex Lie algebra G with multiplicities m_α that are constant on the orbits of the corresponding Weyl group W. Then the function ψ with the above properties exists and is unique.*

The same result is valid also for the nonreduced root system BC_n, if one replaces relations (4) for the roots α and 2α by

$$\psi(k + s\alpha, x) \equiv \psi(k - s\alpha, x)$$

for $(k, \alpha) = 0$ and $s = 1, 2, \ldots, m_\alpha, m_\alpha + 2, \ldots, m_\alpha + 2m_{2\alpha}$.

The proof of this theorem [11] uses the theory of generalized hypergeometric functions developed by G. Heckman and E. Opdam (see [22]).

The following result [11] shows that under some extra assumptions ψ exists only for root systems, thus partially justifying the conjecture mentioned above.

THEOREM 2. *If the set A consists of mutually noncollinear vectors with unit multiplicities, then the function ψ with properties 1) and 2) exists only if $A \cup (-A)$ is the root system of some semisimple Lie algebra.*

Let R_A be the ring of all polynomials $p(k)$ satisfying relation (4):

$$p(k + s\alpha) \equiv p(k - s\alpha) \pmod{(k, \alpha)}$$

for all $\alpha \in A$, $s = 1, \ldots, m_\alpha$.

THEOREM 3. *If the function ψ with properties 1) and 2) does exist, then to every polynomial $p(k) \in R_A$ there corresponds a partial differential operator $L_p(x, \partial/\partial x)$ such that*

$$L_p \psi(k, x) = p(k)\psi(k, x).$$

All such operators form a commutative ring isomorphic to R_A.

Note that $p(k) = -k^2$ belongs to R_A for all A. The corresponding operator has the form

$$(5) \qquad L = -\Delta + \sum_{\alpha \in A} \frac{m_\alpha(m_\alpha + 1)(\alpha, \alpha)}{\sinh^2(\alpha, x)}.$$

When A is the positive part of a root system, any polynomial $p(k)$ invariant under the Weyl group belongs to the ring R_A. According to the Chevalley theorem, the algebra of such polynomials is freely generated by n generators.

It is very important that the ring R_A is much larger than this algebra. Indeed, any polynomial p from the ideal generated by

$$q(k) = \prod_{\alpha \in A} ((k, \alpha)^2 - (\alpha, \alpha)^2) \cdots ((k, \alpha)^2 - m_\alpha^2(\alpha, \alpha)^2),$$

satisfies relations (4) and therefore belongs to R_A.

As a corollary, we see that the corresponding Schrödinger operator (5) is algebraic.

In the rational case, one must replace property 2) in the above construction by the following one:

2') The relation

$$\left(\alpha, \frac{\partial}{\partial k}\right)\left[(\alpha, k)^{-1}\left(\alpha, \frac{\partial}{\partial k}\right)\right]^{s}\psi = 0 \tag{6}$$

holds for all $\alpha \in A$, $s = 0, 1, \ldots, m_\alpha - 1$ and for all k belonging to the corresponding hyperplanes $(k, \alpha) = 0$.

THEOREM 4 [11]. *Let A be a finite set of noncollinear vectors with unit multiplicities. A function ψ with properties 1) and 2') exists if and only if A is the set of normals to the reflection hyperplanes of some Coxeter group.*

Recall that a Coxeter group is a finite group generated by reflections in Euclidean space.

THEOREM 5 [11]. *If A is the set of normals to the reflection hyperplanes of some Coxeter group and the multiplicity function is invariant with respect to this group, then a function ψ with properties 1) and 2') exists and is unique. This function is an eigenfunction of the algebraic Schrödinger operator $L = -\Delta + u(x)$,*

$$u(x) = \sum_{\alpha \in A} \frac{m_\alpha(m_\alpha + 1)(\alpha, \alpha)}{(\alpha, x)^2}. \tag{7}$$

The proof of this theorem uses the so-called *Dunkl operator* [23], which gives an effective procedure to determine the function ψ.

This result was used by Berest and the author to construct new examples of hyperbolic equations satisfying Huygens' Principle [4, 5].

Consider the hyperbolic operator

$$\mathcal{L} = \frac{\partial^2}{\partial t^2} - \frac{\partial^2}{\partial x_1^2} - \cdots - \frac{\partial^2}{\partial x_n^2} - \frac{\partial^2}{\partial y_1^2} - \cdots - \frac{\partial^2}{\partial y_m^2} + u(x_1, \ldots, x_n),$$

where u is given by formula (7) and $N = m + n$ is an arbitrary odd number satisfying the following inequality

$$N \geqslant 3 + 2 \sum_{\alpha \in \mathfrak{R}_+} m_\alpha.$$

THEOREM 6 [4]. *For any Coxeter group and integer-valued invariant function m on the corresponding root system, the equation $\mathcal{L}\varphi = 0$, where the operator $\mathcal{L}$ is described above, satisfies Huygens' principle.*

Note that Stellmacher's equation with the potential $u = -2/x_1^2$ corresponds precisely to the simplest root system of type A_1 with $m_\alpha \equiv 1$. The first new example of Huygens' equation occurs in dimension $N = 9$ and corresponds to the root system of type A_2 with multiplicities $m_\alpha \equiv 1$:

$$\mathcal{L}\varphi = \left(\Box + \frac{2}{x_1^2} + \frac{16(x_1^2 + 3x_2^2)}{(x_1^2 - 3x_2^2)^2}\right)\varphi = 0, \tag{8}$$

where

$$\Box = \frac{\partial^2}{\partial t^2} - \sum_{i=1}^{9} \frac{\partial^2}{\partial x_i^2}.$$

The proof of this result is quite effective in the sense that it gives a procedure for writing out explicit formulas for the fundamental solutions of the corresponding equations (see [4, 5]).

CONJECTURE [4]. *The equations described above together with Stellmacher's and Lagnese's equations solve Hadamard's Problem in the class of equations of the form* (1), *i.e., they exhaust* (*up to certain natural transformations*) *all the equations that satisfy Huygens' principle.*

It turns out (see [6, 5]) that some of the trigonometric algebraic Schrödinger operators (5) also are related to HP, not in Euclidean space, but in spaces with nontrivial metrics, namely *symmetric spaces*. The corresponding equation is the so-called *modified wave equation on the symmetric space* **X**:

$$\varphi_{tt} - \mathcal{L}_X \varphi + c\varphi = 0,$$

where $\mathcal{L}_X$ is the Laplace–Beltrami operator on the symmetric space **X**, c is a certain constant, which, for irreducible spaces, has the form

$$c = \pm|\rho|^2, \qquad \rho = \frac{1}{2} \sum_{\alpha \in \mathcal{R}_+} \mu_\alpha \alpha.$$

Here $\mathcal{R}_+$ is the set of positive restricted roots of the space **X**, the number ρ is their half-sum with multiplicities, and the sign $\pm$ corresponds to the sign of the curvature of **X**: plus for the spaces of compact type, minus for the noncompact type (see e.g. [27]). For reducible symmetric spaces, c is equal to the sum of the constants for the corresponding irreducible components. The value of c is very important: one can show that the following theorem is valid only when the constant c is chosen as described above (see [6]).

THEOREM 7. *Let* **X** *be an odd-dimensional Riemannian symmetric space with the multiplicities* μ_α *of all roots being even. Then the modified wave equation* (9), (10) *on* **X** *satisfies Huygens' principle.*

In the case when **X** is of noncompact type, this result is due to Solomatina [26], Olafsson, and Schlichtkrull [29], and Helgason [28]. The case of spheres was considered by Lax and Phillips in [24]. For compact Lie groups and spaces with complex groups of motion, it was established by Helgason [25]. The general case follows from the explicit formulas for the fundamental solutions of the corresponding modified wave equations found recently by Chalykh and the author [6, 5].

Here is the list of the irreducible symmetric spaces of compact type with even multiplicities (the general case is given by the products of such spaces):
1) odd-dimensional spheres S^{2n+1},
2) simple compact Lie groups,
3) symmetric spaces of type A_{II} in Cartan's notations, i.e.,

$$\mathbf{X} = SU(2n)/Sp(n),$$

4) a special 26-dimensional symmetric space of type E_{IV}: $\mathbf{X} = E_6/F_4$.

The noncompact symmetric spaces dual to them are the following:

1) odd-dimensional Lobachevsky spaces H^{2n+1};

2) symmetric spaces with complex groups of motion, i.e., $\mathbf{X} = G/K$, where G is a simple complex Lie group, regarded as a real one, K is its maximal compact subgroup;

3) symmetric spaces of type A_{II}:

$$\mathbf{X} = SU^*(2n)/Sp(n), \qquad \dim \mathbf{X} = (n-1)(2n+1).$$

Here the group $SU^*(2n)$ is the quaternion analog of $SL(n)$:

$$SU^*(2n) = SL(2n, \mathbb{C}) \cap GL(n, \mathbb{H})$$

and $Sp(n)$ is its intersection with $U(2n)$ (see [27]);

4) special symmetric space of type E_{IV}, dual to $\mathbf{X} = E_6/F_4$, $\dim \mathbf{X} = 26$, and related to Cayley numbers or octonions.

These spaces were considered in our paper [10], where explicit formulas were found for the spherical functions and the inversion of the Abel transform on them. An explanation of their "integrability" is given by the fact that the radial part of the corresponding Laplace–Beltrami operator is conjugate to the algebraic Schrödinger operator (5): the parameters m_α in the potential are equal to half of the root multiplicities μ_α, so when all multiplicities are even the parameters are integers (see [10]). The same circumstance allows us to write out fundamental solutions for the modified wave equations on the spaces listed above explicitly [6, 5] and give an effective proof of HP for them.

It would be interesting to understand whether the other algebraic Schrödinger operators in the trigonometric case (5) are related to HP or not. Another intriguing problem is to investigate, in the theory of Huygens' principle, the possibility of using the corresponding matrix analog of generalized Lamé operators [30] (also see the paper by Krichever in this volume).

Acknowledgements. This work was partially supported by the International Science Foundation and the Russian Fund for Fundamental Research.

References

1. R. Courant and D. Hilbert, *Methods of mathematical physics*, vol. II, Wiley, New York, 1964.
2. J. Hadamard, *Lectures on Cauchy's problem in linear Partial differential equations*, Yale Univ. Press, New Haven, 1923.
3. P. Günther, *Huygens' principle and hyperbolic equations*, Academic Press, Boston, 1988.
4. Yu. Yu. Berest and A. P. Veselov, *Hadamard's problem and Coxeter groups: new examples of the Huygensian equations*, Funktsional. Anal. i Prilozhen. **28** (1994), no. 1, 3–15; English transl. in Functional Anal. Appl. **28** (1994), no. 1.
5. Yu. Yu. Berest and A. P. Veselov, *Huygens' principle and Integrability*, Uspekhi Mat. Nauk **49** (1994), no. 6, 3–75; English transl. in Russian Math. Surveys **49** (1994).
6. O. A. Chalykh and A. P. Veselov, *Integrability and Huygens Principle on the symmetric spaces* (to appear).
7. B. A. Dubrovin, V. B. Matveev, and S. P. Novikov, *Nonlinear equations of KdV type, finite-gap linear operators and Abelian varieties*, Uspekhi Mat. Nauk **51** (1976), no. 1, 51–125; English transl. in Russian Math. Surveys **51** (1976).

8. I. M. Krichever, *Methods of algebraic geometry in the theory of nonlinear equations*, Uspekhi Mat. Nauk **32** (1977), no. 6, 180–208; English transl. in Russian Math. Surveys **32** (1977).
9. O. A. Chalykh and A. P. Veselov, *Commutative rings of partial differential operators and Lie algebras*, Preprint of FIM (ETH, Zürich) (1988); Comm. Math. Phys. **126** (1990), 597–611.
10. O. A. Chalykh and A. P. Veselov, *Integrability in the theory of the Schrödinger operators and Harmonic Analysis*, Comm. Math. Phys. **152** (1993), 29–40.
11. A. P. Veselov, K. L. Styrkas, and O. A. Chalykh, *Algebraic integrability for Schrödinger equation and the groups generated by reflections*, Teoret. Mat. Fiz. **94** (1993), no. 2, 253–275; English transl. in Theoret. and Math. Phys. **94** (1993).
12. J. E. Lagnese and K. L. Stellmacher, *A method of generating classes of Huygens' operators*, J. Math. Mech. **17** (1967), no. 5, 461–472.
13. J. E. Lagnese, *A solution of Hadamard's problem for a restricted class of operators*, Proc. Amer. Math. Soc. **19** (1968), 981–988.
14. R. Schimming, *Korteweg–de Vries Hierarchie und Huygenssches Prinzip*, Sitz. ber. Dresdener Seminar zur Theor. Physik, Nr. 26, 1986.
15. M. Adler and J. Moser, *On a class of polynomials connected with the Korteweg–de Vries equation*, Comm. Math. Phys. **61** (1978), no. 1, 1–30.
16. S. P. Novikov, *Periodic problem for Korteweg–de Vries equation.* I, Funktsional. Anal. i Prilozhen. **8** (1974), no. 3, 54–66; English transl. in Functional Anal. Appl. **8** (1974), no. 3.
17. B. A. Dubrovin, I. M. Krichever, and S. P. Novikov, *Schrödinger equation in magnetic field and Riemann surfaces*, Soviet Math. Dokl. **17** (1976), 947–951.
18. S. P. Novikov and A. P. Veselov, *Two-dimensional Scrödinger operator: inverse scattering transform and evolutionary equations*, Phys. D **18** (1986), 267–273.
19. A. P. Veselov, *Integrable mappings and Lie algebras*, Soviet Math. Dokl. **35** (1987), 211–213.
20. M. A. Olshanetsky and A. M. Perelomov, *Quantum completely integrable systems connected with semisimple Lie algebras*, Lett. Math. Phys. **2** (1977), 7–13.
21. J. Feldman, H. Knörrer, and E. Trubowitz, *There is no two-dimensional analogue of Lamé equation*, Math. Ann. **294** (1992), 295–324.
22. G. J. Heckman, *An elementary approach to the hypergeometric shift operator of Opdam*, Invent. Math. **103** (1991), 341–350.
23. C. F. Dunkl, *Differential-difference operators associated to reflection groups*, Trans. Amer. Math. Soc. **311** (1989), 167–183.
24. P. D. Lax and R. S. Phillips, *An example of Huygens' Principle*, Comm. Pure Appl. Math. **31** (1978), 415–421.
25. S. Helgason, *Wave equations on homogeneous spaces*, Lecture Notes in Math., vol. 1077, Springer-Verlag, Berlin, 1984, pp. 252–287.
26. L. E. Solomatina, *Translation representation and Huygens' principle for an invariant wave equation in a Riemannian symmetric spaces*, Soviet Math. (Iz. VUZ) **30** (1986), 108–111.
27. S. Helgason, *Differential geometry, Lie groups and symmetric spaces*, Academic Press, New York, 1978.
28. _____, *Huygens' Principle for wave equations on Symmetric spaces*, J. Funct. Anal. **107** (1992), 279–288.
29. G. Olafsson and H. Schlichtkrull, *Wave propagation on Riemannian symmetric spaces*, J. Funct. Anal. **107** (1992), 270–278.
30. P. Etingof and K. Styrkas, *Algebraic integrability of Schrödinger operators and representations of Lie algebras*, Preprint (1994), hep-th 9403135 (to appear in Duke Math. J.).

Translated by THE AUTHOR

DEPARTMENT OF MATHEMATICS AND MECHANICS, MOSCOW STATE UNIVERSITY, 119899, MOSCOW, RUSSIA

Other Titles in This Series

(Continued from the front of this publication)

130 **M. M. Lavrent′ev, K. G. Reznitskaya, and V. G. Yakhno,** One-dimensional Inverse Problems of Mathematical Physics
129 **S. Ya. Khavinson,** Two Papers on Extremal Problems in Complex Analysis
128 **I. K. Zhuk et al.,** Thirteen Papers in Algebra and Number Theory
127 **P. L. Shabalin et al.,** Eleven Papers in Analysis
126 **S. A. Akhmedov et al.,** Eleven Papers on Differential Equations
125 **D. V. Anosov et al.,** Seven Papers in Applied Mathematics
124 **B. P. Allakhverdiev et al.,** Fifteen Papers on Functional Analysis
123 **V. G. Maz′ya et al.,** Elliptic Boundary Value Problems
122 **N. U. Arakelyan et al.,** Ten Papers on Complex Analysis
121 **V. D. Mazurov, Yu. I. Merzlyakov, and V. A. Churkin, Editors,** The Kourovka Notebook: Unsolved Problems in Group Theory
120 **M. G. Kreĭn and V. A. Jakubovič,** Four Papers on Ordinary Differential Equations
119 **V. A. Dem′janenko et al.,** Twelve Papers in Algebra
118 **Ju. V. Egorov et al.,** Sixteen Papers on Differential Equations
117 **S. V. Bočkarev et al.,** Eight Lectures Delivered at the International Congress of Mathematicians in Helsinki, 1978
116 **A. G. Kušnirenko, A. B. Katok, and V. M. Alekseev,** Three Papers on Dynamical Systems
115 **I. S. Belov et al.,** Twelve Papers in Analysis
114 **M. Š. Birman and M. Z. Solomjak,** Quantitative Analysis in Sobolev Imbedding Theorems and Applications to Spectral Theory
113 **A. F. Lavrik et al.,** Twelve Papers in Logic and Algebra
112 **D. A. Gudkov and G. A. Utkin,** Nine Papers on Hilbert's 16th Problem
111 **V. M. Adamjan et al.,** Nine Papers on Analysis
110 **M. S. Budjanu et al.,** Nine Papers on Analysis
109 **D. V. Anosov et al.,** Twenty Lectures Delivered at the International Congress of Mathematicians in Vancouver, 1974
108 **Ja. L. Geronimus and Gábor Szegő,** Two Papers on Special Functions
107 **A. P. Mišina and L. A. Skornjakov,** Abelian Groups and Modules
106 **M. Ja. Antonovskiĭ, V. G. Boltjanskiĭ, and T. A. Sarymsakov,** Topological Semifields and Their Applications to General Topology
105 **R. A. Aleksandrjan et al.,** Partial Differential Equations, Proceedings of a Symposium Dedicated to Academician S. L. Sobolev
104 **L. V. Ahlfors et al.,** Some Problems on Mathematics and Mechanics, On the Occasion of the Seventieth Birthday of Academician M. A. Lavrent′ev
103 **M. S. Brodskiĭ et al.,** Nine Papers in Analysis
102 **M. S. Budjanu et al.,** Ten Papers in Analysis
101 **B. M. Levitan, V. A. Marčenko, and B. L. Roždestvenskiĭ,** Six Papers in Analysis
100 **G. S. Ceĭtin et al.,** Fourteen Papers on Logic, Geometry, Topology and Algebra
99 **G. S. Ceĭtin et al.,** Five Papers on Logic and Foundations
98 **G. S. Ceĭtin et al.,** Five Papers on Logic and Foundations
97 **B. M. Budak et al.,** Eleven Papers on Logic, Algebra, Analysis and Topology
96 **N. D. Filippov et al.,** Ten Papers on Algebra and Functional Analysis
95 **V. M. Adamjan et al.,** Eleven Papers in Analysis
94 **V. A. Baranskiĭ et al.,** Sixteen Papers on Logic and Algebra
93 **Ju. M. Berezanskiĭ et al.,** Nine Papers on Functional Analysis
92 **A. M. Ančikov et al.,** Seventeen Papers on Topology and Differential Geometry
91 **L. I. Barklon et al.,** Eighteen Papers on Analysis and Quantum Mechanics

(See the AMS catalog for earlier titles)